MINIMAX MODELS IN THE THEORY OF NUMERICAL METHODS

THEORY AND DECISION LIBRARY

General Editors: W. Leinfellner (*Vienna*) and G. Eberlein (*Munich*)

Series A: Philosophy and Methodology of the Social Sciences

Series B: Mathematical and Statistical Methods

Series C: Game Theory, Mathematical Programming and Operations Research

Series D: System Theory, Knowledge Engineering and Problem Solving

SERIES B: MATHEMATICAL AND STATISTICAL METHODS

VOLUME 21

Scope: The series focuses on the application of methods and ideas of logic, mathematics and statistics to the social sciences. In particular, formal treatment of social phenomena, the analysis of decision making, information theory and problems of inference will be central themes of this part of the library. Besides theoretical results, empirical investigations and the testing of theoretical models of real world problems will be subjects of interest. In addition to emphasizing interdisciplinary communication, the series will seek to support the rapid dissemination of recent results.

The titles published in this series are listed at the end of this volume.

MINIMAX MODELS IN THE THEORY OF NUMERICAL METHODS

ALEKSEI G. SUKHAREV
Moscow State University, Russia

Translated from the Russian by Olga Chuyan

SPRINGER SCIENCE+BUSINESS MEDIA, B.V.

Library of Congress Cataloging-in-Publication Data

Sukharev, A. G. (Alekseĭ Grigor'evich)
[Minimaksnye algoritmy v zadachakh chislennogo analiza. English]
Minimax models in the theory of numerical methods / Aleksei G. Sukharev.
p. cm. -- (Theory and decision library. Series B, Mathematical and statistical methods ; v. 21)
Translation of: Minimaksnye algoritmy v zadachakh chislennogo analiza.
Includes bibliographical references and indexes.
ISBN 978-0-7923-1821-7 ISBN 978-94-011-2759-2 (eBook)
DOI 10.1007/978-94-011-2759-2
1. Numerical analysis. 2. Maxima and minima. I. Title. II. Series.
QA300.S9613 1992
519.4--dc20 92-16531

ISBN 978-0-7923-1821-7

To the memory of my mother

Alexandra I. Sukhareva

and my father

Grigory M. Sukharev

CONTENTS

PREFACE TO THE ENGLISH EDITION

In the Russian edition published in 1989, this book was called "Minimax Algorithms in Problems of Numerical Analysis". The new title is better related to the subject of the book and its style. The basis for every decision or inference concerning the ways to solve a given problem is the computation model. Thus, the computation model is the epicenter of any structure studied in the book. Algorithms are not constructed here, they are rather derived from computation models. Quality of an algorithm depends entirely on consistency of the model with the real-life problem. So, constructing a model is an art, deriving an algorithm is a science.

We study only minimax or, in other words, worst-case computation models. However, one of the characteristic features of the book is a new approach to the notion of the worst-case conditions in dynamic processes. This approach leads to the concept of sequentially optimal algorithms, which play the central role in the book.

In conclusion, I would like to express my gratitude to Prof. Dr. Heinz J. Skala and Dr. Sergei A. Orlovsky for encouraging translation of this book. I also greatly appreciate the highly professional job of Dr. Olga R. Chuyan who translated the book.

Moscow, January 1992 *A.G. Sukharev*

PREFACE

The topics of computational methods efficiency and choice of the most efficient methods for solving a specific problem or a specific class of problems have always played an important role in numerical analysis. Now optimization of the computerized solution process is a major problem of applied mathematics, which stimulates search for new computational methods and ways of their implementation.

In this monograph, the ways of estimating efficiency of computational algorithms and problems of their optimality are studied in the framework of a general computation model. The main elements of the general model determining specific models of computation are: the functional or the operator to be approximated, which corresponds to the problem being solved; the class of functions reflecting the information we have about the problem; the class of algorithms that can be used for solving the problem; the criterion for estimating efficiency of the algorithm; the optimality concept; and, finally, the specific notion of optimality of an algorithm within the framework of the adopted general concept. Using this general computation model allows various problems of numerical analysis to be treated in a unified way and enables us to answer a number of fundamental methodological questions and to establish some general properties of optimal algorithms.

All specific realizations of the general computation model that are dealt with in this monograph are based on the minimax optimality concept, which was used in applied mathematics as long ago as in the 19th century by P.L. Chebyshev. This concept reflects our aim to obtain the best guaranteed result (say, accuracy of the solution) corresponding to the information we have about the problem.

Optimal algorithms for solving problems of numerical analysis have been studied in a great number of papers. There have also been published several monographs which, alongside dealing with specific problems, provide a foundation for a general theory of optimal algorithms.

The subjects of this book are considerably different from the traditional subjects of computational methods. The major difference is the close attention paid to adaptive (sequential) computational algorithms, the process of computations being regarded as a controlled process and the algorithm - as a strategy of control. This approach allows methods of game theory and other methods of operations research and systems analysis to be widely used for constructing optimal algorithms. Using these methods proves to be very fruitful and leads to a number of results connected to both conventional and new optimality concepts.

The ultimate goal of studying the various computation models we deal with in this book is construction of concrete numerical algorithms admitting program implementation. The central role belongs to the concept of a sequentially optimal algorithm, which in many cases reflects the characteristics of real-life computational processes more fully than the traditional optimality concepts. We work out a general scheme for constructing sequentially optimal algorithms and formulate requirements to the computation model allowing application of this scheme to various problems of numerical analysis.

The monograph consists of five chapters. The first chapter plays a special role. In this chapter we introduce the necessary terminology, construct a general computation model, and discuss in detail the arising methodological problems. It is important to note that the terminology related to optimal algorithms is not completely settled yet. For instance, the "terminal operation of an algorithm" introduced in Section 3 of Chapter 1 corresponds to the "algorithm" from the monograph by J.F. Traub and H. Woźniakowski [80] (see the reference list). The results obtained in Chapter 1 are frequently used throughout the book. Thus, in a sense, this chapter has an introductory character.

The subsequent chapters are devoted to construction of optimal methods for solving specific problems of numerical analysis. We deal with problems of integration, recovering functions from their values, search for the global extremum, solving equations and systems of equations, finding maximin, and multi-criterion optimization. Naturally, every problem is investigated under specific (and sometimes rather restrictive) assumptions on the functional and algorithmic classes, the criteria for estimating efficiency of the algorithms, and the optimality concepts being used, that is, within the framework of a specific computation model. The choice of specific models in the monograph is motivated by our wish to demonstrate the intrinsic abilities of the general model as fully as possible, and also to reflect the characteristic features of real-world problems. Some of the results closely related to the subject of the book (such as the results on optimal search for the extremum of a unimodal function) are omitted for the only reason that they have already been presented both in monographs and textbooks.

Every chapter and every section starts with a short introduction, which outlines the range of problems and sometimes contains references.

The reference list is quite extended and contains a number of papers that are not directly referred to in the text. These papers either deal with some optimality concepts and ways of estimating efficiency of computational methods, or are closely related to some specific problems of numerical analysis studied in the book. At the same time, a number of papers directly connected to the subject of the book and included in the annotated bibliography to the monograph by J.F. Traub and H. Woźniakowski [80] are not mentioned for the lack of space.

A few words about our system for referring to material within the text. Within every chapter, theorems, lemmas, and formulas have double numbers; the first number is that of the section, and the second one is that of the theorem, lemma, or formula itself. A reference to material within the same chapter contains just these two numbers; a reference to material in a different chapter names the chapter, for example: Theorem 5.3 of Chapter 1, formula (3.4) of Chapter 2.

The author is very grateful to the faculty of the Operations Research Department of Moscow State University. The creative atmosphere of the department was of great importance for preparing the book.

GLOSSARY OF SYMBOLS

We only list here some notations of the general character, without mentioning commonly used notations and special notations introduced in the book. The latter are either used right after they have been introduced or supplied with the necessary references.

$\stackrel{def}{=}$ – equal by definition (the quantity being defined may occur both on the left- and right-hand sides of the formula)

$\Rightarrow$ – implication (... implies ...)

$\Leftrightarrow$ – equivalence

$A \subset B$ – the set B contains the set A (this does not rule out the case $A=B$)

$\lfloor t \rfloor$ – entire part of t, i.e., the greatest integer that is not greater than t

$\lceil t \rceil$ – the least integer that is not less than t

$x := a$ – operator assigning the value a to the variable x

$\square$ – end of the proof, end of the statement (if it follows from the previous considerations or is given without the proof), and also end of the remark, example, or definition, etc., with a special heading

$\mathbb{R}$ – numerical line (the set of real numbers)

$\mathbb{R}^n$ – n-dimensional coordinate space

$\langle a, b\rangle = \sum_{j=1}^{N} a^j b^j$ – inner product of the vectors $a = (a^1, \ldots, a^N)$ and $b = (b^1, \ldots, b^N)$

$A + B = \{c \mid c = a + b,\ a \in A,\ b \in B\}$

$\|a\|_0 = \max_{i=1,\ldots,n} |a^i|$

$\|a\|_1 = \sum_{i=1}^{n} |a^i|$

$\|a\|_2 = \sqrt{\sum_{i=1}^{n} (a^i)^2}$

$\lim_{\delta \to a+} \phi(\delta)$, $\lim_{\delta \to a-} \phi(\delta)$ – right-hand and left-hand limits of the function ϕ of a numerical variable at the point a

$x_* = \arg\max_{x \in K} f(x)$ – any point of the global maximum of the function f on the set K: $f(x_*) = \max_{x \in K} f(x)$, $x_* \in K$

$\operatorname{Arg}\max_{x \in K} f(x)$ – set of all the points of the global maximum of f on K

CHAPTER 1

GENERAL COMPUTATION MODEL

In this chapter, we discuss the main elements of all the subsequent constructions: functional or operator to be approximated corresponding to the problem being solved; class of algorithms that can be used to solve the problem; criterion for estimating efficiency of algorithms; general concept of optimality; and, finally, specific notion of algorithm's optimality in the framework of the adopted setting. Combination of all these elements forms the so-called *computation model.* We obtain some general results which enable us to solve certain principle methodologial problems and are frequently used in the further constructions.

1. BASIC CONCEPTS

In this section, we briefly describe some elements of the general computation model. We also show how some specific problems of numerical analysis fit into this setting. In the subsequent sections of this chapter, we give a complete description of the general model and provide the necessary concretization of the introduced concepts.

1.1. Scheme of the general computation model

Let F be a set in a linear space over the field of real numbers, and let S be a mapping from F into some metric space B with a metric γ. "The computer" has to solve the problem of constructing the "best approximation" to the element $S(f)\in B$, i.e., an approximation $\alpha(f)\in B$ such that the quantity

$$\varepsilon(\alpha, f)=\gamma(S(f), \alpha(f)) \tag{1.1}$$

is "minimal" (precise definitions of the terms in quotes are given below).

From now on, *"the computer"* is the person who constructs the computation model, organizes the computational process in the framework of this model, and analyzes the results.

The quantity $\varepsilon(\alpha, f)$ is called the *error* (or *accuracy*) of the solution. In what follows, it plays the role of a *criterion for estimating efficiency of algorithms* (we will also discuss other criteria of efficiency, different from (1.1)).

A priori information "the computer" has on the element f (that is, the information obtained at the stage of preliminary investigation before any computations) amounts to the fact that $f \in F$. In the process of solving the problem with the help of an *algorithm* α "the computer" gets additional information and then constructs an approximation $\alpha(f)$ on the basis of this information (see Section 3 for more details).

1.2. Examples

Before we proceed to a more precise description of the general computation model, we illustrate it with some concrete examples.

Let F be a class of real scalar or vector functions defined on a set K of the real Euclidean space $\mathbb{R}^n$. In what follows, we mainly deal with this case.

Numerical integration problem (the problem of computing an integral) in our general setting corresponds to

$$S(f) = \int_K f(x)dx, \quad B = \mathbb{R}. \tag{1.2}$$

Recovery problem (the problem of recovering a function f) corresponds to

$$S(f) = f, \quad B \subset \{\beta \mid \beta : K \to \mathbb{R}\}. \tag{1.3}$$

Global optimization problem (the problem of computing the maximum or supremum of a scalar function f on the set K) corresponds to

$$S(f) = \sup_{x \in K} f(x)dx, \quad B = \mathbb{R}, \tag{1.4}$$

or

$$S(f) = x_*, \quad B = \mathbb{R}^n,$$

where $x_* \in K$ and $f(x_*) = \max_{x \in K} f(x)$ (here we assume that any function $f \in F$ attains its maximum at a unique point).

Finally, the *problem of solving a system of equations*

$$f(x) = 0 \tag{1.5}$$

corresponds to

$$S(f) = x_*, \quad B = \mathbb{R}^n, \tag{1.6}$$

where x_* is the solution of the system. Here we assume that for any n-dimensional vector function $f \in F$ the system (1.5) has a unique solution.

In the future we will pay maximum attention to the problems (1.2)-(1.4) and to the problem of solving the system (1.5) in a setting slightly different from (1.6). However, the definitions and constructions from Chapter 1 also apply to other problems of numerical analysis.

Note that, by tradition, we use the term "problem" to refer to different objects. Thus, we call problem the mapping S (we say "optimization problem", "integration problem", etc.), the combination of S and the class F, and also the concrete element $f \in F$ (in the latter case we imply some specific S).

1.3. *A priori* information on f

We now discuss in more detail some methodological points connected to whether it makes sense to fix a specific functional class F in the computation model. As a rule, "the computer" faces one problem or a series of some concrete problems (here we mean that S and B are fixed and the term "problem" refers to a particular function f). On the face of it, this fully determines the choice of the functional class F: it should consist either of the only function f or of those very functions being considered.

However, this way of defining F would not help us to choose the computational algorithm. Even if "the computer" deals with a unique function f, his choice of the algorithm will be based on some essential and important for him properties of the function, such as continuity, smoothness, monotonicity, convexity, unimodality (having a unique extremum), bounded "rate of change" of the function or its derivatives, etc.

This sort of properties can be detected when studying physical nature of the phenomenon at the stage of construction of its mathematical model. We may know (or compute in the process of preliminary investigation) exact or approximate values of some functional characteristics. We may have a good initial approximation enabling us to obtain the solution through some iterative process. This sort of information allows "the computer" to describe the function or the set of functions in terms of belonging to a specific functional class.

Thus, choice of the functional class is a result of preliminary investigation - the stage at which those characteristics of the problem are detected that will influence our selection of the algorithm for solving the problem.

A more careful preliminary investigation would probably allow us to include the function into a "smaller" class. There is always a trade-off between giving a more complete description of the problem and choosing the algorithm for its solution on the basis of more *a priori* information (thus complicating the choice), i.e., selecting a rather "small" class, on the one hand, and selecting a class for which an efficient computational algorithm could be constructed within acceptable time, on the other hand. Qualification, experience and intuition help to work out a reasonable compromise in every case.

We stress once again that, after the class F has been fixed, the information on the problem to be used when choosing the algorithm for its solution amounts to the fact that $f \in F$.

2. FUNCTIONAL CLASSES UNDER CONSIDERATION

In this section, we give examples of functional classes which are important for both theory and applications and adequately reflect *a priori* information on the problem typical for many situations. These are functional classes determined by quasi-metrics. They are used in many concrete computation models presented in the book. Here we discuss why functional classes determined by quasi-metrics are useful and study their essential properties.

2.1. Formal and informal description of *a priori* information typical for many problems

In practice of computations we encounter many problems with functions of rather complex structure. Every computation of some characteristic of the function (say, its evaluation) may require an expensive physical or computer experiment. Numerous examples of this type are given by computer aided design (see Krasnoshchekov [84], Krasnoshchekov, Morozov and Fedorov [79a, 79b], Krasnoshchekov and Petrov [83], Krasnoshchokov, Petrov and Fedorov [86]) where one evaluation of the (scalar or vector) function describing the system being designed requires full calculation of the corresponding variant of the system. To describe functioning of a complex technological system, we need a whole set of programs. Thus, one function evaluation requires complex and expensive computations. More or less exact computation of the functions' derivatives is often impossible, they even may not exist. Moreover, *a priori* information on such functions is usually very scant.

However, analysis of the real system's properties, as a rule, enables us to justify the assumption that the "rate of change" of the function is bounded and gives numerical estimation of these bounds.

A specific feature of the functions that describe working characteristics of technological objects is their multi-extremality. This property reflects the possibility of essentially different constructive solutions (different designs of the object), each of which can be brought to a local optimum of efficiency by means of "finishing", i.e., slight improvement of construction parameters.

The above informal interpretation of *a priori* information on the function f admits the following formal and rather general representation:

$$|f(u)-f(v)|\le\rho_0(u,v),\quad u,v\in K, \tag{2.1}$$

where ρ_0 is some fixed nonnegative on $K^2=K\times K$ function that characterizes bounds on the "rate of change" of f.

Denote by F_{ρ_0} the class of all functions f satisfying (2.1). It turns out that, even by considering such a general functional class, we arrive at some significant results for various problems of numerical analysis. Moreover, its different realizations can reflect specialities of many real-life problems.

2.2. Representation of a given *a priori* information with the help of a functional class determined by a quasi-metric

We show that, without loss of generality, we can assume that the function on the right-hand side of (2.1) enjoys some additional properties, namely those of a quasi-metric.

A function ρ defined on K^2 is called a *quasi-metric* (has the properties of a quasi-metric) iff

$$\rho(u,u)=0, \tag{2.2}$$

$$\rho(u,v)+\rho(w,v)\ge\rho(u,w), \tag{2.3}$$

$$\rho(u,w)=\rho(w,u), \tag{2.4}$$

$$\rho(u,v)\ge 0 \tag{2.5}$$

for any $u,v,w\in K$ (see Collatz [64]).

Remark. As we know, symmetry (2.4) and positive semi-definiteness (2.5) follow from the properties (2.2) and (2.3). Indeed, from (2.3) for $u=v$ and (2.2) we derive $\rho(w,u)\ge\rho(u,w)$. Since u and w are arbitrary, the opposite inequality is also valid, which yields equation (2.4). Inequality (2.5) follows from (2.2) and (2.3) if we put $u=w$. □

Theorem 2.1 (Sukharev [75]). *Let a functional class F_{ρ_0} be determined by some nonnegative on K^2 function ρ_0 in accordance with* (2.1). *Then there is a quasi-metric ρ defined on K^2 such that $F_\rho=F_{\rho_0}$, where F_ρ is the class of all functions f satisfying the inequality*

$$|f(u)-f(v)|\le\rho(u,v),\quad u,v\in K. \tag{2.6}$$

Proof. Define on K^2 a sequence of functions $\{\rho_i\}_{i=1}^{\infty}$ and a function ρ in the following way:

$$\rho_1(u,u)=0,$$
$$\rho_1(u,v)=\min\{\rho_0(u,v),\rho_0(v,u)\} \quad \text{for} \quad u\neq v,$$
$$\rho_{i+1}(u,w)=\inf_{w\in K}[\rho_i(u,w)+\rho_i(w,u)], \quad i\geq 1,$$
$$\rho(u,v)=\inf_{i\in(0,1,2,\dots)}\rho_i(u,v).$$

We start with proving that ρ is a quasi-metric. It is easily shown by induction that the functions ρ_i for $i\geq 1$ have the properties (2.2), (2.4), and (2.5). Therefore, ρ enjoys the same properties.

Now we show that the sequence $\{\rho_i(u,v)\}_{i=0}^{\infty}$ is monotonic for all $u,v\in K$. We have $\rho_1(u,v)\leq\rho_0(u,v)$ and

$$\rho_{i+1}(u,v)=\inf_{w\in K}[\rho_i(u,w)+\rho_i(w,v)]$$
$$\leq\rho_i(u,u)+\rho_i(u,v)=\rho_i(u,v), \quad i\geq 1.$$

This implies that for any $\varepsilon>0$ there is a number m such that

$$\rho(u,v)+\rho(w,v)\geq\rho_m(u,v)+\rho_m(w,v)-\varepsilon$$
$$\geq\inf_{v\in K}[\rho_m(u,v)+\rho_m(v,w)]-\varepsilon$$
$$=\rho_{m+1}(u,w)-\varepsilon\geq\rho(u,w)-\varepsilon.$$

Since ε is arbitrary, this yields (2.3).

Finally, we have to demonstrate that $F_\rho=F_{\rho_0}$. Since $\rho\leq\rho_0$, (2.6) implies (2.1), i.e., $F_\rho\subset F_{\rho_0}$. We now prove the opposite inclusion. Let f satisfy (2.1). Then it is obvious that

$$|f(u)-f(v)|\leq\rho_1(u,v), \quad u,v\in K.$$

Having assumed that

$$|f(u)-f(v)|\leq\rho_i(u,v), \quad u,v\in K, \tag{2.7}$$

we obtain

$$|f(u)-f(v)|\leq|f(u)-f(w)|+|f(w)-f(v)|\leq\rho_i(u,w)+\rho_i(w,v), \quad w\in K.$$

Therefore,

$$|f(u)-f(v)|\leq\inf_{w\in K}[\rho_i(u,w)+\rho_i(w,v)]=\rho_{i+1}(u,v).$$

Thus, inequality (2.7) has been proved by induction for all $i\geq 0$. Taking infimum with respect to i, we get (2.6). Hence, $F_{\rho_0}\subset F_\rho$, and, therefore, $F_{\rho_0}=F_\rho$, which completes the proof. □

2.3. Properties of functional classes determined by quasi-metrics

We start with a fact which, simple as it is, will play an important role in the subsequent constructions.

Lemma 2.1. *Let ρ be an arbitrary quasi-metric, and let $h_1(x) = -\rho(a, x)$, $a \in K$. Then $\pm h_a \in F_\rho$.*

Proof. Making use of the properties (2.3) and (2.4), we easily obtain

$$|h_a(u) - h_a(v)| = |\rho(a, u) - \rho(a, v)| \leq \rho(u, v),$$

which proves the lemma. □

Lemma 2.1 yields the following statement.

Lemma 2.2. *Let ρ be an arbitrary quasi-metric. Then*

$$\sup_{f \in F_\rho} |f(u) - f(v)| = \rho(u, v).$$

Proof. Apparently, we have

$$\sup_{f \in F_\rho} |f(u) - f(v)| \leq \rho(u, v).$$

On the other hand,

$$\sup_{f \in F_\rho} |f(u) - f(v)| \geq |h_u(u) - h_u(v)| = \rho(u, v),$$

which competes the proof. □

Corollary. *The class F_ρ can not be defined by an inequality*

$$|f(u) - f(v)| \leq \rho'(u, v), \quad u, v \in K,$$

where $\rho'(u, v) < \rho(u, v)$ for at least one pair $(u, v) \in K^2$. □

Using different quasi-metrics, we can define many functional classes which are important for both theory and applications. Before demonstrating this, we establish the following useful fact.

Lemma 2.3. *Let functions ρ_i have the properties of quasi-metrics, and let $k_i \geq 0$, $i = 1, \ldots, s$. Then the functions $\max_{i=1,\ldots,s} (k_i \rho_i)$ and $\sum_{i=1}^{s} k_i \rho_i$ also have the properties of quasi-metrics.*

Proof. We confine ourselves to verifying property (2.3) for the function $\max_{i=1,\ldots,s} (k_i \rho_i)$. We have

$$\max_{i=1,\ldots,s} (k_i \rho_i(u, v)) + \max_{i=1,\ldots,s} (k_i \rho_i(w, v))$$
$$\geq \max_{i=1,\ldots,s} (k_i \rho_i(u, v) + k_i \rho_i(w, v)) \geq \max_{i=1,\ldots,s} (k_i \rho_i(u, w)),$$

thus $\max_{i=1,\ldots,s} k_i \rho_i$ satisfies (2.3). □

Lemma 2.3 makes it easy to check that the functions

$$\rho(u,v)=M\max_{i=1,\dots,n}\{k_i|u^i-v^i|^{\alpha_i}\}, \tag{2.8}$$

$$\rho(u,v)=M\sum_{i=1}^{n}k_i|u^i-v^i|^{\alpha_i} \tag{2.9}$$

have the properties of quasi-metrics for $M\geq 0$, $k_i\geq 0$, $0<\alpha_i\leq 1$, $i=1,\dots,n$; $u=(u^1,\dots,u^n)$, $v=(v^1,\dots,v^n)$. The function

$$\rho(u,v)=M\left(\sum_{i=1}^{n}(u^i-v^i)^2\right)^{1/2} \tag{2.10}$$

also hasthese properties.

With the help of the quasi-metrics (2.8)-(2.10) we can define classes of functions satisfying different variants of the Lipschitz and Hölder conditions. The following lemma establishes this possibility for a variant of the Lipschitz condition for which it may not be quite obvious.

Lemma 2.4. *The class of functions defined on a coordinate parallelepiped*

$$K=\{u=(u^1,\dots,u^n)\mid a^i\leq u^i\leq b^i,\quad i=1,\dots,n\}$$

and satisfying the Lipschitz condition with a constant k_i with respect to the ith variable, $i=1,\dots,n$, coincides with the class F_ρ determined by the quasi-metric (2.9) *with $M=1$, $\alpha_1=\alpha_2=\dots=\alpha_n=1$.*

Proof. Let $f\in F_\rho$, i.e.,

$$|f(u)-f(v)|\leq\sum_{i=1}^{n}k_i|u^i-v^i|,\quad u,v\in K. \tag{2.11}$$

Then for any $u^i,v^i\in[a^i,b^i]$, $j\neq i$, we have

$$|f(w^1,\dots,w^{i-1},u^i,w^{i+1},\dots,w^n)-f(w^1,\dots,w^{i-1},v^i,w^{i+1},\dots,w^n)|$$
$$\leq k_i|u^i-v^i|, \tag{2.12}$$

that is, f satisfies the Lipschitz condition with the constant k_i with respect to the ith variable, $i=1,\dots,n$.

On the other hand, if we assume that (2.12) holds for $i=1,\dots,n$, it is easy to see that

$$\begin{aligned}|f(u)-f(v)|\leq&|f(u^1,u^2,\dots u^{n-1},u^n)-f(v^1,u^2,\dots u^{n-1},u^n)|\\&+|f(v^1,u^2,\dots u^{n-1},u^n)-f(v^1,v^2,\dots u^{n-1},u^n)|\\&+\dots\dots\dots\dots\\&+|f(v^1,v^2,\dots v^{n-1},u^n)-f(v^1,v^2,\dots v^{n-1},v^n)|\\\leq&\sum_{i=1}^{n}k_i|u^i-v^i|,\end{aligned}$$

i.e., (2.11) holds and, therefore, $f \in F_\rho$. □

Assume that a function $\omega(\delta)$ is defined for $\delta \geq 0$ and has the properties of a *modulus of continuity*, i.e., ω is continuous, nondecreasing, $\omega(0)=0$, and $\omega(\delta_1+\delta_2) \leq \omega(\delta_1)+\omega(\delta_2)$. In this case we say that the functional class

$$F_\omega = \{f \mid |f(u)-f(v)| \leq \omega(\|u-v\|), \quad u,v \in K\} \tag{2.13}$$

is determined by the modulus of continuity ω (here $\|\cdot\|$ is some norm in n-dimensional coordinate space). Clearly, all the functions from F_ω are uniformly continuous on K and their moduli of continuity are not greater than ω. Functional classes determined by moduli of continuity are traditional objects of the theory of optimal quadratures, see Babenko [76a, b, 77], Korneichuk [68], Maung Cho Niun and Sharygin [71], Nikolskii [79].

We show that these classes are special cases of functional classes determined by quasi-metrics.

Lemma 2.5. *Let a function $\omega(\delta)$ be defined for all $\delta \geq 0$ and nondecreasing, $\omega(0)=0$, $\omega(\delta_1+\delta_2) \leq \omega(\delta_1)+\omega(\delta_2)$, and let ρ be a quasi-metric. Then the function $\omega(\rho)$ is also a quasi-metric.*

Proof. We have

$$\omega(\rho(u,v))+\omega(\rho(w,v)) \geq \omega(\rho(u,v)+\rho(w,v)) \geq \omega(\rho(u,v)),$$

thus $\omega(\rho)$ satisfies (2.3). The properties (2.2), (2.4), and (2.5) are trivial. □

Since the function $\rho(u,v)=\|u-v\|$, naturally, is a quasi-metric, Lemma 2.5 yields that classes (2.13) are special cases of classes determined by quasi-metrics.

Apparently, functions $\omega(\|u-v\|)$ determining classes (2.13) are invariant with respect to translations as functions of u and v. In general, an arbitrary quasi-metric ρ does not have this property, i.e., the equation $\rho(u+w, v+w)=\rho(u,v)$ may not be valid. This is essential and allows us to define with the help of quasi-metrics important functional classes which cannot be defined using moduli of continuity. Examples are provided by classes of functions satisfying the Lipschitz or Hölder condition with different constants on different subsets of their domain and classes determined by different moduli of continuity on different subsets of their domain. In some cases the corresponding quasi-metric can easily be written explicitly.

We illustrate this with a concrete example.

Lemma 2.6. *Let $K=[0,1]$, $0=a_0<a_1<a_2<\cdots<a_s<a_{s+1}=1$, $k_1 \geq 0, \ldots, k_{s+1} \geq 0$. Then the class of functions f satisfying the Lipschitz condition with the constant k_i on $[a_{i-1}, a_i]$*

$$|f(u)-f(v)| \leq k_i|u-v|, \quad u,v \in [a_{i-1}, a_i], \quad i=1,\ldots,s+1, \tag{2.14}$$

coincides with the class F_ρ determined by the quasi-metric

$$\rho(u,v)=\rho(v,u)=\begin{cases} k_i|u-v| & \text{if } u,v\in[a_{i-1},a_i],\\ k_i(a_i-u)+\sum\limits_{m=i+1}^{j-1} k_m(a_m-a_{m-1})+k_j(v-a_{j-1}) & \\ \qquad \text{if } u\in[a_{i-1},a_i],\ v\in[a_{j-1},a_j],\ i<j. & \end{cases}$$

Proof. It is easily verified that ρ is a quasi-metric. If $f\in F_\rho$, inequalities (2.14) obviously hold.

Assume now that f satisfies inequalities (2.14), and let $u\in[a_{i-1},a_i]$, $v\in[a_{j-1},a_j]$, $i<j$. Then we have

$$|f(u)-f(v)|\leq|f(u)-f(a_i)|+\sum_{m=i+1}^{j-1}|f(a_m)-f(a_{m-1})|$$
$$+|f(v)-f(a_{j-1})|\leq\rho(u,v),$$

that is, $f\in F_\rho$, which completes the proof. □

Another factor that makes it possible to define a wide range of functional classes with the help of quasi-metrics ρ is that the definition of a quasi-metric requires just positive semi-definiteness of ρ rather than positive definiteness.

Let a functional class be determined by the conditions

$$\begin{aligned} &|f(u)-f(v)|\leq\rho_0(u,v), \quad u,v\in K,\\ &|f(u)-f(v)|\leq\alpha_i, \quad \alpha_i\geq 0,\ i\in I. \end{aligned} \tag{2.15}$$

As we know from Theorem 2.1, such a class can be represented in the form of F_ρ, where ρ is some quasi-metric. Moreover, if the function ρ_0 itself is a quasi-metric and the set I is finite, there is an easy constructive way of obtaining ρ. With the help of the second group of conditions in (2.15) we can also specify equality of values of functions from this class at some fixed pairs of points. For this we have to put $\alpha_i=0$. A similar trick makes it possible to specify the fact that the functions should be periodical.

Observe that the class F_ρ contains discontinuous functions if ρ is discontinuous. For instance, if

$$\rho(u,v)=\begin{cases} 0, & u=v,\\ 1, & u\neq v,\end{cases}$$

then, by Lemma 2.1, the class F_ρ contains the discontinuous function $h_a(x)=-\rho(a,x)$.

We have demonstrated that the possibilities of defining functional classes with the help of quasi-metrics are rather wide. Investigation of such classes in the theory of computational methods in this general setting was started in the paper by Sukharev [71]. In some cases it makes sense to consider even less restrictive requirements to the function ρ than (2.2)-(2.5). For instance, Ganshin [76] has extended some results by Sukharev [71] to the case of nonsymmetric functions ρ.

Note that recently much attention has been paid to optimization problems for classes of functions satisfying different forms of the Lipschitz condition (as was mentioned above, such classes are special cases of functional classes determined by quasi-metrics). For functions satisfying the Lipschitz condition various generalizations of the concept of gradient have been introduced (Clarke [83], Dixon [80]), analogues of Lagrange's method of undetermined multipliers have been obtained (Pourciau [77]), numerous computational optimization methods have been constructed (Babii [78], Dixon and Gaviano [80], Danilin and Piyavskii [67], Evtushenko [71], Goldstein [77], Gupal [79], Lbov and Grunov [76], Leonov [70], Piyavskii [67, 72], Podobedov [87], Shubert [72a, b], Sukharev [71, 72, 75, 81a, c], Timonov [77], Ust'uzhaninov [80c]), problems of extension and the best approximation have been studied (Mustăta [77]).

Later we will need the following two lemmas.

Lemma 2.7. *A functional class F_ρ determined by an arbitrary quasi-metric ρ is closed with respect to computations of maximum and minimum.*

Proof. Let $f_1, f_2 \in F_\rho$. Then, for any $u, v \in K$,

$$|\max\{f_1(u), f_2(u)\}| - \max\{f_1(v), f_2(v)\}|$$
$$\leq \max\{|f_1(u) - f_1(v)|, |f_2(u) - f_2(v)|\}| \leq \rho(u, v),$$

i.e., $\max\{f_1, f_2\} \in F_\rho$. In the same way it can be shown that $\min\{f_1, f_2\} \in F_\rho$. □

Lemma 2.8. *A functional class F_ρ determined by an arbitrary quasi-metric ρ is a convex balanced set in the linear space of functions defined on K.*

Proof. Let $f_1, f_2 \in F_\rho$, let $\lambda \in [0, 1]$, and $u, v \in K$. Then

$$|\lambda f_1(u) + (1-\lambda) f_2(u) - \lambda f_1(v) - (1-\lambda) f_2(v)|$$
$$\leq \lambda |f_1(u) - f_1(v)| + (1-\lambda)|f_2(u) - f_2(v)| \leq \rho(u, v),$$

which yields convexity of F_ρ. Verification of the fact that F_ρ is balanced is trivial. □

2.4. Other functional classes

Alongside the classes we have already mentioned, we will deal with other functional classes, for instance, with classes $W_\infty^r([a, b], M)$ consisting of functions f whose $(r-1)$st derivative is absolutely continuous on $[a, b]$ and the L_∞-norm of the rth derivative is bounded by a constant M. For example,

$$W_\infty^1([a, b], M) = \{f \mid |f(u) - f(v)| \leq M|u - v|, \ u, v \in [a, b]\}, \tag{2.16}$$

i.e., the class W^1_∞ coincides with the class of functions satisfying the Lipschitz condition with the constant M on $[a,b]$. We will also work with functional classes of a completely different nature.

3. CLASSES OF DETERMINISTIC ALGORITHMS

In this section, we introduce the general concept of deterministic algorithm for solving the problem of approximation of $S(f)$. We show how special characteristics of the problem, the type of the process of gathering information, and the resources available for its storage and processing determine the class of permissible algorithms.

3.1. Concept of deterministic algorithm

We start with defining deterministic algorithms (or methods, or strategies) for solving the problem of approximation of $S(f)$. Any algorithm can be naturally divided into two stages.

At the *first stage*, we gather information on the specific problem f to be solved. For this we compute values $y_i \in Y_i$ of functionals or operators $x_i \in X_i$, where X_i is a given set of mappings (functionals or operators) defined on F and taking values in a given set Y_i, $i=1,\ldots,N$. Here N is an *a priori* fixed number determined by the computational resources available (later we will also consider the case of N being determined in the process of computations).

At the *second stage*, on the basis of the information gathered at the first stage, we construct an approximation $\beta=\alpha(f)\in B$ to the element $S(t)\in B$.

Formally, a *deterministic algorithm* α is defined as a set of mappings

$$\alpha=(\tilde{x}_1,\ldots,\tilde{x}_N,\tilde{\beta}),$$

where

$$\begin{aligned}
&\tilde{x}_1\equiv x_1: F\to Y_1, \quad x_1\in X_1,\\
&\tilde{x}_2: X_1\times Y_1\to X_2\ni x_2: F\to Y_2,\\
&\tilde{x}_3: X_1\times X_2\times Y_1\times Y_2\to X_3\ni x_3: F\to Y_3,\\
&\cdots\cdots\cdots\cdots\cdots\cdots\cdots\cdots\cdots\cdots \\
&\tilde{x}_N: X_1\times\cdots\times X_{N-1}\times Y_1\times\cdots\times Y_{N-1}\to X_N\ni x_N: F\to Y_N,\\
&\tilde{\beta}: X_1\times\cdots\times X_N\times Y_1\times\cdots\times Y_N\to B.
\end{aligned} \tag{3.1}$$

Alongside α, we also apply the term "algorithm" to the mapping

$$\tilde{x}^i \overset{def}{=} (\tilde{x}_1,\ldots,\tilde{x}_i),$$

corresponding to the first i steps of α, $i \leq N$.

The solution process consists in successive computations of

$$\begin{aligned}
&y_1 = x_1(f),\\
&x_2 = \widetilde{x}_2(x_1, y_1), \quad y_2 = x_2(f),\\
&\dots\dots\dots\dots\dots\dots\dots\dots\\
&x_N = \widetilde{x}_N(x_1, \dots, x_{N-1}, y_1, \dots, y_{N-1}), \quad y_N = x_N(f),\\
&\beta = \widetilde{\beta}(x_1, \dots, x_N, y_1, \dots, y_N).
\end{aligned} \tag{3.2}$$

Then

$$\alpha(f) \stackrel{def}{=} \beta.$$

The computations of $x_i(f)$ are called *informational computations*, $i = 1, \dots, N$. If no explicit assumptions have been done about accuracy of informational computations, they are tacitly assumed to be exact. The computations of $\widetilde{x}_i(x_1, \dots, x_{i-1}, y_1, \dots, y_{i-1})$ are called *algorithmic computations*, $i = 2, \dots, N$. The computation of $\widetilde{\beta}(x_1, \dots, x_N, y_1, \dots, y_N)$ is called *terminal computation* or *terminal operation* of the algorithm[1)]. So the first stage of an algorithm consists in informational and algorithmic computations, and at the second stage the terminal computation is performed. Informational computations are also called tests, experiments, measurements, etc. In what follows, we will often use the terms "step of the algorithm" and "step of the computational process" for the combination of an algorithmic computation and the corresponding informational computation. Thus, at the ith step of the algorithm $x_i = \widetilde{x}_i(x_1, \dots, x_{i-1}, y_1, \dots, y_{i-1})$ and $y_i = x_1(f)$ are computed.

3.2. Some notations and terminology

Now we introduce into consideration some concepts and notations which will be frequently used. For $x_j \in X_j$, $y_j \in Y_j$, $j = 1, \dots, i$, $i \leq N$, denote

$$x^i = (x_1, \dots, x_i), \quad y^i = (y_1, \dots, y_i), \quad z^i = (x_i, y_i),$$
$$F(z^i) = \{f \in F \mid x_j(f) = y_j, \ j = 1, \dots, i\}.$$

All the information on the problem f "the computer" has after the ith step of the computational process amounts to the fact that $f \in F(z^i)$.

The vector z^i is called the *situation after i steps* of the computational process. We say that a situation z^i is *realizable* iff $F(z^i) \neq \emptyset$. A situation $z^i = (x^i, y^i)$ is said to be *realizable for the algorithm α (for the algorithm $\widetilde{x}^i =$*

1) Note that in Traub, Wasilkowski and Woźniakowski [83] and Traub and Woźniakowski [80] the term "algorithm" is used for the computation of $\widetilde{\beta}(x_1, \dots, x_N, y_1, \dots, y_N)$ only.

$(\widetilde{x}_1, \ldots, \widetilde{x}_i))$ iff its components $x_1, y_1, \ldots, x_i, y_i$ can be obtained by applying the first i formulas of (3.2) to some function $f \in F$, $i \leq N$. In this case the vector of the results is also called *realizable for the algorithm* α *(for the algorithm* $\widetilde{x}^i$*)*.

Note that, for simplicity, the domains of the mappings $\widetilde{x}_i$ and $\widetilde{\beta}$ in (3.1) are not minimal possible regarding organization of the computational process. It is easy to see that it would be enough to define the mapping $\widetilde{x}_i$ on the set of situations z^{i-1} realizable for the algorithm $\widetilde{x}^{i-1}$, and the mapping $\widetilde{\beta}$ on the set of situations z^N realizable for the algorithm $\widetilde{x}^N$. When it suits the case, we will define these mappings on "smaller" sets than in (3.1).

3.3. Examples

Now we give some examples of the mappings x_i that appear in (3.1) and (3.2). As a rule, the mapping x_i corresponds to computation of certain characteristics of the function f at the point $x_i \in K$, say, the function value:

$$x_i(f) = f(x_i), \tag{3.3}$$

or the values of the function and its derivatives upto some order:

$$x_i(f) = (f(x_i, f'(x_i), \ldots, f^{(r_i)}(x_i)),$$

or the subdifferential (for convex functions):

$$x_i(f) = \partial f(x_i).$$

We can also consider other cases; for instance, $x_i(f)$ may be the ith coefficient in the expansion of f according to a given system of functions, or

$$x_i(f) = \int_{K_i} f(x)dx, \quad K_i \subset K,$$

or

$$x_i(f) = \max_{x \in K_i} f(x), \quad K_i \subset K.$$

Observe that in the latter case the functional x_i is nonlinear. The case of nonlinear information was studied by Micchelli and Miranker [75] and Traub and Woźniakowski [80]. However, the assumption that the information is linear looks most natural.

The set X_i is defined in accordance with the definition of the mapping x_i. Say, in the case (3.3) we can regard

$$X_1 = X_2 = \cdots = X_N = K$$

as the set of all functionals of the form (3.3) corresponding to the points from the domain K of functions $f \in F$.

Now we proceed to describe sets (classes) of ***permissible*** algorithms. Choice of the class A of permissible algorithms for solving the problem is one of the key points of constructing a computation model. This choice depends on characteristics of the problem, the type of the process of gathering information, and the resources available for its storage and processing.

3.4. Permissible terminal operations

We start with describing the set $\hat{B}_N$ of permissible terminal operations. First of all, $\hat{B}_N$ is a subset of the set $\widetilde{B}_N$ containing all terminal operations of the form (3.1):

$$\hat{B}_N \subset \widetilde{B}_N \stackrel{def}{=} \{\widetilde{\beta} \mid \widetilde{\beta}: X_1 \times \cdots \times X_N \times Y_1 \times \cdots \times Y_n \to B\}.$$

We single out two standard situations:

a) there is only one fixed permissible $\widetilde{\beta}$, i.e.,

$$\hat{B}_N = \{\widetilde{\beta}\}, \quad \widetilde{\beta} \in \widetilde{B}_N; \tag{3.4}$$

b) any terminal operation $\widetilde{\beta}$ of the form (3.1) is permitted, i.e.,

$$\hat{B}_N = B_N. \tag{3.5}$$

The first situation occurs, for instance, in the maximization problem if alongside an approximation to the maximum we have to find the point where the function f takes this value. In this case, having obtained information $y_1 = f(x_1), \ldots, y_N = f(x_N)$, "the computer" will choose $f(x_{i_0}) = \max\limits_{i=1,\ldots,N} f(x_i)$, $1 \le i_0 \le N$, as an approximation to the maximum and x_0 as the maximizing point. Thus, there is only one permissible terminal operation

$$\widetilde{\beta}(z^N) = \max_{i=1,\ldots,N} y_i. \tag{3.6}$$

The second situation occurs, for example, in the problems of integration and optimal recovery of functions we deal with in Chapters 2 and 3.

Intermediate situations are also possible such as

$$\hat{B}_N = \left\{\widetilde{\beta} \mid \widetilde{\beta}(x_1, \ldots, x_N, y_1, \ldots, y_N) = p_0(x_1, \ldots, x_N) + \sum_{i=1}^{N} p_i(x_1, \ldots, x_N) y_i \right\}, \tag{3.7}$$

i.e., only terminal operations linear in $y^N = (y_1, \ldots, y_N)$ are permitted and $Y_1 = \cdots = Y_N = B = \mathbb{R}$.

3.5. Permissible algorithmic computations

Consider now the question of permissible $\tilde{x}^N=(\tilde{x}_1,\dots,\tilde{x}_N)$. Here two extreme cases are possible.

The first case corresponds to the situation where no informational computation can use any information on f obtained through the previous informational computations. This situation occurs, for instance, if "the computer" has either no possibility to store the information in RAM memory or no time for processing the information, or if all the computations have to be performed simultaneously. The latter reason may be caused, for example, by a computer system allowing N simultaneous computations, or by physical or economic necessity to carry out all the computations at the same time, etc.

In this case we have

$$\tilde{x}_i\equiv x_i\in X_i, \quad i=1,\dots,N.$$

Algorithms (or strategies) $(x_1,\dots,x_N)$ are called *nonadaptive*. Denote the set of all such algorithms by

$$X^N \stackrel{def}{=} \prod_{i=1}^{N} X_i \tag{3.8}$$

(here and throughout the book $\prod$ means Cartesian product of sets). The term "nonadaptive" also applies to the algorithms

$$\alpha=(x_1,\dots,x_N,\tilde{\beta})\in A^N \stackrel{def}{=} X^N\times\tilde{B}_N. \tag{3.9}$$

The second extreme case corresponds to the situation where before every informational computation "the computer" knows the results of all the previous informational computations and also has enough memory to store these results and enough time for their processing, i.e., for performing the required algorithmic computation.

In this case, all algorithms composed of mappings of the form (3.1) are permitted. The set of all such algorithms is called the *class of all adaptive algorithms*, and its elements are called *adaptive algorithms*. Denote the sets of all mappings $\tilde{x}_1=x_1,\tilde{x}_2,\dots,\tilde{x}_N$ of the form (3.1) by $\tilde{X}_1=X_1,\tilde{X}_2,\dots,\tilde{X}_N$ respectively, and denote the class of all adaptive algorithms by

$$\tilde{X}^N \stackrel{def}{=} \prod_{i=1}^{N} \tilde{X}_i. \tag{3.10}$$

The term "adaptive" also applies to the algorithms

$$\alpha=(\tilde{x}_1,\dots,\tilde{x}_N,\tilde{\beta})\in\tilde{A}^N \stackrel{def}{=} \tilde{X}^N\times\tilde{B}_N. \tag{3.11}$$

Since nonadaptive algorithms correspond to constant functions from $\tilde{X}^N$, we have

$$X^N\subset\tilde{X}^N, \quad A^N\subset\tilde{A}^N. \tag{3.12}$$

Alongside algorithms from X^N and $\widetilde{X}^N$, algorithms of some "intermediate" types have been considered, such as algorithms with limited memory (Traub [80, 82]), algorithms with delayed information (Beamer and Wilde [69, 71]), block algorithms (Avriel and Wilde [68], Beamer and Wilde [70], Karp and Miranker [68], Shapiro and Wilde [74a, b], Wilde and Beightler [67]), etc.

Suppose that results of informational computations are *delayed* by τ, that is, before the $(i+1)$st informational computation "the computer" learns the result of the $(i-\tau)$th informational computation, $i \geq \tau+1$, and "the computer" has enough time for processing this information. Before the first $\tau+1$ informational computations have been performed, "the computer" gets no information.

In this case, the set of permissible algorithms consists of all the algorithms $\widetilde{x}^N \in \widetilde{X}^N$ such that

$$\widetilde{x}_{i+1}(z^i) = \widetilde{x}_{i+1}(\bar{z}^i), \quad i \leq \tau, \tag{3.13}$$

for any realizable situations z^i, $\bar{z}^i$, and

$$\widetilde{x}_{i+1}(z^i) = \widetilde{x}_{i+1}(\bar{z}^i), \quad i \geq \tau+1, \tag{3.14}$$

for any realizable situations of the form

$$z^i = (x_1, \ldots, x_{i-\tau}, x_{i-\tau+1}, \ldots, x_i, y_1, \ldots, y_{i-\tau}, y_{i-\tau+1}, \ldots, y_i),$$
$$\bar{z}^i = (x_1, \ldots, x_{i-\tau}, \bar{x}_{i-\tau+1}, \ldots, \bar{x}_i, y_1, \ldots, y_{i-\tau}, \bar{y}_{i-\tau+1}, \ldots, \bar{y}_i).$$

Another algorithmic class we describe here is the class of *block* algorithms. The attention they attract is caused by development of multiprocessing computer systems and the theory of parallel computations.

Suppose that all the informational computations are divided into k blocks consisting of $N_1, \ldots, N_k$ computations respectively, $N_1 + \cdots + N_k = N$, and before the $(j+1)$st block "the computer" learns the results of all the computations of the jth block. In practice, say, for a system with r processors, we may have $N_1 = \cdots = N_k = r$.

In this case, the set of permissible algorithms consists of all the algorithms $\widetilde{x}^N \in \widetilde{X}^N$ such that

$$\widetilde{x}_{i+1}(z^i) = \widetilde{x}_{i+1}(\bar{z}^i) \tag{3.15}$$

for any situations of the form

$$z^i = (x_1, \ldots, x_{N_1+\cdots+N_j}, x_{N_1+\cdots+N_j+1}, \ldots, x_i,$$
$$y_1, \ldots, y_{N_1+\cdots+N_j}, y_{N_1+\cdots+N_j+1}, \ldots, y_i),$$
$$\bar{z}^i = (x_1, \ldots, x_{N_1+\cdots+N_j}, \bar{x}_{N_1+\cdots+N_j+1}, \ldots, \bar{x}_i,$$
$$y_1, \ldots, y_{N_1+\cdots+N_j}, \bar{y}_{N_1+\cdots+N_j+1}, \ldots, \bar{y}_i),$$

where

$$N_1 + \cdots + N_j < i < N_1 + \cdots + N_j + N_{j+1}.$$

4. MINIMAX CONCEPT OF OPTIMALITY AND SPECIFIC NOTIONS OF OPTIMALITY

In this section, we introduce several notions of optimality of computational algorithms within the framework of the minimax concept of optimality. We obtain some results that clear up the structure of optimal algorithms and answer some important methodological questions.

4.1. Optimal error algorithms

Chebyshev was probably the first mathematician who studied minimax problems in the theory of computational methods. As was mentioned by Nikolskii [50], the problem of constructing quadrature formulas minimax on some functional classes was posed by A.N. Kolmogorov. The paper by Nikolskii [50] started its investigation. Kiefer [53] was probably the first to consider the problem of constructing a minimax adaptive computational algorithm.

Consider the problem of approximating an operator S on a set F posed in Section 1.1. Let $\alpha \in \widetilde{A}^N$.

We call

$$\sup_{f \in F} \varepsilon(\alpha, f)$$

the *result (accuracy) guaranteed by the algorithm* α *on the class* F, the *worst-case error of* α *on* F, or the *estimate efficiency of* α *on* F.

Let $\hat{A}^N = \hat{X}^N \times \hat{B}^N \subset \widetilde{A}^N = \widetilde{X}^N \times \widetilde{B}^N$ be some set of algorithms with N informational computations (a set of N-step algorithms). An algorithm $\alpha_0 \in \hat{A}^N$ is called *optimal by error* in $\hat{A}^N$ on the class F iff

$$\sup_{f \in F} \varepsilon(\alpha_0, f) = \min_{\alpha \in \hat{A}^N} \sup_{f \in F} \varepsilon(\alpha, f).$$

Let $\varepsilon > 0$[1]. An algorithm $\alpha_\varepsilon \in \hat{A}^N$ is called *ε-optimal by error* in $\hat{A}^N$ on the class F iff

$$\sup_{f \in F} \varepsilon(\alpha_\varepsilon, f) \leq \inf_{\alpha \in \hat{A}^N} \sup_{f \in F} \varepsilon(\alpha, f) + \varepsilon.$$

In the situation where all *a priori* information on the problem f amounts to the fact that $f \in F$, the minimax concept of optimality looks natural and probably most well-found from methodological point of view (see Germeier [71], Luce and Raiffa [57], Moiseyev [79, Chapter 3]). In Section 6 we discuss the opinion sometimes expressed about the minimax approach to organization of computations being "over-cautious".

---

[1] Using here the same letter ε that denotes efficiency criterion will not lead to any confusion.

4.2. Definition of optimal error algorithm reflecting combinatory complexity of algorithms

We now dwell on another informal point in defining optimal algorithms. Having fixed the number of informational computations, or, in other words, ***information complexity*** of the algorithms under consideration (see Nemirovsky and Yudin [79, 83], Traub and Woźniakowski [80]), we defined an optimal error algorithm as an algorithm guaranteeing the best accuracy. We will also deal with algorithms optimal counting informational computations, which are defined in the following way. We fix the required accuracy and call an algorithm guaranteeing this accuracy with the minimal number of informational computations (i.e., with the minimal information complexity) an optimal (counting informational computations) algorithm.

In neither of these two cases the definition of optimality takes into account the resources required for "inner needs" of the algorithms, i.e., the cost of algorithmic and terminal computations, or, in other words, ***combinatory complexity*** of the algorithms under consideration (see Traub and Woźniakowski [80]).

On the face of it, this is not appropriate. However, in computational practice we encounter a lot of complex problems for which every informational computation requires many seconds or even minutes of computer time. In such situations, neglecting the resources required for "inner needs" of the algorithms can often be expedient.

However, even if the situation is different and we can not disregard "inner needs" of the algorithms, obtaining optimal (in the sense of the above definition) and sequentially optimal (in the sense of the definition given below in Section 6) algorithms will still be of use since they will indicate the bounds on "the computer's" abilities. On this basis it is often possible to construct algorithms which are "close" to the optimal ones in some sense and at the same time have acceptable combinatory complexity. Their application gives good results for many practical problems. Examples of constructing this sort of algorithms can be found in all the subsequent chapters.

It would certainly be ideal if our concept of optimality took into account both informational and combinatory complexity of algorithms. However, we do not have many significant results in this area so far. What is worse, if we are really consistent, this way is very unlikely to lead us to a complete computation model. If the model were to reflect combinatory complexity of algorithms, i.e., the cost of computations of $\widetilde{x}_i(z^{i-1})$, $i=2,\ldots,N$, and $\widetilde{\beta}(z^N)$, then we would have to pose the problem of their optimal computation. But in some cases computing $\widetilde{x}_i(z^{i-1})$ amounts to solving an integer optimization problem (as we will see below). It means that we would face the problem of obtaining optimal methods for integer optimization, and so on. In the end, we would come to complicated independent problems of optimal computation of elementary functions, optimal arithmetic, etc., and also the problem of taking into account all individual characteristics of the computer system. Moreover, it would be most difficult to formalize the problem of the cost of research work required to construct optimal algorithms.

4.3. Stochastic concepts of optimality

Alongside the minimax concept, there also exist other optimality concepts. For instance, if the algorithm is meant for solving a series of problems of the same type and we have *a priori* estimates of the frequencies with which these problems appear in the series, then we can try to construct some *a priori* probability distribution on the class of problems and derive algorithms optimal on the average with respect to a certain criterion. This approach, which is called probabilistic, statistical, stochastic, average-case, or Bayesian, is studied, among other papers, in Archetti and Betro [78c, 79, 80], Betro [84], Converse [67], Fine [66], Heyman [68], Kushner [62, 64], Mockus [77, 80], Neimark and Strongin [66], Šaltenis [71], Strongin [78], Timonov [77], Traub, Wasilkowski and Woźniakowski [84], Wasilkowski [85], Wasilkowski and Woźniakowski [84], Woźniakowski [85, 86], and Žilinskas [75, 76, 86]. Most of these papers deal with the problem of search for the extremums of functions of one or more variables. Žilinskas [86] has provided axiomatics for using stochastic models of multi-extremum objective functions in global optimization on the basis of *a priori* information about the objective function that does not include any information on the probability distribution.

4.4. One auxiliary statement

Later we will need the following (almost obvious) lemma which is close to the corresponding statement from Germeier [71]. Let X and Y be sets of arbitrary nature, let $g: X\times Y \to \mathbb{R}$ be a real-valued function on $X\times Y$, and let $\Phi = \{\phi \mid \phi: Y\to X\}$ be the set of all mappings from X into Y. Define the function g on $\Phi\times Y$ by the formula $g(\phi, y) = g(\phi(y), y)$.

Lemma 4.1. *We have*

$$\inf_{\phi\in\Phi}\sup_{y\in Y} g(\phi, y) = \sup_{y\in Y}\inf_{x\in X} g(x, y).$$

Moreover, if there is a function $\phi_0 \in \Phi$ *such that*

$$g(\phi_0(y), y) = \inf_{x\in X} g(x, y), \quad y\in Y,$$

then

$$\sup_{y\in Y} g(\phi_0, y) = \inf_{\phi\in\Phi}\sup_{y\in Y} g(\phi, y).$$

Proof. For arbitrary $\phi \in \Phi$ and $y \in Y$ we have

$$g(\phi, y) = g(\phi(y), y) \geq \inf_{x \in X} g(x, y),$$

which yields

$$\sup_{y \in Y} g(\phi, y) \geq \sup_{y \in Y} \inf_{x \in X} g(x, y)$$

and

$$\inf_{\phi \in \Phi} \sup_{y \in Y} g(\phi, y) \geq \sup_{y \in Y} \inf_{x \in X} g(x, y). \tag{4.1}$$

Define now, for an arbitrary $\varepsilon > 0$, a function $\phi_\varepsilon : Y \to X$ in such a way that, for any $y \in Y$,

$$g(\phi_\varepsilon, y) = g(\phi_\varepsilon(y), y) \leq \inf_{x \in X} g(x, y) + \varepsilon \quad \text{if} \quad \inf_{x \in X} g(x, y) > -\infty$$

and

$$g(\phi_\varepsilon, y) = g(\phi_\varepsilon(y), y) \leq -1/\varepsilon \quad \text{if} \quad \inf_{x \in X} g(x, y) = -\infty.$$

Since $y \in Y$ is arbitrary, we have

$$\sup_{y \in Y} g(\phi_\varepsilon, y) \leq \max \left\{ \sup_{y \in Y} \inf_{x \in X} g(x, y) + \varepsilon, -1/\varepsilon \right\},$$

and, therefore,

$$\inf_{\phi \in \Phi} \sup_{y \in Y} g(\phi, y) \leq \max \left\{ \sup_{y \in Y} \inf_{x \in X} g(x, y) + \varepsilon, -1/\varepsilon \right\}.$$

Letting ε tend to zero, we get

$$\inf_{\phi \in \Phi} \sup_{y \in Y} g(\phi, y) \leq \sup_{y \in Y} \inf_{x \in X} g(x, y). \tag{4.2}$$

Inequalities (4.1) and (4.2) prove the first statement of the lemma. By the definition of ϕ_0 we have

$$\sup_{y \in Y} g(\phi_0(y), y) = \sup_{y \in Y} \inf_{x \in X} g(x, y).$$

This equation combined with the first statement of the lemma yields the second statement of the lemma, which completes the proof. □

4.5. Central terminal operation

In Section 4.1, an algorithm $\alpha_0 = (\tilde{x}_0^N, \tilde{\beta}_0^N) \in \hat{A}^N = \hat{X}^N \times \hat{B}_N$ delivering the minimum with respect to $\tilde{x}^N$ and $\tilde{\beta}$ in the formula

$$\min_{\tilde{x}^N \in \hat{X}^N,\ \tilde{\beta} \in \hat{B}_N} \sup_{f \in F} \varepsilon((\tilde{x}^N, \tilde{\beta}), f)$$

was called optimal by error in $\hat{A}^N$ on the functional class F. It is easy to see that, in general, optimality of $(\tilde{x}_0^N, \tilde{\beta}_0)$ does not imply that $\tilde{\beta}_0$ delivers the minimum in the formula $\min_{\widetilde{\beta} \in \hat{B}_N} \sup_{f \in F} \varepsilon((\widetilde{x}^N, \widetilde{\beta}), f)$ for $\widetilde{x}^N \neq \widetilde{x}_0^N$, even if for some $\widetilde{\beta}$ the algorithm $(\widetilde{x}^N, \widetilde{\beta})$ is optimal by error in $\hat{A}^N$. In this sense, $\widetilde{\beta}_0$ is not a universal optimal terminal operation.

Assume that $\hat{B}_N = \widetilde{B}_N$, i.e., all terminal operations are permitted. In this case, there exists a universal optimal terminal operation (and also a universal ε-optimal terminal operation for any $\varepsilon > 0$). Theorem 4.1 gives a strict definition of this term. We now construct this terminal operation.

Suppose that the informational computations have been completed and we have a situation z^N. This means that all the information on the problem f now amounts to the fact that $f \in F(z^N)$. Apparently, the best guaranteed accuracy in this case is

$$\varepsilon_N(z^N) = \inf_{\beta \in B} \sup_{f \in F(z^N)} \gamma(S(f), \beta). \tag{4.3}$$

Assume that infimum in (4.3) is attained for any realizable z^N. Denote by $\widetilde{\beta}_*(z^N)$ the element of B delivering this infimum (if there are several such elements, then we can fix any of them and denote it by $\widetilde{\beta}_*(z^N)$). Thus, for any realizable z^N we have

$$\sup_{f \in F(z^N)} \gamma(S(f), \beta_*(z^N)) = \inf_{\beta \in B} \sup_{f \in F(z^N)} \gamma(S(f), \beta). \tag{4.4}$$

The terminal operation $\widetilde{\beta}_*$ is called the *central terminal operation*[1)].

The analogue of the central terminal operation in the situation where the infimum in (4.3) is not attained is the concept of *ε-central terminal operation* $\widetilde{\beta}_*^\varepsilon$ such that

$$\sup_{f \in F(z^N)} \gamma(S(f), \widetilde{\beta}_*^\varepsilon(z^N)) \leq \inf_{\beta \in B} \sup_{f \in F(z^N)} \gamma(S(f), \beta) + \varepsilon$$

for any realizable z^N.

To justify the usage of the term "central terminal operation", recall that $r = \inf_{\beta \in B} \sup_{s \in S} \gamma(s, \beta)$ is called the *Chebyshev radius* of the set S in the metric space B with the metric γ. If there exists a point β_* such that $\sup_{s \in S} \gamma(s, \beta_*) = r$, then it is called the *Chebyshev center* of the set S. Loosely speaking, r is the radius of the smallest ball containing S, and β_* is the center of this ball. Note that β_* may not be uniquely determined.

On completion of the informational computations, the image $S(F(z^N))$ of the set $F(z^N)$ under the mapping S is the "uncertainty set": all its el-

---

[1)] We will use the same notation $\widetilde{\beta}_*$ for the central terminal operations for different N. This will not lead to any confusion, though N is not indicated explicitly in this notation. The number of arguments of the function f will always be clear from the context.

ements (and only they) can appear to be the element $S(f)$ being approximated (provided the information "the computer" has about f amounts to the fact that $f \in F(z^N)$). Rewriting (4.3) in the form

$$\varepsilon_N(z^N) = \inf_{\beta \in B} \sup_{s \in S(F(z^N))} \gamma(s, \beta),$$

we see that $\varepsilon_N(z^N)$ is the Chebyshev radius of the set $S(F(z^N))$ and $\widetilde{\beta}_*(z^N)$ is its Chebyshev center. That is why $\widetilde{\beta}_*$ is called the central terminal operation.

The following theorem establishes universal optimality of the central terminal operation.

Theorem 4.1. *For any* $\widetilde{x}^N \in \widetilde{X}^N$,

$$\min_{\widetilde{\beta} \in \widetilde{B}_N} \sup_{f \in F} \varepsilon((\widetilde{x}^N, \widetilde{\beta}), f) = \sup_{f \in F} \varepsilon((\widetilde{x}^N, \widetilde{\beta}_*), f),$$

where $\widetilde{\beta}_*$ *is the central terminal operation.*

Proof. Fix an arbitrary $\widetilde{x}^N \in \widetilde{X}^N$. To prove the theorem, it suffices to justify the following chain of equations:

$$\begin{aligned}
\inf_{\widetilde{\beta} \in \widetilde{B}_N} \sup_{f \in F} \varepsilon((\widetilde{x}^N, \widetilde{\beta}), f) &= \inf_{\widetilde{\beta} \in \widetilde{B}_N} \sup_{f \in F} \gamma(S(f), \widetilde{\beta}(z^N)) \\
&= \inf_{\widetilde{\beta} \in \widetilde{B}_N} \sup_{z^N \in \{z^N\}_{\widetilde{x}^N}} \sup_{f \in F(z^N)} \gamma(S(f), \widetilde{\beta}(z^N)) \\
&= \sup_{z^N \in \{z^N\}_{\widetilde{x}^N}} \inf_{\beta \in B} \sup_{f \in F(z^N)} \gamma(S(f), \beta) \\
&= \sup_{z^N \in \{z^N\}_{\widetilde{x}^N}} \sup_{f \in F(z^N)} \gamma(S(f), \widetilde{\beta}_*(z^N)) \\
&= \sup_{f \in F} \gamma(S(f), \widetilde{\beta}_*(z^N)) = \sup_{f \in F} \varepsilon((\widetilde{x}^N, \widetilde{\beta}_*), f).
\end{aligned}$$

The symbol $z^N = (x_1, \ldots, x_N, y_1, \ldots, y_N))$ on the right-hand side of the first equation denotes the vector obtained by formulas (3.2) for fixed $\widetilde{x}^N$ and f; thus, this equation makes use only of the definition (1.1) of the criterion ε and takes into account the definition of $\alpha(f)$ by formulas (3.2).

The symbol $\{z^N\}_{\widetilde{x}^N}$ denotes the set of all z^N realizable for the algorithm $\widetilde{x}^N$ (see Section 3.2). So, the second equation is based on the obvious formula

$$F = \bigcup_{z^N \in \{z^N\}_{\widetilde{x}^N}} F(z^N). \tag{4.5}$$

The third equation follows from Lemma 4.1 applied to the function $g(\beta, z^N) = \sup_{f \in F(z^N)} \gamma(S(f), \beta)$. The fourth equation follows from (4.4). The fifth equation follows from (4.5); the vector z^N on its right-hand side has the same sense as on the right-hand side of the first equation. Finally, the sixth equation, as well as the first one, makes use only of the definition of the criterion ε. □

Corollary. *If an algorithm* $(\widetilde{x}_0^N, \widetilde{\beta}_0)$ *is optimal by error in* $\hat{X}^N \times \hat{B}_N$ *on the class* F *and* $\widetilde{\beta}_*$ *is the central terminal operation, then the algorithm* $(\widetilde{x}_0^N, \widetilde{\beta}_*)$ *is optimal by error in* $\hat{X}^N \times \widetilde{B}_N$ *on the class* F.

Proof. In the following chain of equations

$$\inf_{\widetilde{x}^N \in \hat{X}^N} \inf_{\widetilde{\beta} \in \widetilde{B}_N} \sup_{f \in F} \varepsilon((\widetilde{x}^N, \widetilde{\beta}), f) = \sup_{f \in F} \varepsilon((\widetilde{x}_0^N, \widetilde{\beta}_0), f)$$
$$= \inf_{\widetilde{\beta} \in \widetilde{B}_N} \sup_{f \in F} \varepsilon((\widetilde{x}_0^N, \widetilde{\beta}), f) = \sup_{f \in F} \varepsilon((\widetilde{x}_0^N, \widetilde{\beta}_*), f)$$

the first two equations reflect optimality of the algorithm $(\widetilde{x}_0^N, \widetilde{\beta}_0)$, and the third equation follows from Theorem 4.1. Equality of the leftmost and the rightmost expressions proves the corollary. □

4.6. Optimality of a linear terminal operation

Let S be a functional, and let $\gamma(S(f), \beta) = |S(f) - \beta|$. Suppose that we use a nonadaptive algorithm $x^N = (x_1, \dots, x_N) \in X^N$, where $x_1, \dots, x_N$ are fixed functionals. Then terminal operations can be regarded as functionals θ of the variables $y_1 = x_1(f), \dots, y_N = x_N(f)$; $\theta(y_1, \dots, y_N) = \widetilde{\beta}(x_1, \dots, x_N, y_1, \dots, y_N)$.

A terminal operation θ_0 is called an *optimal terminal operation for fixed* $x_1, \dots, x_N$ iff

$$\sup_{f \in F} |S(f) - \theta_0(x_1(f), \dots, x_N(f))| = \min_{\theta \in \Theta} \sup_{f \in F} |S(f) - \theta(x_1(f), \dots, x_N(f))|, \tag{4.6}$$

where Θ is the set of all numerical functions θ of the variables

$$(y_1, \dots, y_N) \in \{(y_1, \dots, y_N) \mid y_1 = x_1(t), \dots, y_N = x_N(t),\ f \in F\}.$$

In view of Theorem 4.1, the central terminal operation $\widetilde{\beta}_*(x_1, \dots, x_N, y_1, \dots, y_N)$ regarded as a function of the variables $y_1, \dots, y_N$ for fixed $x_1, \dots, x_N$ is optimal in the sense of the above definition. However, the optimal terminal operation for fixed $x_1, \dots, x_N$ may not be unique. This brings up the problem of choosing a function θ_0 of the simplest possible form among all the optimal terminal operations. The best candidate satisfying the property of simplicity is a linear function. It turns out that there is an important class of problems for which a linear optimal terminal operation does exist. This is the contents of the following theorem generalizing a similar result from Smolyak [65] (see also Bakhvalov [71]).

Theorem 4.2 (Sukharev [86]). *Let F be a convex set in a linear space, and let $S, x_1, \ldots, x_N$ be linear functionals. Then among all the optimal terminal operations for fixed $x_1, \ldots, x_N$ there is a linear function $\theta_0(y_1, \ldots, y_N) = p_0 + p_1 y_1 + \cdots + p_N y_N$. Moreover, if, in addition, the set F is balanced, then $p_0 = 0$.*

Proof. Put

$$Y = \{y = (y_0, y_1, \ldots, y_N) \mid y_0 = S(f), y_1 = x_1(f), \ldots, y_N = x_N(f),\ f \in F\}.$$

Convexity of the set F and linearity of the functionals $S, x_1, \ldots, x_N$ imply convexity of the set $Y \subset \mathbb{R}^{N+1}$. For an arbitrary function $\theta \in \Theta$, we have

$$\sup_{f \in F} |S(f) - \theta(x_1(f), \ldots, x_N(f))| = \sup_{(y_0, y_1, \ldots, y_N) \in Y} |y_0 - \theta(y_1, \ldots, y_N)|$$

$$= \sup_{(y_1, \ldots, y_N) \in \pi(Y)} \sup_{y_0 \in \sigma(y_1, \ldots, y_N)} |y_0 - \theta(y_1, \ldots, y_N)|,$$

where $\sigma(y_1, \ldots, y_N) = \{y_0 \mid (y_0, y_1, \ldots, y_N) \in Y\}$ is the interval with the extreme points

$$a(y_1, \ldots, y_N) = \inf_{(y_0, y_1, \ldots, y_N) \in Y} y_0,$$

$$b(y_1, \ldots, y_N) = \sup_{(y_0, y_1, \ldots, y_N) \in Y} y_0,$$

and $\pi(Y) = \{(y_1, \ldots, y_N) \mid \sigma(y_1, \ldots, y_N) \neq \emptyset\}$ is the projection of the set Y on the subspace of the variables $y_1, \ldots, y_N$. Therefore, in view of Lemma 4.1, we have

$$\inf_{\theta \in \Theta} \sup_{f \in F} |S(f) - \theta(x_1(f), \ldots, x_N(f))|$$

$$= \inf_{\theta \in \Theta} \sup_{(y_1, \ldots, y_N) \in \pi(Y)} \sup_{y_0 \in \sigma(y_1, \ldots, y_N)} |y_0 - \theta(y_1, \ldots, y_N)|$$

$$= \sup_{(y_1, \ldots, y_N) \in \pi(Y)} \inf_{r \in \mathbb{R}} \sup_{y_0 \in \sigma(y_1, \ldots, y_N)} |y_0 - r|$$

$$= \sup_{(y_1, \ldots, y_N) \in \pi(Y)} \inf_{r \in \mathbb{R}} \max\{b(y_1, \ldots, y_N) - r, r - a(y_1, \ldots, y_N)\}$$

$$= \sup_{(y_1, \ldots, y_N) \in \pi(Y)} \frac{1}{2}(b(y_1, \ldots, y_N) - r, r - a(y_1, \ldots, y_N)) \stackrel{def}{=} d. \tag{4.7}$$

If $d = +\infty$, then any function $\theta \in \Theta$ is optimal. Let $d < +\infty$. In this case, to make sure that a linear optimal terminal operation exists, it suffices to point out a function $\theta_0(y_1, \ldots, y_N) = p_0 + p_1 y_1 + \cdots + p_N y_N$ such that

$$\sup_{f \in F} |S(f) - \theta_0(x_1(f), \ldots, x_N(f))| \leq d$$

or, in other words,

$$|y_0 - (p_0 + p_1 y_1 + \cdots + p_N y_N)| \leq d \text{ for all } (y_0, y_1, \ldots, y_N) \in Y. \tag{4.8}$$

Consider the sets $Y_1 = Y - (d, 0, \ldots, 0)$, $Y_2 = Y + (d, 0, \ldots, 0)$ (see Fig. 1). Let $y = (y_0, y_1, \ldots, y_N) \in \operatorname{ri} Y_1 \cap \operatorname{ri} Y_2$ (here, as usual, $\operatorname{ri} A$ denotes the relative interior of the set A). Then $(y_0+d, y_1, ..., y_N)\in \operatorname{ri} Y$ and $(y_0-d, y_1, ..., y_N)\in \operatorname{ri} Y$. Apparently, the line passing through these two points belongs to the affine hull of the set Y. The points $(y_0+d+\varepsilon/2, y_1, \dots, y_N)$ and $(y_0-d-\varepsilon/2, y_1, \dots, y_N)$ lie on the same line. Hence, for a small $\varepsilon>0$, they belong to the set Y, and, therefore,

$$b(y_1, \dots, y_N)-a(y_1, \dots, y_N)\geq (y_0+d+\varepsilon/2)-(y_0-d-\varepsilon/2)=2d+\varepsilon,$$

which contradicts the last equation in (4.7). Thus, $\operatorname{ri} Y_1 \cap \operatorname{ri} Y_2=\emptyset$.

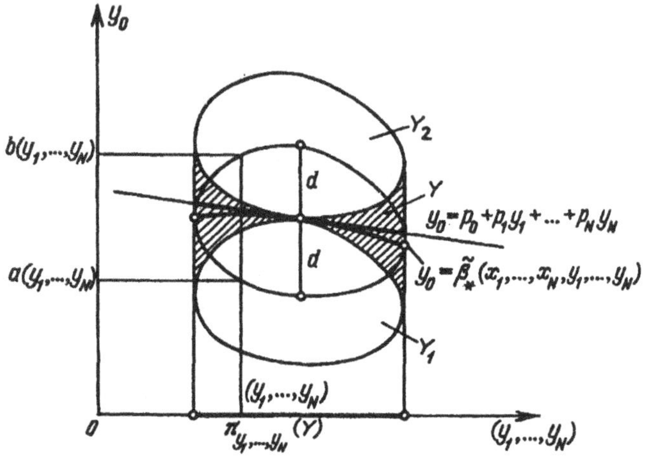

Fig. 1. To the concept of optimal terminal operation for fixed $x_1, \dots, x_N$

Due to the separation theorem (see, e.g., Sukharev, Timokhov and Fedorov [86]), in this case there is a hyperplane $\langle c, y\rangle=\xi$, $c=(c_0, c_1, \dots, c_N)$, separating the sets Y_1 and Y_2 properly, so that

$$\langle c, y\rangle \leq \xi \leq \langle c, y''\rangle \quad \text{for all} \quad y' \in Y_1, \; y'' \in Y_2 \tag{4.9}$$

and

$$\langle c, \bar{y}\rangle < \langle c, \bar{y}''\rangle \quad \text{for some} \quad \bar{y}' \in Y_1, \; \bar{y}'' \in Y_2. \tag{4.10}$$

Assume that $c_0=0$. Consider an arbitrary vector $y'=(y_0', y_1', \dots, y_N')\in Y$ and the vector $y''=(y_0'+2d, y_1', \dots, y_N')\in Y_2$. We have $\langle c, y'\rangle=\langle c, y''\rangle$. Hence, due to (4.9), $\langle c, y'\rangle=\xi$ for all $y'\in Y_1$. Similarly, $\langle c, y''\rangle=\xi$ for all $y''\in Y_2$, which contradicts (4.10). The contradiction shows that $c_0\neq 0$.

Let, for definiteness, $c_0>0$. Putting $p_1=-c_1/c_0$, $\dots$, $p_N=-c_N/c_0$, $p_0=\xi/c_0$ and making use of the definitions of Y_1 and Y_2, we can rewrite (4.9) as

$$y_0-d-\sum_{i=1}^{N} p_i y_i \leq p_0 \leq y_0+d-\sum_{i=1}^{N} p_i y_i \quad \text{for all} \quad y=(y_0, y_1, \dots, y_N)\in Y,$$

which proves (4.8).

To complete the proof, it now suffices to make sure that $p_0 = 0$ if F is balanced. In this case the set Y is also balanced. Let $(y_1, \ldots, y_N) \in \pi(Y)$. Then the points $(a(y_1, \ldots, y_N), y_1, \ldots, y_N)$ and $(b(y_1, \ldots, y_N), y_1, \ldots, y_N)$ belong to the closure $\bar{Y}$ of Y. It is easy to see that $\bar{Y}$ is also balanced. Therefore, $(-a(y_1, \ldots, y_N), -y_1, \ldots, -y_N) \in \bar{Y}$. As is known, convexity of Y implies convexity of $\bar{Y}$ (see, e.g., Sukharev, Timokhov and Fedorov [86]), and so

$$\frac{1}{2}(b(y_1, \ldots, y_N), y_1, \ldots, y_N) + \frac{1}{2}(-a(y_1, \ldots, y_N), -y_1, \ldots, -y_N)$$
$$= \left(\frac{b(y_1, \ldots, y_N) - a(y_1, \ldots, y_N)}{2}, 0, \ldots, 0\right) \in \bar{Y},$$

and, due to symmetry of $\bar{Y}$

$$\left(\frac{a(y_1, \ldots, y_N) - b(y_1, \ldots, y_N)}{2}, 0, \ldots, 0\right) \in \bar{Y}.$$

By the definition of the functions a and b,

$$b(y_1, \ldots, y_N) - a(y_1, \ldots, y_N)$$
$$\geq \frac{b(y_1, \ldots, y_N) - a(y_1, \ldots, y_N)}{2} - \frac{b(y_1, \ldots, y_N) - a(y_1, \ldots, y_N)}{2}$$
$$= b(y_1, \ldots, y_N) - a(y_1, \ldots, y_N).$$

Now, recalling the definition of d in (4.7) and the facts that $(y_1, \ldots, y_N) \in \pi(Y)$ is arbitrary and Y is symmetric, we get successively that $b(0, \ldots, 0) - a(0, \ldots, 0) = 2d$ and $b(0, \ldots, 0) = -a(0, \ldots, 0) = d$.

If $y_1 = \cdots = y_N = 0$, then the inequality (4.8) takes the form $|y_0 - p_0| \leq d$ for any $y_0 \in (a(0, \ldots, 0), b(0, \ldots, 0)) = (-d, d)$, which is possible only for $p_0 = 0$. □

In Fig. 1, the heavy lines show the graphs of the function $y_0 = \widetilde{\beta}_*(x_1, \ldots, x_N, y_1, \ldots, y_N)$ corresponding to the central terminal operation ($x_1, \ldots, x_N$ are fixed) and the function $y_0 = p_0 + p_1 y_1 + \cdots + p_N y_N$ corresponding to the linear optimal terminal operation for fixed $x_1, \ldots, x_N$. Clearly, any function $y_0 = \theta_0(y_1, \ldots, y_N)$ whose graph lies entirely within the shaded area in Fig. 1 determines an optimal terminal operation for fixed $x_1, \ldots, x_N$, since $|y_0 - \theta_0(y_1, \ldots, y_N)| \leq d$ for any $(y_0, y_1, \ldots, y_N) \in Y$.

4.7. Comments on the concept of optimal terminal operation

The definition (4.6) of optimal terminal operation for fixed $x_1, \ldots, x_N$ can easily be extended to the case of an arbitrary $x^N = (x_1, \ldots, x_N) \in X^N$.

Namely, a terminal operation $\widetilde{\beta}_{\text{opt}}$ such that, for any $x^N \in X^N$,

$$\sup_{f\in F}\gamma(S(f),\widetilde{\beta}_{\text{opt}}(x_1,\ldots,x_N,x_1(f),\ldots,x_N(f)))$$

$$= \min_{\widetilde{\beta}\in\widetilde{B}_N}\sup_{f\in F}\gamma(S(f),\widetilde{\beta}(x_1,\ldots,x_N,x_1(f),\ldots,x_N(f))) \tag{4.11}$$

could be called a *terminal operation optimal in the set* $\widetilde{B}_N$.

However, this definition would only be adequate for the setting where only nonadaptive algorithms $(x^N,\widetilde{\beta})\in A^N$ are used (see (3.9)). The point is that an optimal terminal operation defined in the above way does not have the property of universal optimality, in contrast with the central terminal operation (see Section 4.5). The following example demonstrates that, in general, optimality of an algorithm $(\widetilde{x}_0^N,\widetilde{\beta}_0)$ does not imply optimality of the algorithm $(\widetilde{x}_0^N,\widetilde{\beta}_{\text{opt}})$ (cf. the corollary of Theorem 4.1).

Example. Let $F=\{0,2,4,6,8,10,12,14\}$, $S(f)=f$, $B=\mathbb{R}$ (recall that the operator S takes values from the metric space B with a metric γ), $N=2$, $X_1=X_2=\{a_1,\ldots,a_7\}$, where

$$a_i(f)=\begin{cases} 0, & f<2i-1,\\ 1, & f>2i-1,\end{cases} \qquad i=1,\ldots,7.$$

Put $\widetilde{\beta}_{\text{opt}}(a_4,a_6,1,0)=\widetilde{\beta}_{\text{opt}}(a_6,a_4,0,1)=10$. For all the other $z^2=(x_1,x_2,y_1,y_2)$, put $\widetilde{\beta}_{\text{opt}}(x_1,x_2,y_1,y_2)=\widetilde{\beta}_*(x_1,x_2,y_1,y_2)$, where $\widetilde{\beta}_*$ is the central terminal operation. It is easily verified that $\widetilde{\beta}_{\text{opt}}$ is an optimal terminal operation in the sense of the definition (4.11). For instance, we have

$$\sup_{f\in F}\gamma(S(f),\widetilde{\beta}_{\text{opt}}(a_4,a_6,a_4(f),a_6(f)))$$

$$= \min_{\widetilde{\beta}\in\widetilde{B}_N}\sup_{f\in F}\gamma(S(f),\widetilde{\beta}(a_4,a_6,a_4(f),a_6(f)))=3,$$

where the supremum on the left-hand side of the second equation is attained at $f=0$ and $f=6$.

Consider an adaptive algorithm $(\widetilde{x}_1^0,\widetilde{x}_2^0)$ defined in the following way: $\widetilde{x}_1^0=a_4$, $\widetilde{x}_2^0(a_4,0)=a_2$, $\widetilde{x}_2^0(a_4,1)=a_6$. Apparently, the algorithm $(\widetilde{x}_1^0,\widetilde{x}_2^0,\widetilde{\beta}_*)$ guarantees accuracy 1 (see Section 4.1) and is optimal by error in $\widetilde{A}^2$ (see (3.11)) on the class F, whereas the accuracy guaranteed by the algorithm $(\widetilde{x}_1^0,\widetilde{x}_2^0,\widetilde{\beta}_{\text{opt}})$ is worse (for this algorithm, the worst-case error is 2 and the maximum is attained at $f=8$), and, therefore, the algorithm $(\widetilde{x}_1^0,\widetilde{x}_2^0,\widetilde{\beta}_{\text{opt}})$ is not optimal. □

The formula for the best guaranteed result (accuracy) in an algorithmic class $\hat{A}^N$

$$\inf_{\alpha\in\hat{A}^N}\sup_{f\in F}\varepsilon(\alpha,f) \tag{4.12}$$

contains the infimum over the set $\hat{A}^N$. In many cases, say, if $\hat{A}^N=\widetilde{A}^N$ (see (3.11)), or if $\hat{A}^N$ is defined either by conditions (3.13) and (3.14) or by condition (3.15), the set $\hat{A}^N$ has a complex functional structure.

We pose the problem of representing (4.12) in the form of a multiple minimax, or, more precisely, in the form of successive computations of infimums and supremums of the criterion ε over sets whose structure is much simpler than that of $\hat{A}^N$. The solution to this problem given in Sections 4.8–4.11 is based on Lemma 4.1. This solution will enable us to clarify some important theoretical questions.

4.8. Best guaranteed accuracy in the class of nonadaptive algorithms

Consider the class $\hat{A}^N$ of nonadaptive algorithms (see (3.9)) under the assumption that all terminal operations $\widetilde{\beta}\in\widetilde{B}_N$ are permitted. Recall that $\varepsilon_N(z^N)$ is defined by (4.3).

Theorem 4.3.

$$\inf_{\alpha\in A^N}\sup_{f\in F}\varepsilon(\alpha,f)=\inf_{x^N\in X^N}\sup_{y^N\in\{y^N|F(z^N)\neq\emptyset\}}\varepsilon_N(z^N)$$
$$=\inf_{x^N\in X^N}\sup_{y^N\in\{y^N|F(z^N)\neq\emptyset\}}\inf_{\beta\in B}\sup_{f\in F(z^N)}\gamma(S(f),\beta).$$

Proof. Apparently, we have

$$\inf_{\alpha\in A^N}\sup_{f\in F}\varepsilon(\alpha,f)=\inf_{x^N\in X^N}\inf_{\widetilde{\beta}\in\widetilde{B}_N}\sup_{y^N\in\{y^N|F(z^N)\neq\emptyset\}}\sup_{f\in F(z^N)}\gamma(S(f),\widetilde{\beta}(z^N)).$$

Fix $x^N\in X^N$ and apply Lemma 4.1 to the function

$$g(\widetilde{\beta},y^N)=\sup_{f\in F(z^N)}\gamma(S(f),\widetilde{\beta}(z^N)).$$

To complete the proof, it now suffices to take the infimum with respect to $x^N\in X^N$ of the both sides of the obtained equation. □

4.9. Best guaranteed accuracy in the class of adaptive algorithms

Consider the class $\widetilde{A}^N$ of adaptive algorithms (see (3.11)) under the assumption that all terminal operations $\widetilde{\beta}\in\widetilde{B}_N$ are permitted. Put

$$\varepsilon_N=\inf_{x_1\in X_1}\sup_{y_1\in\{y_1|F(z^1)\neq\emptyset\}}\cdots\inf_{x_N\in X_N}\sup_{y_N\in\{y_N|F(z^N)\neq\emptyset\}}\inf_{\beta\in B}\sup_{f\in F(z^N)}\gamma(S(f),\beta)$$
$$=\inf_{x_1\in X_1}\sup_{y_1\in\{y_1|F(z^1)\neq\emptyset\}}\inf_{x_N\in X_N}\sup_{y_N\in\{y_N|F(z^N)\neq\emptyset\}}\varepsilon_N(z^N). \tag{4.13}$$

Theorem 4.4.

$$\inf_{\alpha\in\widetilde{A}^N}\sup_{f\in F}\varepsilon(\alpha,f)=\varepsilon_N.$$

Proof. In this proof, we denote by $\widetilde{x}^{i+1}(y^i)$ the vector obtained by formulas (3.2):

$$\widetilde{x}^{i+1}(y^i)=(x_1,x_2,\dots,x_{i+1}),$$

where $x_1=\widetilde{x}_1$ is the first component of $\widetilde{x}^{i+1}$, $x_2=\widetilde{x}_2(x_1,y_1),\dots,x_{i+1}=\widetilde{x}_{i+1}(x_1,\dots,x_i,y_1,\dots,y_i)$, and the vector y^i is assumed to be realizable for the algorithm $\widetilde{x}^{i+1}$.

We now write the following chain of equations proving the theorem (and explained below):

$$\begin{aligned}
\inf_{\alpha\in\widetilde{A}^N}\sup_{f\in F}\varepsilon(\alpha,f) &= \inf_{\widetilde{x}^N}\inf_{\widetilde{\beta}}\sup_{y^N}\sup_{f\in F(\widetilde{x}^N(y^{N-1}),y^N)}\gamma(S(f),\widetilde{\beta}(\widetilde{x}^N(y^{N-1}),y^N))\\
&= \inf_{\widetilde{x}^N}\sup_{y^N}\inf_{\beta\in B}\sup_{f\in F(\widetilde{x}^N(y^{N-1}),y^N)}\gamma(S(f),\widetilde{\beta}(z^N))\\
&= \inf_{\widetilde{x}^{N-1}}\inf_{\widetilde{x}_N}\sup_{y^{N-1}}\sup_{y_N}\inf_{\beta\in B}\sup_{f\in F(\widetilde{x}^N(y^{N-1}),y^N)}\gamma(S(f),\widetilde{\beta}(z^N))\\
&= \inf_{\widetilde{x}^{N-1}}\sup_{y^{N-1}}\inf_{\widetilde{x}_N\in X_N}\sup_{y_N}\inf_{\beta\in B}\sup_{f\in F(\widetilde{x}^{N-1}(y^{N-2}),x_N,y^N)}\gamma(S(f),\widetilde{\beta}(z^N))\\
&=\dots=\varepsilon_n.
\end{aligned}$$

In the first equation, the infimums are taken over the set of all $\widetilde{X}^N\times\widetilde{B}^N=\widetilde{A}^N$, and the supremum is taken over the set of all y^N realizable for the algorithm $\widetilde{x}^N$. Apparently, $F=\cup_{y^N}F(\widetilde{x}^N(y^{N-1}),y^N)$, which justifies replacing the supremum with respect to $f\in F$ by two successive supremums.

To prove the second equation in this chain, we fix $\widetilde{x}^N$, apply Lemma 4.1 to the function

$$g(\widetilde{\beta},y)=\sup_{f\in F(\widetilde{x}^N(y^{N-1}),y^N)}\gamma(S(f),\widetilde{\beta}(\widetilde{x}^N(y^{N-1}),y^N))$$

and take the infimum with respect to $\widetilde{x}^N$ of the both sides of the obtained equation.

All the rest equations in the chain are proved in the same way. □

4.10. Case of fixed terminal operation

Suppose that the terminal operation $\widetilde{\beta}$ is fixed and thus the choice of an algorithm $\alpha=(\widetilde{x}^N,\widetilde{\beta})$ is reduced to the choice of $\widetilde{x}^N$. This is the case when the class of permissible terminal operations consists of one operation $\widetilde{\beta}$ (see (3.4)). Moreover, in the case where arbitrary terminal operations are permitted (see (3.5)) and the infimum in (4.3) is attained for every realizable z^N, that is, the central terminal operation $\widetilde{\beta}_*$ exists, we can also fix $\widetilde{\beta}=\widetilde{\beta}_*$ (if (4.4) does not determine the central terminal operation uniquely, we can fix one). The results of Section 4.5 show that in this case, by fixing $\widetilde{\beta}$ in the above way, we also reduce the choice of an optimal algorithm α to the choice of an optimal $\widetilde{x}^N$.

Assume that the terminal operation $\widetilde{\beta}$ has been fixed in this or that way. Then it is convenient to define the criterion ε, already defined by (1.1) on the set $\widetilde{A}^N \times F$ (see (3.11)), also on the sets $X^N \times F$ (see (3.8)) and $\widetilde{X}^N \times F$ (see (3.10)). Put

$$\varepsilon(x^N, f) = \gamma(S(f), \widetilde{\beta}(x^N, x_1(f), \ldots, x_N(f))),$$
$$\varepsilon(\widetilde{x}^N, f) = \varepsilon(x^N, f). \tag{4.14}$$

In the latter formula, the vector x^N is determined by the algorithm $\widetilde{x}^N$ and the function f in accordance with (3.2).

We now reformulate Theorems 4.3 and 4.4 in such a way that they include the case (3.4) of a unique permissible terminal operation as well as the case (3.5). In order to do this, we extend the notation $\varepsilon_N(z^N)$ introduced by (4.3) to the case (3.4). Namely, put

$$\varepsilon_N(z^N) = \sup_{f \in F(z^N)} \gamma(S(f), \widetilde{\beta}(z^N)), \tag{4.15}$$

where $\widetilde{\beta}$ is the fixed terminal operation. If in the case (3.5) the central terminal operation exists, then (4.15) coincides with (4.3) since we have agreed to fix the central terminal operation $\widetilde{\beta}_*$ as $\widetilde{\beta}$ in this situation. It is easy to see that

$$\varepsilon_N(z^N) = \sup_{f \in F(z^N)} \varepsilon(x^N, f). \tag{4.16}$$

Using the introduced notations, we can present the best guaranteed accuracy in the class X^N of nonadaptive algorithms (see (3.8)) as $\inf\limits_{x^N \in X^N} \sup\limits_{f \in F} \varepsilon(x^N, f)$. Theorem 4.3 can now be reformulated in the following way.

Theorem 4.5.

$$\inf_{x^N \in X^N} \sup_{f \in F} \varepsilon(x^N, f) = \inf_{x^N \in X^N} \sup_{y^N \in \{y^N | F(z^N) \neq \emptyset\}} \varepsilon_N(z^N). \qquad \square$$

The analogue of Theorem 4.4 reads as follows.

Theorem 4.6.

$$\inf_{\widetilde{x}^N \in \widetilde{X}^N} \sup_{f \in F} \varepsilon(\widetilde{x}^N, f) = \varepsilon_N.$$

Proof. In the case (3.5) the theorem coincides with Theorem 4.4, and in the case (3.4) its proof is just the same, the only difference is that we need not take the infimums with respect to β. □

4.11. Best guaranteed accuracy in the classes of algorithms with delayed information and block algorithms

The following two theorems are proved in the same way as Theorem 4.4.

Theorem 4.7. *Let $\hat{X}^N$ be the set of algorithms with delayed information defined by* (3.13) *and* (3.14). *Then*

$$\inf_{\tilde{x}^N \in \hat{X}^N} \sup_{f \in F} \varepsilon(\tilde{x}^N, f) = \inf_{x_1,\dots,x_{r+1}} \sup_{y_1} \inf_{x_{r+2}} \sup_{y_2} \dots \inf_{x_N} \sup_{y_{N-r},\dots,y_N} \varepsilon_N(z^N),$$

where the infimums and supremums are taken over such sets that the vector z^N is realizable. □

Theorem 4.8. *Let $\hat{X}^N$ be the set of block algorithms defined by* (3.15). *Then*

$$\inf_{\tilde{x}^N \in \hat{X}^N} \sup_{f \in F} \varepsilon(\tilde{x}^N, f)$$

$$= \inf_{x_1,\dots,x_{N_1}} \sup_{y_1,\dots,y_{N_1}} \dots \inf_{x_{N_1+\dots+N_{k-1}+1},\dots,x_N} \sup_{y_{N_1+\dots+N_{k-1}+1},\dots,y_N} \varepsilon_N(z^N),$$

where the infimums and supremums are taken over such sets that the vector z^N is realizable. □

4.12. Setting with unfixed N

Alongside the setting with a fixed number of informational computations N we have considered above, we shall also deal with its modification for the case of unfixed N. Let

$$\tilde{x} = (\tilde{x}_1, \tilde{x}_2, \dots, \tilde{x}_N, \dots) \tag{4.17}$$

be a sequence of mappings of the form (3.1). Denote by $\tilde{X}$ the set of all such sequences, which will now play the role of algorithms.

For estimating efficiency of algorithms $\tilde{x} \in \tilde{X}$ we will use the criterion (4.14), assuming that it is defined for all N. Thus, if N informational computations have been performed (where N is not fixed in advance), then the terminal operation will be some operation $\tilde{\beta}$ fixed in advance for every N.

We say that the algorithm $\widetilde{x}$ defined by (4.17) *converges* on the functional class F iff

$$\sup_{f \in F} \varepsilon(\widetilde{x}^N, f) \to 0 \quad \text{as} \quad N \to \infty;$$

we call $\widetilde{x}$ *asymptotically optimal in* $\widetilde{X}$ on the class F iff

$$\sup_{f \in F} \varepsilon(\widetilde{x}^N, f) \sim \varepsilon_N;$$

and we call it *optimal by order in* $\widetilde{X}$ on the class F iff

$$\sup_{f \in F} \varepsilon(\widetilde{x}^N, f) \asymp \varepsilon_N,$$

where $\widetilde{x}^N = (\widetilde{x}_1, \ldots, \widetilde{x}_N)$ and ε_N is defined by (4.13) and (4.15).

The notation $a_N \sim b_N$ means that two sequences $\{a_N\}$ and $\{b_N\}$ converging to zero as $N \to \infty$ are *equivalent*, i.e., $\lim_{N \to \infty} a_N / b_N = 1$ (here we agree that $0/0 = 1$).

The notation $a_N \asymp b_N$ means that two sequences $\{a_N\}$ and $\{b_N\}$ converging to zero as $N \to \infty$ *have the same order of convergence*, i.e., there are N_0, $c_1 > 0$ and $c_2 > 0$ such that for any $N \geq N_0$

$$c_1 |b_N| \leq |a_N| \leq c_2 |b_N|.$$

Suppose we have a sequence of algorithms $\{\widetilde{x}(N)\}$, $\widetilde{x}(N) \in \widetilde{X}^N$. If

$$\sup_{f \in F} \varepsilon(\widetilde{x}^N, f) \sim \varepsilon_N \quad \text{or} \quad \sup_{f \in F} \varepsilon(\widetilde{x}^N, f) \asymp \varepsilon_N,$$

then the sequence $\{\widetilde{x}(N)\}$ is called *asymptotically optimal* or *optimal by order* respectively. However, it should be emphasized that it may turn out to be impossible to construct an asymptotically optimal or optimal by order algorithm of the form (4.17) using the elements of this sequence.

4.13. Optimal (counting informational computations) algorithms

Let us now fix the accuracy ε of the solution that the algorithm has to guarantee. Denote

$$N_\varepsilon = \min\{N \mid \varepsilon_N \leq \varepsilon\}. \tag{4.18}$$

An algorithm α_0 is called *optimal (counting informational computations)* on the class F among all the adaptive algorithms guaranteeing the prescribed accuracy ε iff

$$\alpha_0 \in \widetilde{A}^{N_\varepsilon} \quad \text{and} \quad \sup_{f \in F} \varepsilon(\alpha_0, f) \leq \varepsilon. \tag{4.19}$$

Remark. All the concepts introduced above (asymptotically optimal algorithm, algorithm optimal by order, and algorithm optimal counting informational computations) can be defined for an arbitrary subset of the set of all adaptive algorithms as well as for the whole set. □

4.14. One-step optimal algorithms

The concept of one-step optimal algorithm in the framework of the minimax approach and other optimality models was studied, for instance, in Gross and Johnson [59] (see also Bellman and Dreyfus [62]) for the problem of solving equations and in Chernous'ko [70b], Strongin[78], Šaltenis [71], and Žilinskas [75] for the optimization problem.

Let an algorithm $\tilde{x}$ be defined by (4.17), and let its accuracy be estimated using the criterion (4.14). In addition, we have to define the *stopping criterion*.

As a rule, an algorithm stops either after N steps, where the number N is prescribed in advance, or after having guaranteed a prescribed accuracy ε, i.e., after j steps, where

$$\varepsilon_{j-1}(z^{j-1}) > \varepsilon \quad \varepsilon_j(z^j) \leq \varepsilon, \tag{4.20}$$

see (4.16). Clearly, the actual accuracy of the solution for a given function f may appear to be much better then the accuracy guaranteed *a posteriori*.

An algorithm $\tilde{x}$ is called *one-step optimal* in $\tilde{X}$ on the functional class F iff

$$\min_{x \in X_1} \sup_{f \in F} \varepsilon(x, f) = \sup_{f \in F} \varepsilon(x_1, f), \tag{4.21}$$

where $x_1 = \tilde{x}_1$, and for any realizable situation $z^i = (x^i, y^i)$, $i \geq 1$,

$$\min_{x \in X_{i+1}} \sup_{f \in F(z^i)} \varepsilon(x^i, x, f) = \sup_{f \in F(z^i)} \varepsilon(x^i, x^{i+1}, f), \tag{4.22}$$

where $x_{i+1} = \tilde{x}_{i+1}(z^i)$.

Thus, a one-step optimal algorithm guarantees the maximal possible improvement of accuracy of the solution at every step.

Despite their intuitive attractiveness and simplicity of the definition, using one-step optimal algorithms does not necessarily lead to success. For example, suppose that we are seeking the extremum of a unimodal function (see Wilde [64]) with the help of algorithms that use only function values. Assume that, after i steps, the point of the extremum has been localized within an interval of the length l and the function has already been evaluated at a point lying at the distance δ from one of the endpoints.

Consider the algorithm that at every step evaluates the function at the point lying within the localization interval and symmetric to the point of the previous evaluation within this interval. It is easy to see that this algorithm is one-step optimal. After its $(i+n)$th step, the length of the localization interval will be $l - n\delta$, i.e., the improvement can be arbitrary small provided that δ is small enough. In contrast with this one-step optimal algorithm, applying optimal nonadaptive (Shapiro [84], Wilde [64]) and especially optimal adaptive (Chernous'ko [70a], Gal [71], Johnson [55], Kiefer [53], Witzgall [72]) algorithms gives far better improvement.

Nevertheless, there is a number of reasons that arouse interest in constructing one-step optimal algorithms in context of different computation models. An important stimulant is the possibility to modify the computation model in the process of computations. For instance, the estimates of some "global" parameters determining the functional class can be improved in the computational process (see Section 4.2 of Chapter 2).

For this or that reason, we often have to decide on the number of informational computations and the required absolute accuracy of the solution in the process of computations, which is another motivation for using one-step optimal algorithms. Our choice can also be influenced by lack of *a priori* information on complexity of function evaluations (which makes *a priori* estimation of the required resources impossible), by necessity of finding the extremum with a prescribed relative accuracy, etc. Finally, a strong motivation for using one-step optimal algorithms is created by simplicity of their construction and implementation.

5. COMPARISON OF THE BEST GUARANTEED RESULTS FOR ADAPTIVE AND NONADAPTIVE ALGORITHMS

In this section, we obtain sufficient conditions for coincidence of the best guaranteed results in the classes of nonadaptive and all adaptive algorithms.

5.1. Statement of the problem

Let the terminal operation $\widetilde{\beta}$ be fixed (see Section 4.10). Since the set X^N of nonadaptive algorithms is a subset of the set $\widetilde{X}^N$ of all adaptive algorithms (see (3.8), (3.10) and (3.12)), the best guaranteed result (or accuracy) in the algorithmic class $\widetilde{X}^N$ is not worse than in X^N:

$$\inf_{\widetilde{x}^N \in \widetilde{X}^N} \sup_{f \in F} \varepsilon(\widetilde{x}^N, f) \leq \inf_{x^N \in X^N} \sup_{f \in F} \varepsilon(x^N, f), \tag{5.1}$$

where the criterion ε is defined by (4.14).

For some problems the inequality (5.1) is strict, the left-hand side being considerably less than the right-hand one. Say, for the problem of finding zero of a strictly monotonic function with values of different signs at the endpoints of the segment $[0, 1]$ and for the problem of search for the extremum of a function unimodal on $[0, 1]$, the accuracy obtained by adaptive algorithms depends on N exponentially (see Johnson [55], Kiefer [53, 57], Traub and Woźniakowski [80]).

On the other hand, for some other problems we have equality in (5.1):

$$\inf_{\widetilde{x}^N \in \widetilde{X}^N} \sup_{f \in F} \varepsilon(\widetilde{x}^N, f) = \inf_{x^N \in X^N} \sup_{f \in F} \varepsilon(x^N, f). \tag{5.2}$$

This means that the possibility of using all adaptive algorithms rather than only nonadaptive ones does not lead to any improvement of the guaranteed result. However, later we will see that this by no means makes adaptive algorithms inexpedient. Application of adaptive algorithms to specific problems can produce dramatic effect even if (5.2) holds. This idea will be developed in Section 6 where we introduce the concept of sequentially optimal algorithm and also in the subsequent chapters where we construct sequentially optimal algorithms for various problems of numerical analysis and discuss the results of their application. The fact that (5.2) is valid has great practical as well as theoretical value since it often simplifies considerably the procedure of constructing sequentially optimal algorithms.

The first result of this kind was probably obtained by Kiefer [57] for integration of monotonic functions. Then the problem was studied in Bakhvalov [71], Gal and Micchelli [80], Plaskota [86], Sukharev [71], Zaliznyak and Ligun [78]. For instance, Bakhvalov [71] has derived sufficient conditions for (5.2) to hold in the case where $S, x_1 \in X_1, \ldots, x_N \in X_N$ are linear functionals. In Sukharev [71] (5.2) was for the first time proved for a certain nonlinear functional.

Here we establish a general fact (Theorem 5.1) which has useful applications and enables us to obtain a result analogous to that of Bakhvalov [71] under less restrictive assumptions (Theorem 5.2).

5.2. Main theorem

We put (see (4.14)–(4.16))

$$\varepsilon_N(x^N, f) \overset{def}{=} \varepsilon_N(x^N, f) = \sup_{g \in F(x^N, x^N(f))} \varepsilon(x^N, g), \tag{5.3}$$

where $x^N(f) = (x_1(f), \ldots, x_N(f))$ and, in accordance with the notation introduced in Section 3.2, $F(x^N, x^N(f)) = \{g \in F \mid x^N(g) = x^N(f)\}$.

Theorem 5.1 (Sukharev [85]). *If the function $\varepsilon_N(x^N, f)$ has a generalized saddle point on $X^N \times F$, i.e.,*

$$\inf_{x^N \in X^N} \sup_{f \in F} \varepsilon_N(x^N, f) = \sup_{f \in F} \inf_{x^N \in X^N} \varepsilon_N(x^N, f), \tag{5.4}$$

then (5.2) *is valid, that is, the best guaranteed results in the classes of nonadaptive and all adaptive algorithms coincide.*

Proof. Let (5.4) hold, and let v be the value of both its left- and right-hand sides. Due to (5.3), we have

$$\sup_{f \in F} \varepsilon_N(x^N, f) = \sup_{f \in F} \varepsilon(x^N, f),$$

hence

$$\inf_{x^N \in X^N} \sup_{f \in F} \varepsilon(x^N, f) = v. \tag{5.5}$$

Equation (5.4) implies that for any $\delta>0$ there is a function $f_\delta \in F$ such that

$$\inf_{x^N \in X^N} \varepsilon(x^N, f_\delta) \geq v-\delta,$$

which yields

$$\sup_{f\in F(x^N, x^N(f_\delta))} \varepsilon(x^N, f) = \varepsilon_N(x^N, f_\delta) \geq v-\delta, \quad x^N \in X^N. \tag{5.6}$$

Let $\widetilde{x}^N \in \widetilde{X}^N$ be an arbitrary adaptive algorithm, and let a vector $x_\delta^N \in X^N$ be determined by the algorithm $\widetilde{x}^N$ and the function f_δ according to (3.2). Then, due to the definition (4.14) and formula (5.6), we have

$$\sup_{f\in F} \varepsilon(\widetilde{x}^N, f) \geq \sup_{f\in F(x_\delta^N, x_\delta^N(f_\delta))} \varepsilon(x_\delta^N, f) \geq v-\delta$$

(indeed, for any function $f\in F(x_\delta^N, x_\delta^N(f_\delta))$ the algorithm $\widetilde{x}^N$ determines the same vector x_δ^N). Since $\widetilde{x} \in \widetilde{X}^N$ and $\delta>0$ are arbitrary,

$$\inf_{\widetilde{x}^N \in \widetilde{X}^N} \sup_{f\in F} \varepsilon(\widetilde{x}^N, f) \geq v. \tag{5.7}$$

Now (5.2) follows from (5.1), (5.5) and (5.7), which completes the proof. □

Theorem 5.1 immediately yields the following important result (first obtained by Gal and Micchelli [80] under more restrictive assumptions).

Corollary. *If there is a function $f_0 \in F$ such that*

$$\varepsilon_N(x^N, f_0) \geq \varepsilon_N(x^N, f), \quad x^N \in X^N, \quad f\in F, \tag{5.8}$$

then (5.2) is valid, that is, the best guaranteed results in the classes of non-adaptive and all adaptive algorithms coincide.

Proof. Due to (5.8),

$$\sup_{f\in F} \inf_{x^N\in X^N} \geq \inf_{x^N\in X^N} \varepsilon_N(x^N, f_0) \geq \inf_{x^N\in X^N} \sup_{f\in F} \varepsilon_N(x^N, f).$$

This yields (5.4) since, as we know,

$$\sup_{f\in F} \inf_{x^N\in X^N} \varepsilon_N(x^N, f) \leq \inf_{x^N\in X^N} \sup_{f\in F} \varepsilon_N(x^N, f). \qquad \square$$

5.3. Approximation of a linear functional on a convex set using linear information

Let S be a functional, $B=\mathbb{R}$,

$$\gamma(S(f), \beta) = |S(f)-\beta|, \tag{5.9}$$

and let arbitrary terminal operations be permitted. Suppose we have fixed the terminal operation equal to the central terminal operation $\widetilde{\beta}_*$ defined by (4.4), i.e., in (4.14) we have

$$\widetilde{\beta} = \widetilde{\beta}_*. \tag{5.10}$$

Then, by the definitions (5.3) and (4.3),

$$\begin{aligned}\varepsilon_N(x^N,f)&=\inf_{\beta\in\mathbb{R}}\sup_{g\in F(x^N,x^N(f))}|S(g)-\beta|\\&=\frac{1}{2}\left(\sup_{g\in F(x^N,x^N(f))}S(g)-\inf_{g\in F(x^N,x^N(f))}S(g)\right)\\&=\frac{1}{2}\sup_{g_1,g_2\in F(x^N,x^N(f))}(S(g_1)-S(g_2)).\end{aligned}\tag{5.11}$$

When determining whether $\varepsilon_N(x^N,f)$ has a saddle point or not, the following fact (established by Gal and Micchelli [80] under more restrictive assumptions) may prove helpful. In its formulation, the sets $Y_1,\ldots,Y_N$ are implied to be linear spaces.

Lemma 5.1. *If* (5.9) *and* (5.10) *hold, F is convex, $x_i\in X_i$ are linear operators, $i=1,\ldots,N$, and S is a linear functional, then the function $\varepsilon_N(x^N,f)$ is convex on F with respect to f.*

Proof. Let $f_1,f_2\in F$. By virtue of (5.11), for any $\delta>0$ there are f_i' and f_i'' such that

$$x^N(f_i')=x^N(f_i'')=x^N(f_i),\ \varepsilon_N(x^N,f_i\leq\frac{1}{2}(S(f_i'')-S(f_i'))+\delta,\ i=1,2.$$

Observing that, due to linearity of x^N,

$$x^N(\xi f_1''+(1-\xi)f_2'')=x^N(\xi f_1'+(1-\xi)f_2')=x^N(\xi f_1+(1-\xi)f_2),$$

we get for an arbitrary $\xi\in[0,1]$

$$\begin{aligned}\varepsilon_N(x^N,\xi f_1+(1-\xi)f_2)&\geq\frac{1}{2}S(\xi f_1''+(1-\xi)f_2'')-\frac{1}{2}S(\xi f_1'+(1-\xi)f_2')\\&=\frac{1}{2}(S(f_1'')-S(f_1'))+\frac{1}{2}(1-\xi)(S(f_2'')-S(f_2'))\\&\geq\xi\varepsilon_N(x^N,f_1)+(1-\xi)\varepsilon_N(x^N,f_2)-\delta.\end{aligned}$$

Since $\delta>0$ is arbitrary, this proves the lemma. □

Now, using the corollary of Theorem 5.1, we prove the main result of this section.

Theorem 5.2. *Assume that* (5.9) *and* (5.10) *hold, F is a convex centrally symmetric set with the center f_0, the sets X_i consist of linear operators, $i=1,\ldots,N$, and the functional S is also linear. Then* (5.2) *is valid, that is, the best guaranteed results in the classes of nonadaptive and all adaptive algorithms coincide, and so every algorithm optimal in X^N is also optimal in $\widetilde{X}^N$.*

Proof. Fix arbitrary $f\in F$, $x^N\in X^N$. Let

$$g_1,g_2\in F(x^N,x^N(f)).\tag{5.12}$$

Then $2f_0 - g_1 \in F$ and $2f_0 - g_2 \in F$ since F is symmetric and

$$f_1 \stackrel{def}{=} \frac{1}{2}(g_1 + 2f_0 - g_2) \in F, \quad f_2 \stackrel{def}{=} \frac{1}{2}(g_2 + 2f_0 - g_1) \in F$$

since F is convex. Moreover, we have $x^N(f_1) = x^N(f_2) = x^N(f_0)$. Taking (5.11) into account, we get

$$\varepsilon_N(x^N, f_0) \geq \frac{1}{2}(S(f_1) - S(f_2)) = \frac{1}{2}(S(g_1) - S(g_2)).$$

Since g_1 and g_2 are arbitrary functions satisfying (5.12), due to (5.11) we have

$$\varepsilon_N(x^N, f_0) \geq \varepsilon_N(x^N, f).$$

To complete the proof, it now suffices to apply the corollary of Theorem 5.1. □

5.4. Global optimization problem

Suppose we have to solve the global optimization problem (1.4), i.e.,

$$S(f) = \sup_{x \in K} f(x).$$

The question about coincidence of the best guaranteed results in the classes of nonadaptive and all adaptive algorithms for this problem was probably considered for the first time in Sukharev [71]. Later the same methods were used to answer this question for more general functional classes in Zaliznyak and Ligun [78] (see also Plaskota [86]).

Here we will only establish one simple result with the help of argument different from those used in Zaliznyak and Ligun [78] and Sukharev [71].

Let $x_i(f) = f(x_i)$, $x_i \in K$, $i = 1, \ldots, N$, and let only one terminal operation

$$\widetilde{\beta}(z^N) = \max_{i=1,\ldots,N} y_i$$

be permitted (see (3.6)). Then, due to (4.14),

$$\varepsilon_N(x^N, f) = \sup_{x \in K} f(x) - \max_{i=1,\ldots,N} f(x_i). \tag{5.13}$$

Theorem 5.3. *Let the functional class F contain the constants $c \in \mathbb{R}$ and satisfy the following conditions:*

$$f \in F \Rightarrow f + c \in F, \quad c \in \mathbb{R}, \tag{5.14}$$

$$f \in F \Rightarrow f_+ \stackrel{def}{=} \max\{f, 0\} \in F. \tag{5.15}$$

Then for the global optimization problem with the criterion (5.13) *equation* (5.2) *is valid, that is, the best guaranteed results in the classes of nonadaptive and all adaptive algorithms coincide.*

Proof. Let $x^N \in X^N = K^N$ and $g \in F$. Then, due to (5.14) and (5.15),

$$\bar{g} \overset{def}{=} \left(g - \max_{i=1,\ldots,N} g(x_i)\right)_+ \in F, \quad x^N(\bar{g}) = x^N(0) = 0$$

and, in view of (5.3),

$$\varepsilon_N(x^N, g) = \varepsilon_N(x^N, \bar{g}) \leq \varepsilon_N(x^N, 0).$$

Fixing an arbitrary function $f \in F$ and taking in the above inequality the supremum with respect to $d \in F(x^N, x^N(f))$, we get $\varepsilon_N(x^N, f) \leq \varepsilon_N(x^N, 0)$. Application of the corollary of Theorem 5.1 completes the proof. □

Corollary. *For the problem of search for the global extremum of a function from the functional class F_ρ determined by an arbitrary quasi-metric ρ (see Section 2), the best guaranteed results in the classes of nonadaptive and all adaptive algorithms coincide, i.e., (5.2) is valid, and so every algorithm optimal in X^N is also optimal in $\widetilde{X}^N$.*

Proof. Clearly, the class F_ρ contains the constants, satisfies condition (5.14) and, due to Lemma 2.7, also satisfies condition (5.15). □

Observe in conclusion that Theorems 5.2 and 5.3 were proved with the help of the corollary of Theorem 5.1. However, in some situations coincidence of the worst-case errors in the classes of nonadaptive and all adaptive algorithms can be proved with the help of Theorem 5.1, whereas a universal "worst" function f_0 does not exist (i.e., the corollary of Theorem 5.1 does not work). For example, this is the case with the integration problem for the classes of monotonic functions and Lipschitz functions defined on a segment of the real line and having fixed and different values at the endpoints of the segment (see Sukharev and Chuyan [84]).

6. SEQUENTIALLY OPTIMAL ALGORITHMS

In this section, we introduce and discuss the central concept of the book – the concept of sequentially optimal algorithm.

6.1. Examples demonstrating necessity of further specification of the optimality concept

We start with giving some examples of optimal algorithms for different problems of numerical analysis and different functional classes and constructing situations in which these algorithms prescribe strategies that are far from the best. Here we have to run ahead and use the results on optimality of these algorithms that will be proved in the subsequent chapters (Examples 1–3).

Example 1. Let F be the class of functions satisfying the Lipschitz condition on [0, 1]] with a given constant M. Assume that every informational computation (both in this example and in the next three examples) is evaluation of the function at some point. According to Theorem 2.1 and formula (3.2) from Chapter 2, the compound rectangle quadrature formula with the nodes $1/(2N), 3/(2N), \ldots, (2N-1)/(2N)$ is optimal. Due to Theorem 1.3 of Chapter 2, this quadrature is an optimal by error nonadaptive algorithm of numerical integration. When applying this algorithm, we approximate $\int_0^1 f(x)\,dx$ by $\frac{1}{N}\sum_{i=1}^{N} f\left(\frac{2i-1}{2N}\right)$. In view of Theorem 5.2 and Lemma 2.8, this algorithm is also optimal in the class of all adaptive algorithms. Since the order of the nodes is of no importance, assume that f is successively evaluated at the nodes

$$\frac{1}{2N}, \frac{2N-1}{2N}, \frac{3}{2N}, \frac{5}{2N}, \ldots, \frac{2N-3}{2N} \tag{6.1}$$

and the first two steps give the results

$$f\left(\frac{1}{2N}\right)=M\frac{1}{2N}, \quad f\left(\frac{2N-1}{2N}\right)=M\frac{2N-1}{2N}. \tag{6.2}$$

It is easy to see that in this case we have $f(x)=Mx$ for $x\in[1/2(2N), (2n-1)/(2N)]$. This means that the optimal algorithm suggests that we perform the rest $N-2$ evaluations at the points where the values of the function are already known, which apparently makes no sense. □

Example 2. Consider the problem of global optimization for the functional class from Example 1. The optimal nonadaptive algorithm for solving this problem consists in exhaustive search over the lattice (6.1); this algorithm is also optimal in the class of all adaptive algorithms (see Remarks 2 and 3 after Theorem 2.1, Chapter 4). If the results (6.2) are obtained, then, as well as in Example 1, it makes no sense to perform the rest $N-2$ informational computations prescribed by the optimal algorithm. □

Example 3. Let F be the class of functions nondecreasing on [0, 1] and having fixed values at the endpoints: $f(0)=0$, $f(1)=1$. As is shown in Section 3.6 of Chapter 2 (Theorem 3.4), in this case the compound trapezoid quadrature formula with the nodes 0, $1/(N+1)$, $2/(N+2), \ldots, N/(N+1)$, 1 is an optimal by error nonadaptive algorithm of numerical integration. It is also shown there that this algorithm is optimal in the class of all adaptive algorithms. Assume that

$$f\left(\frac{1}{N+1}\right)=1. \tag{6.3}$$

Clearly, in this case $f(x)=1$ for $x \in [1/(N+1), 1]$ and it makes no sense to perform the rest informational computations at the points $2/(N+2), \ldots, N/(N+1)$. □

Before giving the last example and discussing the necessity of further specification of the optimality concept, note that Examples 1–3 allow us to make yet another important conclusion. They show that in favourable situations, for instance, in the situations (6.2) and (6.3), adaptive algorithms can guarantee a better accuracy than nonadaptive ones, even though the best guaranteed results in the classes of adaptive and nonadaptive algorithms on the whole functional class F coincide (i.e., (5.2) holds).

Example 4. Consider the problem of maximizing a unimodal on $[0, 1]$ function. It is well-known (see Johnson [55], Kiefer [53], Traub and Woźniakowski [80], Wilde [64]) that the N-step ε-optimal by error adaptive algorithm for $N=4$ requires evaluations of f at the points $x_1=2/5$, $x_2=3/5$, and

$$x_3 = \begin{cases} 1/5 & \text{if } f(2/5) > f(3/5), \\ 4/5 & \text{if } f(2/5) < f(3/5), \\ 1/5 \text{ or } 4/5 & \text{if } f(2/5) = f(3/5); \end{cases}$$

finally, if, say, $f(3/5) < f(2/5) < f(1/5)$, then $x_4 = 1/5 + \varepsilon$. This algorithm guarantees the accuracy $1/5+\varepsilon$. However, if after two steps it turns out that

$$f(2/5) = f(3/5), \tag{6.4}$$

it is possible to guarantee the accuracy $1/10+\varepsilon$; we just have to put $x_3=1/2$ and $x_4=1/2+\varepsilon$. □

The above examples show that the definition of an optimal algorithm given in Section 4 does not reflect some important aspects of organizing the computational process. Although an optimal error algorithm guarantees the best possible accuracy for "the worst" function from the class F, in general it makes no use of the favourable (not the worst for "the computer") characteristics of the function that may be revealed in the solution process. For instance, functions having the properties (6.2)–(6.4) in Examples 1–4 are obviously not the worst ones (actually, they are the best ones).

Sometimes examples of this sort arouse criticism of the minimax approach (principle of the best guaranteed result, see Germeier [71]). It is accused of being "over-cautious" and orientated to the least favourable situation that but seldom occurs in practice. However, the actual reason why optimal in the framework of the minimax approach algorithms do not always make proper use of information on the problem obtained in the solution process is, in fact, lack of consistency in applying the principle of the best guaranteed result when defining optimal algorithms. Examples 1–4 bring us to the conclusion that the optimality concept which was formulated in Section 4 needs further specification. Below we give such a specification: we single out an important subset of the set of all optimal algorithms, namely the set of sequentially optimal algorithms.

6.2. Concept of sequentially optimal algorithm

If we are really consistent in following the principle of the best guaranteed result, our aim should not be just constructing ***any*** optimal algorithm, but rather obtaining the optimal algorithm that at every step[1)] makes the best use of the information accumulated through the previous steps (no matter if it is "worst" or not) in order to achieve either the best possible accuracy with a given number of steps or the maximal reduction of the number of steps required to guarantee a prescribed accuracy. We call such algorithms sequentially optimal (formal definitions are given below).

The concept of sequentially optimal algorithm was introduced in Sukharev [72] (in that paper the term "best strategy" was used). Note that, for the optimization problem, Chernous'ko [70a] and Witzgall [72] have actually constructed sequentially optimal algorithms for certain functional classes, though their aim was just to obtain an optimal algorithm within the minimax setting. Ideologically, the concept of a sequentially optimal algorithm is closely related to schemes of sequential analysis (see Bellman [57], Bellman and Dreyfus [62], Wald [47, 50], Mikhalevich [65]) based on A.A. Markov's ideas.

The first sequentially optimal algorithms for recovering functions were probably constructed in Sukharev [76, 77, 78], and those for numerical integration – in Sukharev [79b, c].

To construct a sequentially optimal algorithm is a considerably more difficult problem than to construct just an optimal algorithm. Below we will see that its solution actually calls for constructing at every step an algorithm optimal on some subclass of the original functional class, determined by the information accumulated through the previous steps.

Observe that for some problems the concept of optimal algorithm actually serves the practical purpose of optimization of the computational process as perfectly as the concept of sequentially optimal algorithm. Say, for the problem of search for the extremum of a unimodal function, the only possible case of favourable behaviour of the function is exact equality between the function value computed at the present step and one of the function values computed at the previous steps. In a sense, this must be an extraordinary situation.

---

1) Recall that the term "step of an algorithm" refers to a pair of an algorithmic and an informational computation (see Section 3.1).

6.3. Formal definitions

Assume that the terminal operation $\widetilde{\beta}$ is fixed ($\widetilde{\beta}$ is either the only permissible or the central terminal operation); this means that we will use the criterion (4.14). For $i \geq 1$, put

$$\varepsilon(x^i, \widetilde{x}_{i+1}, \ldots, \widetilde{x}_N, f) = \varepsilon(x^i, x_{i+1}, \ldots, x_N, f), \tag{6.5}$$

where $x_{i+1}, \ldots, x_N$ for fixed x^i are determined by the mappings $\widetilde{x}_{i+1}, \ldots, \widetilde{x}_N$ and the function f according to (3.2). We will call the set of mappings $(\widetilde{x}_{i+1}, \ldots, \widetilde{x}_N)$ (as well as $(\widetilde{x}_1, \ldots, \widetilde{x}_N)$) an algorithm. Recall that $\widetilde{X}^N$ denotes the class of all adaptive algorithms.

Definition 1. An algorithm $\widetilde{x}_0^N = (\widetilde{x}_1^0, \ldots, \widetilde{x}_N^0) \in \widetilde{X}^N$ is called *sequentially optimal (by error)* on the class F iff it is optimal by error in $\widetilde{X}^N$ on the class F and for any realizable situation z^i (see Section 3.2), $1 \leq i \leq N-1$,

$$\sup_{f \in F(z^i)} \varepsilon(x^i, \widetilde{x}_{i+1}^0, \ldots, \widetilde{x}_N^0, f) = \min_{\widetilde{x}_{i+1}, \ldots, \widetilde{x}_N} \sup_{f \in F(z^i)} \varepsilon(x^i, \widetilde{x}_{i+1}, \ldots, \widetilde{x}_N, f), \tag{6.6}$$

where the minimum is taken over all possible sets of mappings $(\widetilde{x}_{i+1}, \ldots, \widetilde{x}_N)$ of the form (3.1). □

In full accordance with (4.13), (4.15) and (4.16), we introduce the following notation:

$$\begin{aligned} \varepsilon_N(z^i) = & \inf_{x_{i+1} \in X_{i+1}} \sup_{y_{i+1} \in \{y_{i+1} | F(z^{i+1}) \neq \emptyset\}} \cdots \\ & \inf_{x_N \in X_N} \sup_{y_N \in \{y_N | F(z^N) \neq \emptyset\}} \sup_{f \in F(z^N)} \gamma(S(f), \widetilde{\beta}(z^N)) \\ = & \inf_{x_{i+1} \in X_{i+1}} \sup_{y_{i+1} \in \{y_{i+1} | F(z^{i+1}) \neq \emptyset\}} \cdots \\ & \inf_{x_N \in X_N} \sup_{y_N \in \{y_N | F(z^N) \neq \emptyset\}} \sup_{f \in F(z^N)} \varepsilon_N(z^N). \end{aligned} \tag{6.7}$$

Applying Theorem 4.6 to the functional class $F(z^i)$ and the criterion $\varepsilon(x^i, \widetilde{x}_{i+1}, \ldots, \widetilde{x}_N, f)$, we obtain

$$\inf_{\widetilde{x}_{i+1}, \ldots, \widetilde{x}_N} \sup_{f \in F(z^i)} \varepsilon(x^i, \widetilde{x}_{i+1}, \ldots, \widetilde{x}_N, f) = \varepsilon_N(z^i). \tag{6.8}$$

Comparison of (6.6), (6.7), and (6.8) shows that Definition 1 is equivalent to the following definition.

Definition 2. An algorithm $\widetilde{x}_0^N = (\widetilde{x}_1^0, \ldots, \widetilde{x}_N^0) \in \widetilde{X}^N$ is called *sequentially optimal (by error)* on the class F iff the outermost infimum in (4.13) for ε_N is attained at $x_1 = \widetilde{x}_1^0$, and the outermost infimum in (6.7) for $\varepsilon_N(z^i)$ is attained at $x_{i+1} = \widetilde{x}_{i+1}^0(z^i)$ for any realizable situation z^i, $1 \leq i \leq N-1$. □

Similarly to the notion of ε-optimal algorithm, we can define the notion of ε-sequentially optimal algorithm.

Naturally, a sequentially optimal algorithm need not be unique (and is not unique in many real-life problems, as we will see below). In this case, our choice of a specific sequentially optimal algorithm in every situation should be based on the characteristic features of the problem being solved. This also goes for sequentially optimal (counting informational computations) algorithms, which are defined in the next section.

6.4. Sequentially optimal (counting informational computations) algorithms

We now fix the accuracy ε we have to guarantee when solving the problem. Similarly to the definition (4.18) of N_ε, define for an arbitrary realizable situation z^i

$$N_*(z') \overset{def}{=} \min\{N \mid \varepsilon_{i+N}(z^i) \le \varepsilon\}, \tag{6.9}$$

where $\varepsilon_{i+N}(z^i)$ is defined by (6.7).

Assume that the stopping criterion (4.20) is used, i.e., the process stops after j steps in a situation z^i if (see (4.15) and (4.16))

$$\varepsilon_{j-1}(z^{j-1}) > \varepsilon, \quad \varepsilon_j(z^j) \le \varepsilon.$$

Assume that, as before, the terminal operation $\widetilde{\beta}$ is fixed ($\widetilde{\beta}$ is either the only permissible or the central terminal operation).

An algorithm $\widetilde{x}_0 = (\widetilde{x}_1^0, \widetilde{x}_2^0, \dots)$ is called *sequentially optimal (counting informational computations)* on the class F for the fixed accuracy ε iff it is optimal on F (counting informational computations), i.e.,

$$\widetilde{x}_0 \in \widetilde{X}^{N_\varepsilon}, \quad \sup_{f \in F} \varepsilon(\widetilde{x}_0, f) \le \varepsilon, \tag{6.10}$$

and for any realizable situation z^i, $i \ge 1$,

$$\sup_{f \in F(z^i)} \varepsilon(x^i, \widetilde{x}_{i+1}^0, \dots, \widetilde{x}_{i+N_\varepsilon(z^i)}^0, f) \le \varepsilon. \tag{6.11}$$

Clearly, if $\{z^i\}$ is a sequence of situations obtained when solving some problem with the help of a sequentially optimal algorithm, then

$$N_\varepsilon(z^i) \le N_\varepsilon - 1, \quad N_\varepsilon(z^{i+1}) \le N_\varepsilon(z^i) - 1, \quad i \ge 1. \tag{6.12}$$

Later we will show that sequentially optimal algorithms often prove much more efficient than optimal algorithms (we could also make this conjecture on the basis of Examples 1–3). In such cases, it is not enough to find *a priori* estimates of the algorithms' efficiency on the functional classes (these estimates could even create the false impression of intractability of the problem) and construct one of the optimal algorithms on their basis. And it is certainly not enough to construct optimal by order or asymptotically optimal algorithms.

7. STOCHASTIC ALGORITHMS

Stochastic (probabilistic, randomized, statistical) algorithms are widely used for solving various problems of numerical analysis, such as numerical integration (see Bakhvalov [59, 61, 64, 73], Yermakov [71], Sobol [73]), solution of equations (see Vasil'ev, P.P. [84]), and optimization (see Anderssen [72], Archetti and Betro [78b], Brooks [52], Chichinadze [83], Gaviano [75], Karmanov [86], McCormick [72], Nemirovsky and Yudin [79], Rastrigin [68, 73, 74], Sysoyev and Petrov [76], Ust'uzhaninov [80, 83, 85], Zhigliavskii [85]), including problems of experimental design (see Nalimov and Chernova [65], Fedorov [71], Finney [60]) where randomized procedures are also used. Algorithms like those studied by Niederreiter and Curley [79], Niederreiter and Peart [86], Sobol [69, 73, 79, 82] are also closely related to stochastic algorithms. However, problems of estimating efficiency of stochastic algorithms and constructing optimal stochastic algorithms are posed and solved much less often (see, e.g., Aird and Rice [77], Anderssen [72], Anderssen and Bloomfield [75], Bakhvalov [59, 61, 64, 73], Gal and Micchelli [80], Nemirovsky and Yudin [79], Sukharev [71, 81a, c], Ust'uzhaninov [78, 80a, b, c, 81, 83, 85], Vasil'ev, P.P. [84]).

It is important to note that a precondition to applying stochastic algorithms, comparing them in terms of efficiency and constructing optimal stochastic algorithms is "the computer's" willingness to use averaged criteria of efficiency. This is usually the case when randomization procedures are frequently used in the computational process.

In this section, we describe different classes of stochastic algorithms and define several specific optimality concepts for stochastic algorithms in the framework of the minimax approach. But, in order to clarify the question about expediency of stochastic algorithms, we first discuss some aspects of applying deterministic algorithms.

7.1. Solution process as an antagonistic game

We now describe the solution process as a multistep antagonistic game between "the computer" (minimizing player) and "nature" (maximizing player). To be definite, assume that "the computer" can use any algorithm from $\widetilde{A}^N$ (see (3.11)). In this case the game looks as follows.

At the first step, the minimizing player chooses $x_1 \in X_1$; then, knowing x_1, the maximizing player chooses $y_1 \in \{y_1 \mid F(z^1) \neq \emptyset\}$. At the second step, the minimizing player chooses $x_2 \in X_2$ knowing x_1 and y_1, and the maximizing player chooses $y_2 \in \{y_2 \mid F(z^2) \neq \emptyset\}$ knowing x_1, x_2 and y_1. In this way N steps are performed. Finally, at the $(N+1)$st step the minimizing player chooses $\beta \in B$ knowing z^N and the maximizing player chooses $f \in F(z^N)$

knowing Z^N and β. The gain of the maximizing player (the loss of the minimizing player) is equal to $\gamma(S(f),\beta)$. The game can be schematized as the following sequence of alternative choices made by the players:

$$x_1, y_1, x_2, y_2, \dots, x_N, y_N, \beta, f. \tag{7.1}$$

It is easy to see that this is a game with complete information. However, in general, every player in every position has an infinite (and even uncountable) set of alternatives, and so the well-known in the game theory Zermelo theorem, which states that every game with complete information has a saddle point (see Kuhn [53], Zermelo [13]), does not apply to this game directly.

Nevertheless, the game (7.1) does have a saddle point, provided that all the infimums and supremums in formula (4.13) for ε_N are attained so (4.13) can be rewritten as

$$\varepsilon_N = \min_{x_1 \in X_1} \max_{y_1 \in \{y_1 | F(z^1) \neq \emptyset\}} \dots \min_{x_N \in X_N} \max_{y_N \in \{y_N | F(z^N) \neq \emptyset\}} \min_{\beta \in B} \max_{f \in F(z^N)} \gamma(S(f),\beta). \tag{7.2}$$

Indeed, by virtue of Theorem 4.4, the minimizing player can guarantee that his/her loss will not exceed ε_N by applying any optimal strategy (existence of an optimal (and even of a sequentially optimal) strategy follows from the assumption that the minimums in (7.2) are attained). On the other hand, the maximizing player can guarantee that his/her gain will not be less than ε_N by using a strategy such that at every step the alternative he/she chooses delivers the corresponding maximum in (7.2).

If "the computer" can apply only nonadaptive algorithms, that is, the set of strategies of the minimizing player is A^N (see (3.9)), the game can be represented by the following sequence of alternative choices by the players:

$$x^N, y^N, \beta, f. \tag{7.3}$$

Under the assumption that the maximums and minimums in the formula

$$\min_{x^N \in X^N} \max_{y^N \in \{y^N | F(z^N) \neq \emptyset\}} \min_{\beta \in B} \max_{f \in F(z^N)} \gamma(S(f),\beta) \tag{7.4}$$

are attained (see Theorem 4.3), the game (7.3) has a saddle point (the proof is similar to that for scheme (7.1)).

If only one terminal operation $\widetilde{\beta}$ is permitted, we have to omit the minimum with respect to $\beta \in B$ in (7.2) and (7.4), replace $\gamma(S(f),\beta)$ by $\gamma(S(f),\widetilde{\beta}(z^N))$, and also omit β in (7.1), (7.3).

The games corresponding to the settings where "the computer" applies algorithms with delayed information or block algorithms could be analyzed in a similar way.

In some cases, the fact that the game describing the solution process has a saddle point helps us construct optimal and sequentially optimal deterministic computational algorithms. A typical example of such construction is given in Sections 2 and 3 of Chapter 3.

At the same time, when considering stochastic computational algorithms, the approach to the solution process as to a game with complete information may not reflect the real state of affairs adequately and may lead to the wrong conclusion about inexpediency of applying stochastic algorithms.

We could come to this conclusion using the following argument: a game with complete information has a saddle point in pure strategies, and, hence, the best guaranteed result in mixed strategies (stochastic algorithms) coincides with the best guaranteed result in pure strategies (deterministic algorithms), and an optimal deterministic algorithm is also optimal in the class of all stochastic algorithms.

The mistake of this argument is that the best guaranteed result in the class of stochastic algorithms for the original problem should be defined in the same way as the best guaranteed result in mixed strategies for the corresponding antagonistic game.

We illustrate this with a simple example. Assume that "the computer" can apply only nonadaptive algorithms $x^N \in X^N$ and that some terminal operation $\tilde{\beta}$ is fixed. In this case, the corresponding game with complete information is even simpler than the game (7.3) and can be represented by the following sequence of alternative choices made by the players:

$$x^N, f. \tag{7.5}$$

The payoff function in this game is the criterion $\varepsilon(x^N, f)$ defined by (4.14).

Nonadaptive stochastic algorithms (nonadaptive mixed strategies) are determined by probability measures $\sigma^N \in {}^N$, where ^{N} is the set of all probability measures on some σ-algebra of the space of elementary events X^N. Application of a nonadaptive stochastic algorithm σ^N consists in random choice of some $x^N \in X^N$ in accordance with the probability distribution σ^N and subsequent computation of $y_1 = x_1(f), \dots, y_N = x_N(f)$, $\beta = \beta(x_1, \dots, x_n, y_1, \dots, y_N)$ (like in the deterministic setting).

The best guaranteed result in mixed strategies for the game (7.5) is

$$\inf_{\sigma^N \in {}^N} \int_{X^N} \sup_{f \in F} \varepsilon(x^N, f)\sigma^N\{dx^N\} = \inf_{x^N \in X^N} \sup_{f \in F} \varepsilon(x^N, f). \tag{7.6}$$

However, the original statement of the problem gives no reason to think that "nature" will "learn" the concrete realization of the random choice made by "the computer". Therefore, in the original setting the best result "the computer" can guarantee using stochastic algorithms should be defined by the formula

$$\inf_{\sigma^N \in {}^N} \sup_{f \in F} \int_{X^N} \varepsilon(x^N, f)\sigma^N\{dx^N\}. \tag{7.7}$$

Note that the integrands in (7.6), (7.7) and all the rest formulas in this section are assumed to be integrable. Now we can proceed to define various optimality concepts for stochastic algorithms.

7.2. Optimal by error nonadaptive stochastic algorithms

An algorithm $\sigma_0^N \in {}^N$ is called *optimal by error on F in the class of nonadaptive stochastic algorithms* ^{N} iff

$$\sup_{f\in F} \int_{X^N} \varepsilon(x^N, f)\sigma_0^N\{dx^N\} = \min_{\sigma^N \in {}^N} \sup_{f\in F} \int_{X^N} \varepsilon(x^N, f)\sigma^N\{dx^N\}. \tag{7.8}$$

In the case where arbitrary terminal operations are permitted and "the computer" can randomize the choice of β, a stochastic nonadaptive algorithm is determined by a pair $(\sigma^N, \tilde{\tau})$, where $\sigma^N \in {}^N$,

$$\tilde{\tau}\colon \{z^N \mid F(z^N) \neq \emptyset\} \to \ , \tag{7.9}$$

and is the set of all probability measures on some σ-algebra of the space of elementary events B. Denote the set of all mappings of the form (7.9) by $\tilde{\ }$.

Application of a nonadaptive stochastic algorithm $(\sigma^N, \tilde{\tau}) \in {}^N \times \tilde{\ }$ to a function $f \in F$ consists in random choice of some $x^N \in X^N$ in accordance with the probability distribution σ^N, computation of $y^N = x^N(f) = (x_1(f), \dots, x_N(f))$, determination of $\tau = \tilde{\tau}(z^N) \in$, and, finally, random choice of some $\beta \in B$ in accordance with the probability distribution τ.

The result guaranteed by the algorithm $(\sigma^N, \tilde{\tau})$ on F is equal to

$$\sup_{f\in F} \int_{X^N} \sigma^N\{dx^N\} \int_B \tilde{\tau}(x^N, x^N(f))\{d\beta\}\gamma(S(f), \beta).$$

Now it is natural to give the following definition. An algorithm $(\sigma_0^N, \tilde{\tau}_0) \in {}^N \times \tilde{\ }$ is called *optimal by error on F in the class of nonadaptive stochastic algorithms* ${}^N \times \tilde{\ }$ iff

$$\sup_{f\in F} \int_{X^N} \sigma_0^N\{dx^N\} \int_B \tilde{\tau}(x^N, x^N(f))\{d\beta\}\gamma(S(f), \beta)$$

$$= \min_{(\sigma^N, \tilde{\tau}) \in {}^N \times \tilde{\ }} \sup_{f\in F} \int_{X^N} \sigma^N\{dx^N\} \int_B \tilde{\tau}(x^N, x^N(f))\{d\beta\}\gamma(S(f), \beta).$$

Similarly to Theorem 4.3, it can be verified that the best guaranteed result on F for algorithms from ${}^N \times \tilde{\ }$ satisfies the equation

$$\inf_{(\sigma^N, \tilde{\tau}) \in {}^N \times \tilde{\ }} \sup_{f\in F} \int_{X^N} \sigma^N\{dx^N\} \int_B \tilde{\tau}(x^N, x^N(f))\{d\beta\}\gamma(S(f), \beta)$$

$$= \inf_{\sigma^N \in {}^N} \sup_{f_1\in F} \int_{X^N} \sigma^N\{dx^N\} \inf_{\tau \in} \sup_{f_2 \in F(x^N, x^N(f_1))} \int_B \tau\{d\beta\}\gamma(S(f_2), \beta). \tag{7.10}$$

7.3. Optimal by error adaptive stochastic algorithms

An *adaptive stochastic algorithm* (in case the terminal operation $\tilde{\beta}$ is fixed) is defined as a set of mappings

$$\tilde{\sigma}^N = (\tilde{\sigma}_1, \tilde{\sigma}_2, \ldots, \tilde{\sigma}_N), \tag{7.11}$$

where

$$\tilde{\sigma}_1 \equiv \sigma_1 \in {}_1, \quad \tilde{\sigma}_{i+1}: \{z^i \mid F(z^i) \neq \emptyset\} \rightarrow {}_{i+1}, \quad i \geq 1, \tag{7.12}$$

and ${}_i$ is the set of all probability measures on some σ-algebra of the space of elementary events X_i, $i \geq 1$ (certainly, we could also define every mapping $\tilde{\sigma}_{i+1}$, $i \geq 1$, on the whole set $X_1 \times \cdots \times X_i \times Y_1 \times \cdots \times Y_i$, like in the deterministic setting). Denote the set of all algorithms $\tilde{\sigma}^N$ of the form (7.11), (7.12) by $\tilde{}^N$.

Application of an adaptive stochastic algorithm $\tilde{\sigma}^N$ to a function $f \in F$ consists in random choice of some $x_1 \in X_1$ in accordance with the probability distribution σ_1, computation of $y_1 = x_1(f)$, determination of $\sigma_2 = \tilde{\sigma}_2(z^1)$, random choice of $x_2 \in X_2$ in accordance with the probability distribution σ_2, and so on.

The result guaranteed on F by the algorithm $\tilde{\sigma}^N$ is equal to

$$\sup_{f \in F} \int_{X_1} \tilde{\sigma}_1\{dx_1\} \int_{X_2} \tilde{\sigma}_2(x_1, x_1(f))\{dx_2\} \ldots$$
$$\int_{X_N} \tilde{\sigma}_N(x^{N-1}, x^{N-1}(f))\{dx_N\} \varepsilon(x^N, f),$$

where $x^{N-1}(f) = (x_1(f), \ldots, x_{N-1}(f))$ and the criterion $\varepsilon(x^N, f)$ is defined by (4.14).

An algorithm $\tilde{\sigma}_0^N = (\tilde{\sigma}_1^0, \ldots, \tilde{\sigma}_N^0) \in \tilde{}^N$ is called *optimal by error on F in the class of adaptive stochastic algorithms* $\tilde{}^N$ iff

$$\sup_{f \in F} \int_{X_1} \tilde{\sigma}_1\{dx_1\} \int_{X_2} \tilde{\sigma}_2^0(x_1, x_1(f))\{dx_2\} \ldots$$
$$\int_{X_N} \tilde{\sigma}_N^0(x^{N-1}, x^{N-1}(f))\{dx^N\} \varepsilon(x^N, f)$$
$$= \min_{\tilde{\sigma}^N \in \tilde{}^N} \sup_{f \in F} \int_{X_1} \tilde{\sigma}_1\{dx_1\} \int_{X_2} \tilde{\sigma}_2(x_1, x_1(f))\{dx_2\} \ldots$$
$$\int_{X^N} \tilde{\sigma}_N^0(x^{N-1}, x^{N-1}(f))\{dx^N\} \varepsilon(x^N, f).$$

Similarly to Theorem 4.6, it can be obtained that the best guaranteed result on F for algorithms from $\widetilde{\ }^N$ satisfies the equation

$$\inf_{\widetilde{\sigma}\in \widetilde{\ }^N} \sup_{f\in F} \int_{X_1} \widetilde{\sigma}_1\{dx_1\} \int_{X_2} \widetilde{\sigma}_2(x_1, x_1(f))\{dx_2\} \ldots$$

$$\int_{X_N} \widetilde{\sigma}_N(x^{N-1}, x^{N-1}(f))\{dx_N\}\varepsilon(x^N, f)$$

$$= \inf_{\sigma_1\in \ _1} \sup_{f_1\in F} \int_{X_1} \sigma_1\{dx_1\} \inf_{\sigma_2\in \ _2} \sup_{f_2\in F(x_1,x_1(f_1))} \int_{X_2} \sigma_2\{dx_2\} \ldots$$

$$\inf_{\sigma_N\in \ _N} \sup_{f_N\in F(x^{N\ 1}, x^{N\ 1}(f_{N1}\ ))} \int_{X_N} \sigma_N\{dx_N\}\varepsilon(x^N, f_N). \tag{7.13}$$

In the case where arbitrary terminal operations are permitted and "the computer" can randomize the choice of $\beta_{\widetilde{\ }}$ an adaptive stochastic algorithm is determined by a pair $(\widetilde{\sigma}^N, \widetilde{\tau}) \in \widetilde{\ }^N \times \widetilde{\ }$ (see (7.9)), and the best result guaranteed by this algorithm is equal to

$$\sup_{f\in F} \int_{X_1} \widetilde{\sigma}_1\{dx_1\} \int_{X_2} \widetilde{\sigma}_2(x_1, x_1(f))\{dx_2\} \ldots$$

$$\int_{X_N} \widetilde{\sigma}_N(x^{N-1}, x^{N-1}(f))\{dx_N\} \int_B \widetilde{\tau}(x^N, x^N(f))\{d\beta\}\gamma(S(f), \beta).$$

For this case we can also introduce the concept of optimal algorithm (analogous to the optimality concepts introduced above) and obtain the formula for the best guaranteed result on F for algorithms from $\widetilde{\ }^N \times \widetilde{\ }$.

We now make an important comment on the definition of adaptive stochastic algorithm. This definition implies local randomization of alternative choice, that is, random choice of one of the possible alternatives is supposed to be performed in every realizable situation (in every position, it terms of the game theory) z^i. In the game theory, such algorithms are called *strategies of behaviour* (see Krasnoshchekov, Morozov and Fedorov [79a]). It is also possible to use global randomization, that is, to randomize choice of adaptive algorithms from the class $\widetilde{A}^N$ (or from $\hat{X}^N$ if the terminal operation is fixed).

As is known in the game theory (see Krasnoshchekov, Morozov and Fedorov [79a]), the local and global randomization methods are equivalent for games with full memory and a finite set of alternatives in every position for every player. Equivalence here means that for every pair of the players' strategies obtained by one of the two methods there is a pair of strategies obtained by the other method and guaranteeing the same average payoffs.

It is natural to expect that in our setting, though sets of alternatives are infinite in every position, the local and global randomizatiom methods will also be equivalent (under certain technical assumptions) since the corresponding game is a game with full memory. This equivalence has been proved for the optimization problem (see Nemirovsky and Yudin [79, Chapter 1, Theorem 3.4]).

7.4. One-step optimal stochastic algorithms

We now formulate the concept of one-step optimal stochastic algorithm, still assuming that all the integrands are integrable. Let the terminal operation be fixed, and let an algorithm

$$\tilde{\sigma}=(\tilde{\sigma}_1,\tilde{\sigma}_2,\dots,\tilde{\sigma}_N,\dots) \tag{7.14}$$

be a sequence of mappings from (7.12). Denote the set of all such algorithms by $\widetilde{\ }$. We will estimate their efficiency using the criterion obtained by averaging the criterion (4.14). The algorithm (7.14), as well as deterministic algorithms, stops either after a given number of steps N or after a prescribed accuracy ε has been guaranteed (see (4.20)).

An algorithm $\tilde{\sigma}$ is called *one-step optimal on F in the algorithmic class* $\widetilde{\ }$ iff

$$\sup_{f\in F}\int_{X_1} \varepsilon(x,f)\sigma_1\{dx\}=\min_{\sigma\in\ _1}\sup_{f\in F}\int_{X_1}\varepsilon(x,f)\sigma\{dx\}, \tag{7.15}$$

where $\sigma_1=\tilde{\sigma}_1$, and for any realizable situation $z^i=(x^i,y^i)$, $i\geq 1$,

$$\sup_{f\in F(z^i)}\int_{X_{i+1}} \varepsilon(x^i,x,f)\sigma_{i+1}\{dx\}=\min_{\sigma\in\ _{i+1}}\sup_{f\in F(z^i)}\int_{X_{i+1}}\varepsilon(x^i,x,f)\sigma\{dx\}, \tag{7.16}$$

where $\sigma_{i+1}=\tilde{\sigma}_{i+1}(z^i)$.

Note that the definition of one-step optimal stochastic algorithm would not be adequate if we changed the order of taking the supremum and integration in (7.15) and (7.16). It would mean that we regard "nature", which "chooses" f, as a thinking adversary who can "learn" the concrete realization of the random variable x_{i+1}. This approach would directly lead us to the conclusion that a one-step optimal deterministic algorithm is also one-step optimal among all stochastic algorithms.

We also note that, for any $\sigma_{i+1}\in\ _{i+1}$, the worst-case *a posteriori* accuracy after the $(i+1)$st step

$$\sup_{y_{i+1}\in\{y_{i+1}|F(z^{i+1})\neq\emptyset\}}\ \sup_{f\in F(z^{i+1})}\varepsilon(x^{i+1},f)=\sup_{f\in F(z^i)}\varepsilon(x^{i+1},f)$$

for any $\sigma_{i+1}\in\ _{i+1}$ on the average is not better than the average *a priori* accuracy guaranteed before the $(i+1)$st step (the latter is given by the both left- and right-hand sides of (7.16)) since

$$\sup_{f\in F(z^i)}\int_{X_{i+1}}\varepsilon(x^{i+1},f)\sigma_{i+1}\{dx_{i+1}\}\leq\int_{X_{i+1}}\sup_{f\in F(z^i)}\varepsilon(x^{i+1},f)\sigma_{i+1}\{dx_{i+1}\}.$$

7.5. Sequentially optimal stochastic algorithms

Now we define the concept of sequentially optimal by error stochastic algorithm, assuming that the terminal operation $\widetilde{\beta}$ is fixed and, as before, all the integrands are integrable. The definition given below is analogous to Definition 2 from Section 6. We could also give a definition analogous to Definition 1 from the same section and prove the equivalence of these two definitions of sequentially optimal by error stochastic algorithms using representation (7.13) and the corresponding representations of the best guaranteed results on classes $F(z^i)$.

A stochastic algorithm $\widetilde{\sigma}_0^N = (\widetilde{\sigma}_1^0, \dots, \widetilde{\sigma}_N^0) \in {}^{\sim N}$ is called *sequentially optimal by error on* F iff the outermost infimum in the formula of the best guaranteed result on F

$$\inf_{\sigma_1 \in \ _1} \sup_{f_1 \in F} \int_{X_1} \sigma_1\{dx_1\} \inf_{\sigma_2 \in \ _2} \sup_{f_2 \in F(x_1, x_1(f_1))} \int_{X_2} \sigma_2\{dx_2\} \dots$$

$$\inf_{\sigma_N \in \ _N} \sup_{f_N \in F(x^{N\ 1}, x^{N\ 1}(f_{N\ 1}))} \int_{X_N} \sigma_N\{dx_N\} \varepsilon(x^N, f_N)$$

is attained at $\sigma_1 = \widetilde{\sigma}_1^0$, and the outermost infimum in the formula of the best guaranteed result on $F(z^i)$

$$\inf_{\sigma_{i+1} \in \ _{i+1}} \sup_{f_{i+1} \in F(z^i)} \int_{X_{i+1}} \sigma_{i+1}\{dx_{i+1}\} \dots$$

$$\inf_{\sigma_N \in \ _N} \sup_{f_N \in F(x^{N\ 1}, x^{N\ 1}(f_{N\ 1}))} \int_{X_N} \sigma_N\{dx_N\} \varepsilon(x^N, f_N)$$

is attained at $\sigma_{i+1} = \widetilde{\sigma}_{i+1}^0$ for any realizable situation z^i, $1 \le i \le N-1$.

CHAPTER 2

NUMERICAL INTEGRATION

In this chapter, we consider numerical integration of functions of one or more variables. We derive optimal quadrature formulas, optimal adaptive, nonadaptive, and sequentially optimal integration algorithms for various functional classes. We analyze the influence of computational errors on the accuracy of the solution. We also deal with the problem of program implementation of the algorithms derived.

1. OPTIMAL QUADRATURES FOR FUNCTIONAL CLASSES DETERMINED BY QUASI-METRICS

In this section, we obtain a number of general rezults on optimal nodes and weights of quadrature formulas for functional classes determined by quasi-metrics.

1.1. Statement of the problem

Let K be a measurable set of finite Lebesgue measure in n-dimensional Euclidean space $\mathbb{R}^n$. Consider a function $\rho(u,v)$ defined on $K \times K$, integrable with respect to u for any fixed v, and satisfying the properties of a *quasi-metric*, see (2.2)-(2.5), Chapter 1. Throughout this chapter we denote by F_ρ the class of measurable functions on K satisfying the inequality

$$|f(u)-f(v)| \leq \rho(u,v), \quad u,v \in K.$$

Thus, in addition to the assumptions of Section 2.2 of Chapter 1, functions from F_p are assumed to be measurable.

Suppose that the information on the function f obtained through the process of computations consists of N function values

$$x_i(f)=f(x_i), \quad x_i \in K, \ i=1,\dots,N,$$

so in (3.1), Chapter 1, we have

$$X_i=\dots=X_N=K.$$

Methods of approximating integrals $\int_K f(x)dx$ with sums $\sum_{i=1}^N p_i f(x_i)$ are a traditional subject of investigation. Integration formulas of such a form (and sometimes also functions $\sum_{i=1}^N p_i f(x_i)$) are called *quadrature formulas*, or *quadratures*. The points $x_1, \dots, x_N$ are called *nodes*, and the numbers $p_1, \dots, p_N$ are called *weights* or *coefficients* of the quadrature formula.

Optimal weights of a quadrature formula on the class F for fixed nodes $x_1, \dots, x_N$ are coefficients $p_1^0(x^N), \dots, p_N^0(x^N)$ such that

$$\sup_{f\in F}\left|\int_K f(x)dx - \sum_{i=1}^N p_i^0(x^N)f(x_i)\right| = \min_{p^N\in\mathbb{R}^N}\sup_{f\in F}\left|\int_K f(x)dx - \sum_{i=1}^N p_i f(x_i)\right|, \tag{1.1}$$

where $x^N = (x_1, \dots, x_N)$ and $p^N = (p_1, \dots, p_N)$.

Optimal nodes on the class F are points $x_1^0, \dots, x_N^0 \in K$ such that

$$\min_{p^N\in\mathbb{R}^N}\sup_{f\in F}\left|\int_K f(x)dx - \sum_{i=1}^N p_i f(x_i^0)\right|$$
$$= \min_{x^N\in K^N}\min_{p^N\in\mathbb{R}^N}\sup_{f\in F}\left|\int_K f(x)dx - \sum_{i=1}^N p_i f(x_i)\right|, \tag{1.2}$$

where K^N is the Nth Cartesian power of the set K.

A quadrature formula with optimal nodes and weights is called *optimal*. The problem of obtaining optimal quadrature formulas was closely considered by many authors, see lists of references in Nikolskii [79], Traub and Woźniakowski [80], and Zhensykbayev [81].

Our aim is to estimate the error of a given quadrature formula on the class $F = F_\rho$, i.e., to obtain a method of evaluating

$$\sup_{f\in F}\left|\int_K f(x)dx - \sum_{i=1}^N p_i f(x_i)\right|$$

for given p^N and x^N. We also want to find optimal weights $p_1^0(x^N), \dots, p_N^0(x^N)$ for given nodes x^N and, finally, to describe the way of selecting optimal nodes.

1.2. Auxiliary statements

We now obtain some auxiliary results which will enable us to solve the posed problems. Throughout this section we suppose that $F = F_p$. In particular, we have

$$F(z^i) = \{f \in F_\rho \mid f(x_j) = y_j,\ j = 1, \dots, i\},$$

where $z^i=(x^i,y^i)$. Let, for $1\leq i\leq N$,

$$\phi_{1i}(x)=\max_{j=1,\dots,i}\{y_j-\rho(x,x_j)\}, \quad \phi_{2i}(x)=\min_{j=1,\dots,i}\{y_j+\rho(x,x_j)\}, \tag{1.3}$$

$$\Phi_1(z^i)=\int_K \phi_{1i}(x)dx, \quad \Phi_2(z^i)=\int_K \phi_{2i}(x)dx,$$

$$\Psi_1(z^i,p^i)=\langle p^i,y^i\rangle-\Phi_1(z^i), \quad \Psi_2(z^i,p^i)=\langle p^i,y^i\rangle-\Phi_2(z^i). \tag{1.4}$$

Note that it is only for simplicity that we do not list $z^i=(x_1,\dots,x_i,y_1,\dots,y_i)$ among the arguments and the subscripts of the functions $\phi_{1i}(x)$ and $\phi_{2i}(x)$. However, the subscript i reminds the reader of their dependence on z^i.

To begin with, we prove three simple lemmas.

Lemma 1.1. *If $|y_\mu-y_\nu|\leq(x_\mu,x_\nu)$ for $\mu,\nu=1,\dots,i$, then $\phi_{1i}\in F(z^i)$ and $\phi_{1i}\in F(z^i)$.*

Proof. The assumption of the lemma yields $y_\nu-\rho(x_\mu,x_\nu)\leq y_\mu$. Hence, $\phi_{1i}(x_\mu)=\max_{\nu=1,\dots,i}\{y_\nu-\rho(x_\mu,x_\nu)\}\leq y_\mu$. On the other hand, we have $\phi_{1i}(x_\mu)\geq y_\mu-\rho(x_\mu,y_\mu)=y_\mu$. Thus, $\phi_{1i}(x_\mu)=y_\mu$, $\mu=1,\dots,i$. Since $\phi_{1i}\in F$ due to Lemmas 2.1 and 2.7 of Chapter 1, we have $\phi_{1i}\in F(z^i)$. In the same way we prove that $\phi_{2i}\in F(z^i)$. □

Lemma 1.2. *If $f\in F(z^i)$, then $\phi_{1i}(x)\leq f(x)\leq\phi_{2i}(x)$, $x\in K$.*

Proof. Since $f\in F(z^i)$, we have

$$-\rho(x,x_j)\leq f(x)-f(x_j)=f(x)-y_j\leq\rho(x,x_j), \quad j=1,\dots,i.$$

Therefore,

$$\max_{j=1,\dots,i}\{y_j-\rho(x,x_j)\}\leq f(x)\leq\min_{j=1,\dots,i}\{y_j+\rho(x,x_j)\}. \qquad \square$$

Lemmas 1.1 and 1.2 yield that ϕ_{1i} provides a sharp lower envelope for functions from $F(z^i)$, and ϕ_{2i} provides a sharp upper envelope for those functions.

Lemma 1.3. 1) *The set $l(x^i)=\{y^i \mid F(z^i)\neq 0\}$ admits the representation*

$$l(x^i)=\{y^i \mid |y_\mu-y_\nu|\leq\rho(x_\mu,x_\nu),\ \mu,\nu=1,\dots,i\}.$$

2) *The set $l(x^i)$ is convex.*

Proof. 1) The inclusion

$$l(x^i)\subset\{y^i \mid |y_\mu-y_\nu|\leq\rho(x_\mu,x_\nu),\ \mu,\nu=1,\dots,i\}$$

is obvious. The opposite inclusion follows from Lemma 1.1. This proves the first statement of the lemma.

2) Let $u^i = (u_1, \ldots, u_i) \in l(x^i)$, $v^i = (v_1, \ldots, v_i) \in l(x^i)$, $0 \le \lambda \le 1$. Then for $\mu, \nu = 1, \ldots, i$ we have
$|[\lambda u_\mu + (1-\lambda)v_\mu] - [\lambda u_\nu + (1-\lambda)v_\nu]| \le \lambda |u_\mu - u_\nu| + (1-\lambda)|v_\mu - v_\nu| \le \rho(x_\mu, x_\nu)$,
i.e., $\lambda u^i + (1-\lambda)v^i \in l(x^i)$, which completes the proof. □

Let m take the values 1, 2. Denote by $K_{mi}(z^N)$, $i = 1, \ldots, N$, an arbitrary collection of sets forming a partition of K, i.e.,

$$\cup_{i=1}^N K_{mi}(z^N) = K, \quad K_{mi}(z^N) \cap K_{mj}(z^N) = 0 \quad \text{if} \quad i \ne j,$$

and satisfying the following conditions:

$$\begin{aligned} &K_{mi}(z^N) \quad \text{is measurable}, \quad i = 1, \ldots, N; \\ &y_i + (-1)^m \rho(x, x_i) = \phi_{mN}(x) \quad \text{if} \quad x \in K_{mi}(z^N). \end{aligned} \tag{1.5}$$

Observe that in the case under consideration (when ρ is a quasi-metric) this partition may not be unique, as well as the measures of the sets forming the partition. Denote the Lebesgue measure of a set A by $\mu(a)$ and consider the set

$$A_m(z^N) = \{g = (g_1, \ldots, g_N) \mid g_i = \mu(K_{mi}(z^N)), \ g = 1, \ldots, N\}, \quad m = 1, 2.$$

It is clear from the definition that every element g of $A_m(z^N)$ corresponds to some partition (1.5) such that the components of the vector g are the Lebesgue measures of the sets $K_{mi}(z^N)$ forming the partition.

Lemma 1.4. *The sets $A_1(z^N)$ and $A_2(z^N)$ are convex and closed.*

Proof. Since we assume z^N to be fixed, we do not explicitly point out the dependence of K_{mi}, A_m and the sets introduced below on z^N. For any pairwise different $i_1, \ldots, i_k \in \{1, \ldots, N\}$, $1 \le k \le N$, consider the set

$$\begin{aligned} B_{i_1,\ldots,i_k} = \{x \in K \mid\ & y_i - \rho(x, x_i) = \phi_{1N}(x) > y_j - \rho(x, x_j), \\ & i \in \{i_1, \ldots, i_k\}, \ j \in \{1, \ldots, N\} \setminus \{i_1, \ldots, i_k\}\}. \end{aligned}$$

Consider for $\alpha = (\alpha^1, \ldots, \alpha^N) \in A_1$ a partition K_{1i}^α of K satisfying conditions (1.5) such that $\alpha^i = \mu(K_{1i}^\alpha)$, $i = 1, \ldots, N$. Put for $k > 2$

$$B_{i_1,\ldots,i_k}^{\alpha i} = K_{1i}^\alpha \cap B_{i_1,\ldots,i_k} = \mu(B_{i_1,\ldots,i_k}^{\alpha i}) < i \in \{i_1, \ldots, i_k\}.$$

Let $\beta = (\beta^1, \ldots, \beta^N) \in A_1$, $0 \le \lambda \le 1$, and let the notation used for α be retained for β. Let

$$\gamma_{i_1,\ldots,i_k}^i = \lambda \alpha_{i_1,\ldots,i_k}^i + (1-\lambda)\beta_{i_1,\ldots,i_k}^i, \quad i \in \{i_1, \ldots, i_k\}.$$

We can partition $B_{i_1,\ldots,i_k}$ into sets $B_{i_1,\ldots,i_k}^{\gamma i}$ such that $\mu(B_{i_1,\ldots,i_k}^{\gamma i}) = \gamma_{i_1,\ldots,i_k}^i$, $i \in \{i_1, \ldots, i_k\}$. Now it is clear that the sets

$$K_{1i}^\gamma \stackrel{def}{=} \bigcup_{i \in \{i_1,\ldots,i_k\};\ k \ge 2} B_{i_1,\ldots,i_k}^{\gamma i} \cup B_i, \quad i = 1, \ldots, N,$$

satisfy conditions (1.5). Putting $\gamma^i = \mu(K_{1i}^\gamma)$, $\gamma = (\gamma^1, \ldots, \gamma^N)$ we have $\gamma \in A_1$.

On the other hand, for $i=1,\dots,N$ we have

$$\gamma^i=\mu(K_{1i}^{\gamma})=\sum_{i\in\{i_1,\dots,i_k\};\ k\geq 2}\gamma^i_{i_1,\dots,i_k}+\mu(B_i)$$

$$=\lambda\left[\sum_{i\in\{i_1,\dots,i_k\};\ k\geq 2}\alpha^i_{i_1,\dots,i_k}+\mu(B_i)\right]$$

$$+(1-\lambda)\left[\sum_{i\in\{i_1,\dots,i_k\};k\geq 2}\beta^i_{i_1,\dots,i_k}+\mu(B_i)\right]$$

$$=\lambda\alpha^i+(1-\lambda)\beta^i.$$

Therefore, $\gamma=\lambda\alpha-(1-\alpha)\beta\in A_1$, which proves the convexity of A_1.

Now we prove that A_1 is closed. Let $\alpha=(\alpha^1(t),\dots,\alpha^N(t))\in A_1$, and let $\alpha(t)$ tend to α as $t\to\infty$. Define $\alpha^i_{i_1,\dots,i_k}(t)$, $i\in\{i_1,\dots,i_k\}$, $k\geq 2$, by analogy with the definition of $\alpha^i_{i_1,\dots,i_k}$ in the first part of the proof. Select $t_r\to\infty$ such that $\alpha^i_{i_1,\dots,i_k}(t_r)\to\alpha^i_{i_1,\dots,i_k}$ as $r\to\infty$ for any permissible $i_1,\dots,i_k,i,k$. Let sets $B^{\alpha i}_{i_1,\dots,i_k}$ form a partition of $B_{i_1,\dots,i_k}$ such that $\mu(B^{\alpha i}_{i_1,\dots,i_k})=\alpha_{i_1,\dots,i_k}$, $i\in\{i_1,\dots,i_k\}$, $k\geq 2$. Then partitioning K into sets

$$K^{\alpha}_{1i}=\bigcup_{i\in\{i_1,\dots,i_k\};\ k\geq 2}B^{\alpha i}_{i_1,\dots,i_k}\cup B_i,\quad i=1,\dots,N,$$

corresponds to α, i.e., $\mu(K^{\alpha}_{1i})=\alpha^i$, $i=1,\dots,N$. For the proof we note that

$$\mu(K^{\alpha}_{1i})=\sum_{i\in i_1,\dots,i_k;\ k\geq 2}\alpha^i_{i_1,\dots,i_k}+\mu(B_i)$$

$$=\lim_{r\to\infty}\left[\sum_{i\in i_1,\dots,i_k;\ k\geq 2}\alpha^i_{i_1,\dots,i_k}(t_r)+\mu(B_i)\right]$$

$$=\lim_{r\to\infty}\alpha^i(t_r)=\alpha^i,\quad i=1,\dots,N.$$

Hence, $\alpha\in A_1$; therefore, A_1 is closed.

The fact that A_2 is convex and closed is established in the same way. □

Recall that the set

$$\partial_{y^N}\Phi(y^N)=\{g\in\mathbb{R}^N \mid \Phi(y^N+\Delta y)-\Phi(y^N)\geq\langle g,\Delta y\rangle,\ \forall\ \Delta y\in\mathbb{R}^N\}$$

is called the subdifferential of the convex function Φ: $\mathbb{R}^N\to\mathbb{R}$ at the point y^N, and the set

$$\partial_{y^N}\Psi(y^N)=\{g\in\mathbb{R}^N \mid \Psi(y^N+\Delta y)-\Psi(y^N)\leq\langle g,\Delta y\rangle,\ \forall\ \Delta y\in\mathbb{R}^N\}$$

is called the subdifferential of the concave function Ψ: $\mathbb{R}^N\to\mathbb{R}$ at y^N.

Lemma 1.5. *The function $\Phi_1(z^N)$ is convex on $\mathbb{R}^N$ with respect to y^N; the function $\Phi_2(z^N)$ is concave on $\mathbb{R}^N$ with respect to y^N. Moreover,*

$$\partial_{y^N}\Phi_1(y^N)=A_1(z^N)\ \text{ and }\ \partial_{y^N}\Phi_2(y^N)=A_2(z^N).$$

Proof. The convexity of Φ_1 is obvious. For any partition $K_{1i}(z^N)$, $i=1,\dots,N$, of K satysfying conditions (1.5) and for any $\Delta y=(\Delta y_1,\dots,\Delta y_N)$ we obtain

$$\begin{aligned}
&\Phi_1(x^N,y^N+\Delta y)-\Phi_1(x^N,y^N)\\
&\quad=\int_K\left[\max_{j=1,\dots,N}\{y_j+\Delta y_j-\rho(x,x_j)\}-\max_{j=1,\dots,N}\{y_j-\rho(x,x_j)\}\right]dx\\
&\quad=\sum_{i=1}^N\int_{K_{1i}(z^N)}\left[\max_{j=1,\dots,N}\{y_j+\Delta y_j-\rho(x,x_j)\}-y_j+\rho(x,x_j)\right]dx\\
&\quad\geq\sum_{i=1}^N\int_{K_{1i}(z^N)}\Delta y_i\,dx=\sum_{i=1}^N\Delta y_i\mu(K_{1i}(z^N)),
\end{aligned}$$

hence

$$\partial_{y^N}\Phi_1(z^N)\supset A_1(z^N). \tag{1.6}$$

Let

$$K_{11}^0(z^N)=\{x|\ y_1-\rho(x,x_1)=\Phi_{1N}(x)\},$$

$$K_{1i}^0(z^N)=\{x|\ y_i-\rho(x,x_i)=\Phi_{1N}(x)\}\setminus\bigcup_{j=1}^{i-1}K_{1j}^0(z^N),\quad i=1,\dots,N.$$

Clearly, the sets $K_{1i}^0(z^N)$, $i=1,\dots,N$, for any y^N constitute a partition of K satisfying conditions (1.5), and for any $\Delta y\in\mathbb{R}^N$ we have

$$\mu(K_{1i}^0(x^N,y^N+\delta\Delta y))\to\mu(K_{1i}^{\Delta y}(z^N))\quad\text{as}\quad\delta\to 0.$$

Furthermore, the sets $K_{1i}^{\Delta y}(z^N)$, $i=1,\dots,N$, also constitute a partition of K satisfying conditions (1.5). Thus, for any $\Delta y\in\mathbb{R}^N$, $\delta>0$

$$\begin{aligned}
&\sup_{g\in\partial_{y^N}\Phi_1(y^N)}\langle g,\delta\Delta y\rangle\leq\Phi_1(x^N,y^N+\delta\Delta y)-\Phi_1(x^N,y^N)\\
&=\int_K\left[\max_{j=1,\dots,N}\{y_j+\delta\Delta y_j-\rho(x,x_j)\}-\max_{j=1,\dots,N}\{y_j-\rho(x,x_j)\}\right]dx\\
&=\sum_{j=1}^N\int_{K_{1i}^0(x^N,y^N+\delta\Delta y)}\left[y_i+\delta\Delta y_i-\rho(x,x_i)-\max_{j=1,\dots,N}\{y_j-\rho(x,x_j)\}\right]dx
\end{aligned}$$

$$\leq \delta \sum_{j=1}^{N} \Delta y_i \mu(K_{1i}^{0}(x^N, y^N + \delta\Delta y))$$

$$= \delta \sum_{j=1}^{N} \Delta y_i \mu(K_{1i}^{\Delta y}(z^N)) + \delta \sum_{j=1}^{N} \Delta y_i \left[\mu(K_{1i}^{0}(x^N, y^N + \delta\Delta y))\mu(K_{1i}^{\Delta y}(z^N))\right]$$

$$= \delta \sum_{j=1}^{N} \Delta y_i \mu(K_{1i}^{\Delta y}(z^N)) + o(\delta).$$

Therefore, for any $\Delta y \in \mathbb{R}^N$

$$\sup_{g \in \partial_{y^N}\Phi_1(y^N)} \langle g, \Delta y\rangle \leq \langle g^{\Delta y}, \Delta y\rangle + o(\delta)/\delta,$$

where $g^{\Delta y} = (\mu(K_{11}^{\Delta y}(z^N))) \in A_1(z^N)$, hence

$$\sup_{g \in \partial_{y^N}\Phi_1(y^N)} \langle g, \Delta y\rangle \leq \sup_{g \in A_1(z^N)} \langle g, \Delta y\rangle.$$

Since the opposite inequality holds due to (1.6), we get for any $\Delta y \in \mathbb{R}^N$

$$\sup_{g \in \partial_{y^N}\Phi_1(y^N)} \langle g, \Delta y\rangle = \sup_{g \in A_1(z^N)} \langle g, \Delta y\rangle.$$

According to Lemma 1.1 from Pshenichnyi [69], this implies that $\partial_{y^N}\Phi_1(z^N) = A_1(z^N)$ since the sets $\partial_{y^N}\Phi_1(z^N)$ and $A_1(z^N)$ are convex and closed due to Theorem 11 from Pshenichnyi [69], and Lemma 1.4 respectively. In the same way we obtain the equation

$$\partial_{y^N}\Phi_2(z^N) = A_2(z^N). \qquad \square$$

1.3. Error estimate for a given quadrature formula

We now proceed to the first of the problems posed in Section 2.1 – the problem of error estimation for a quadrature formula with given nodes x^N and weights p^N. Using Lemmas 1.1, 1.2 and notations (1.3), (1.4), we obtain

$$\max_{f \in F} \left| \int_K f(x)dx - \sum_{i=1}^{N} p_i f(x_i) \right| = \max_{y^N \in l(x^N)} \max_{f \in F(z^N)} \left| \int_K f(x)dx - \langle p^N, y^N\rangle \right|$$

$$= \max_{y^N \in l(x^N)} \max \left\{ \langle p^N, y^N\rangle - \int_K \phi_{1N}(x)dx, \int_K \phi_{1N}(x)dx - \langle p^N, y^N\rangle \right\}$$

$$= \max\left\{ \max_{y^N \in l(x^N)} \Psi_1(z^N, p^N), \max_{y^N \in l(x^N)} \Psi_1(z^N, p^N) \right\}. \tag{1.7}$$

Thus, the problem amounts to maximizing two concave functions on the convex (due to Lemma 1.3) set $l(x^N)$, i.e., to solving two convex programming problems. The following theorem provides necessary and sufficient conditions of the extremum for such problems.

Theorem 1.1 (Sukharev [81d]). *A vector $y^N(m)=(y_1(m),\dots,y_N(m))$ maximizes the function $\Psi_m(x^N,y^N,p^N)$ on the set $l(x^N)$ for fixed x^N and p^N if and only if there exist sets $K_{mi}(x^N,y^N(m))$, $i=1,\dots,N$, satisfying conditions* (1.5) *for $y^N=y^N(m)$ and numbers λ_m^{jk}, $j,k=1,\dots,N$, such that*

$$(-1)^m\left[\mu(K_{1i}^0(x^N,y^N(m)))-p_i\right]=\sum_{j,k=1}^{N}\lambda_m^{jk}g_i^{jk},$$

$$\lambda_m^{jk}(y_j(m)-y_k(m)-\rho(x_j,x_k))=0,\quad i,j,k=1,\dots,N,$$

where

$$g_j^{jk}=\begin{cases}1, & i=j\neq k,\\ -1, & i=k\neq j,\\ 0 & \textit{otherwise}.\end{cases}$$

Proof. Recall that $\Psi(y)=\max_{y'\in A}\Psi(y')$ for a convex set A and a concave function Ψ if and only if

$$\Gamma(y)^*\cap[-\partial_y\Psi(y)]\neq 0, \tag{1.8}$$

where $\Gamma(y)^*$ is the cone dual to the cone $\Gamma(y)$ of permissible directions at the point y with respect to the set A, see Pshenichnyi [69].

We describe the cones $\Gamma(y^N)$ and $\Gamma(y^N)^*$ for $A=l(x^n)$. By Lemma 1.3,

$$l(x^N)=\{y^N=(y_1,\dots,y_N)\mid y_j-y_k\le\rho(x_j,x_k),\quad j,k=1,\dots,N\}.$$

Therefore

$$\Gamma(y^N)=\{\Delta y=(\Delta y_1,\dots,\Delta y_N)\mid \Delta y_j-\Delta y_k\le 0\quad\text{if}\quad y_j-y_k=\rho(x_j,x_k)\}.$$

It follows by applying the Farkas lemma that

$$\Gamma(y^N)^*=\left\{-\sum_{j,k=1}^{N}\lambda^{jk}g^{jk}\;\middle|\;\lambda^{jk}(y_j-y_k-\rho(x_j,x_k))=0,\right.$$
$$\left.\lambda^{jk}\le 0,\quad j,k=1,\dots,N\right\}, \tag{1.9}$$

where $g^{jk}=(g_1^{jk},\dots,g_N^{jk})$.

Since Φ_1 is convex and Φ_2 is concave with respect to y^N, the functions Ψ_1 and Ψ_2 are concave with respect to y^N. Hence, by Lemma 1.5,

$$\partial_{y^N}\Psi_m(z^N,p^N)=\{g=(g_1,\dots,g_N)\mid g_i=(-1)^m[\mu(K_{mi}(z^N))=p_i],$$
$$i=1,\dots,N\ m=1,2.\}. \tag{1.10}$$

The theorem now follows directly from (1.8)–(1.10). □

The optimality conditions provided by Theorem 1.1 are useful for solving problems connected with estimating errors of quadrature formulas, in particular, problems of deriving optimal quadrature formulas (maybe under some restrictions on the weights). However, those conditions do not lead directly to any numerical procedure. We will not use Theorem 1.1 in our presentation.

The proof of Theorem 1.2, answering the question about optimal quadrature weights, is based on the necessary and sufficient condition of the maximum for the problem of unconstrained optimization of a concave function Ψ, i.e., for the problem of computing $\Psi(y)=\max_{y'\in A}\Psi(y')$. This condition, which is a special case of (1.8) for $A=\mathbb{R}^N$, has the form

$$0\in\partial_y\Psi(y). \tag{1.11}$$

1.4. Search for optimal weights and optimal nodes of a quadrature

The following theorem answers the question about optimal weights $p_1^0(x^N),\dots,p_N^0(x^N)$ of a quadrature formula with given nodes x^N, and its corollary enables us to solve the problem of finding optimal nodes in a number of cases. Some related results under more restrictive assumptions were obtained by Babenko [76a, b, 77], Korneichuk [68], Maung Cho Niun and Sharygin [71].

Let $\theta^N=(\theta,\dots,\theta)$, where $\theta\in\mathbb{R}$ is any real number. It is easy to show that the set of all partitions of K of the form $K_{mi}(x^N,\theta^N)$, $i=1,\dots,N$, depends neither on θ nor on m. Denote an arbitrary partition of this form by $K_i(x^N)$, $i=1,\dots,N$. In other words, $K_1(x^N),\dots,K_N(x^N)$ is an arbitrary collection of measurable pairwise disjoint sets whose union is K satisfying the condition

$$\rho(x,x_i)=\min_{j=1,\dots,N}\rho(x,x_j) \quad \text{for} \quad x\in K_i(x^N),\ i=1,\dots,N.$$

Such partitions are called *Voronoi-Dirichlet ρ-partitions.* In the case under consideration (where ρ is a quasi-metric), the sets $K_i(x^N)$ forming a Voronoi-Dirichlet partition and their measures $\mu(K_i(x^N))$, $i=1,\dots,N$, in general are not determined uniquely.

Theorem 1.2 (Sukherav [81d]). *The quadrature formula with the weights $p_i^0(x^N)=\mu(K_i(x^N))$, $i=1,\dots,N$, is optimal on F_ρ among all quadratures with fixed nodes $x_1,\dots,x_N$, and its guaranteed result (worst-case error) is equal to*

$$\int_K \min_{i=1,\dots,N}\rho(x,x_i)dx.$$

Note that, in general, due to non-uniqueness of the measures $\mu(K_i(x^N))$, optimal weights are not determined uniquely either.

Proof. For every $p^N \in \mathbb{R}^N$ we have

$$\max_{f\in F}\left|\int_K f(x)dx - \sum_{i=1}^N p_i f(x_i)\right| \geq \max_{y^N\in l(x^N)} \Psi_1(z^N, p^N)$$

$$\geq \Psi_1(x^N,\theta^N,p^N) = \theta\sum_{i=1}^N p_i - \int_K \max_{i=1,\ldots,N}\{\theta-\rho(x,x_i)\}dx$$

$$=\theta\left(\sum_{i=1}^N p_i - \mu(K)\right) + \int_K \min_{i=1,\ldots,N}\rho(x,x_i)dx$$

$$\geq \int_K \min_{i=1,\ldots,N}\rho(x,x_i)dx. \tag{1.12}$$

Here, the first inequality holds due to (1.7), the second one holds because $\theta^N \in l(x^N)$, and, finally, the third one holds since $\theta\in\mathbb{R}$ can be selected in such a way that $\theta\left(\sum_{i=1}^N p_i - \mu(K)\right)\geq 0$.

For $y^N=\theta^N$ and $p^N=p_0^N(x^N)=(p_1^0(x^N),\ldots,p_N^0(x^N))$, formula (1.10) takes the form

$$\partial_{y^N}\Psi_m(x^N,\theta^N,p_0^N(x^N))$$
$$=\{g=(g_1,\ldots,g_N)|g_i=(-1)^m(\mu(K_i(x^N))-p_i^0(x^N)),\ i=1,\ldots,N\},\quad m=1,2.$$

Every element g of this set corresponds to some Voronoi-Dirichlet partition. The definition of optimal weights yeilds that, for optimal weights $p_0^N(x^N)$, the function Ψ_m satisfies optimality condition (1.11) if $y^N=\theta^N$ and $p^N = p_0^N(x^N)$, i.e.,

$$0\in\partial_{y^N}\Psi_m(x^N,\theta^N,p_0^N(x^N)),\quad m=1,2.$$

Therefore, observing that $\theta^N\in l(x^N)$, we get

$$\Psi_m(x^N,\theta^N,p_0^N(x^N)) = \max_{y^N\in\mathbb{R}^N}\Psi_m(x^N,y^N,p_0^N(x^N))$$

$$=\max_{y^N\in l(x^N)}\Psi_m(x^N,y^N,p_0^N(x^N)),\quad m=1,2. \tag{1.13}$$

Since

$$\sum_{i=1}^N p_i^0(x^N) = \sum_{i=1}^N \mu(K_i(x^N)) = \mu(K),$$

chain of equations (1.12) yeilds that, for $p^N=p_0^N(x^N)$,

$$\Psi_1(x^N,\theta^N,p_0^N(x^N)) = \int_K \min_{i=1,\ldots,N}\rho(x,x_i)dx.$$

Similary,

$$\Psi_2(x^N,\theta^N,p_0^N(x^N))=\int_K \min_{i=1,\dots,N}\rho(x,x_i)dx.$$

Hence, taking into account (1.7) and (1.13), we have

$$\max_{f\in F}\left|\int_K f(x)dx-\sum_{i=1}^N p_i^0(x^N)f(x_i)\right|=\int_K \min_{i=1,\dots,N}\rho(x,x_i)dx. \tag{1.14}$$

The theorem now follows directly from (1.12) and (1.14). □

This theorem leads to a method of obtaining optimal nodes. We formulate it as a corollary of the theorem.

Corollary. *Obtaining optimal nodes of a quadrature formula amounts to finding points* $x_1^0,\dots,x_N^0$ *from* K *such that*

$$\int_K \min_{i=1,\dots,N}\rho(x,x_i^0)dx=\min_{x^N\in K^N}\int_K \min_{i=1,\dots,N}\rho(x,x_i)dx. \tag{1.15}$$

The best guaranteed result on F_ρ *in the class of all quadratures is*

$$m_N \overset{\text{def}}{=} \inf_{x^N\in K^N}\int_K \min_{i=1,\dots,N}\rho(x,x_i)dx.$$ □

Theorem 1.3. *The quadrature formula with the nodes* $p_i^0(x^N)$, $i=1,\dots,N$, *is an optimal terminal operation for fixed* $x_1,\dots,x_N\in K$. *An optimal quadrature formula, i.e., a quadrature formula with nodes* $x_0^N=(x_1^0,\dots,x_N^0)$ *satisfying condition* (1.15) *and the weights* $p_i^0(x_0^N)$, $i=1,\dots,N$, *is an optimal error algorithm on* F_ρ *in the class of all nonadaptive algorithms* A^N. *The best guaranteed result is equal to* m_N.

Proof. For any fixed $x_1,\dots,x_N\in K$,

$$\min_{\tilde\beta\in\tilde B_N}\sup_{f\in F}\left|\int_K f(x)dx-\tilde\beta(x_1,\dots,x_N,f(x_1),\dots,f(x_N))\right|$$

$$=\min_{\theta\in\Theta}\sup_{f\in F}\left|\int_K f(x)dx-\theta(f(x_1),\dots,f(x_N))\right|$$

$$=\sup_{f\in F}\left|\int_K f(x)dx-\sum_{i=1}^N p_i^0(x^N)f(x_i)\right|=\int_K \min_{i=1,\dots,N}\rho(x,x_i)dx, \tag{1.16}$$

where $\tilde B_N$ is the set of all possible terminal operations and Θ is the set of all numerical functions θ of the variables $(y_1,\dots,y_N)\in\{(y_1,\dots,y_N)\mid y_1=$

$f(x_1),\ldots, y_N=f(x_N), f\in F\}$. In (1.16), both the left-hand and the right-hand sides of the first equation represent the same quantity in different notations. The second equation follows from Lemma 2.8 and Theorem 4.2 of Chapter 1 and from Theorem 1.2. Indeed, Lemma 2.8 of Chapter 1 yeilds that the functional class $F=F_\rho$ is convex and balanced, hence, due to Theorem 4.2 of Chapter 1, among optimal terminal operations with fixed $x_1,\ldots,x_N$ there is a quadrature formula. Finally, Theorem 1.2 implies that this quadrature formula has the weights $p_i^0(x^N)$, $i=1,\ldots,N$. The third equation also follows from Theorem 1.2.

The second equation in (1.16) is just the first statement of the theorem. The second statement follows by minimizing all parts of (1.16) with respect to $x^N=(x_1,\ldots,x_N)\in K^N$. □

The problem (1.15) is related to discrete geometry problems. For instance, if ρ is the Euclidean metric, the problem (1.15) is closely connected to the problem of obtaining an optimal covering of the set K with balls of the same radius and to the problem of obtaining a least density covering of n-dimensional space with unit balls. We shall discuss this question in greater detail in the next section.

It is but seldom that we are able to obtain the explicit solution of (1.15), see Section 2. If a nonadaptive algorithm with a fixed number of nodes is meant to be used many times, it will probably make sense to solve (1.15) numerically.

In the case where N and the space dimension n are small, it is possible to obtain an approximate solution through the following iterative process. Let

$$\eta(x^N)\stackrel{def}{=}\int_K \min_{i=1,\ldots,N} \rho(x,x_i)dx=\sum_{i=1}^{N}\int_{K_i(x^N)}\rho(x,x_i)dx,$$

$$D_i(x^N)\stackrel{def}{=}\int_{K_i(x^N)}\rho(x,x_i)dx.$$

The $(k+1)$th iteration consists in minimizing the function η with respect to the vector variable $x_{m(k)}$ on the segment $[x_{m(k)}(k),a(k)]$, where the vector $x^N(k)=(x_1(k),\ldots,x_N(k))$ is the result of the kth iteration, the number $m(k)\in\{1,\ldots,N\}$ is defined by the equation

$$D_{m(k)}(x^N(k))=\max_{i=1,\ldots,N} D_i(x^N(k)),$$

and

$$a(k)=\arg\max_{x\in K_{m(k)}(x^N(k))}\rho(x,x_{m(k)}(k)).$$

If $n=2$, the set K is the unit square and ρ is the Euclidean metric, the above algorithm provides the following approximate values of the best guaranteed accuracy: $m_2 \approx 0.297$ $\left(\text{in fact, } m_2 = \frac{1}{6}\left(\frac{\sqrt{5}}{4} + \ln\left(\frac{\sqrt{5}+1}{2}\right)\right) + \frac{1}{24}\left(\sqrt{5} - \frac{1}{2}\ln\left(\frac{1}{\sqrt{5}+2}\right)\right) = 0.2966\ldots\right)$, $m_3 \approx 0.237$, $m_4 \approx 0.187$, $m_5 \approx 0.175$.

Theorem 1.2 enables us to compare numerically different "uniform" (in some sense) distributions of the nodes $x_1, \ldots, x_N$ in the set K with respect to the criterion $\eta(x^N)$.

Compare, for instance, the following three distributions (here, as before, $n=2$, K is the unit square and ρ is the Euclidean metric).

The first one is the lattice Γ_1^N, see Remark 6 after Theorem 2.3 of the next section. For $n=2$, the basic coordinate frame of this lattice forms a regular triangle. The length of its side is selected in such a way that the unit square contains at least N nodes of the lattice and if the side is longer, the unit square contains less than N nodes.

The second distribution is the "cubic" lattice from Aird and Rice [77]. For $N = m^m$ (where m is integer), this lattice coincides with the ordinary cubic lattice.

Finally, the third distribution is obtained by choosing points from LP_τ-sequence, see Sobol [69, Chapter 6, Section 3].

The results of computations are listed in Table 1, which contains approximate values of $\eta(x^N) = \int\limits_K \min\limits_{i=1,\ldots,N} \rho(x, x_i)dx$. The table shows that, starting from $N=7$, the quadrature formulas using the lattice Γ_1^n are most effective. To compare with, note that the average of 50 values $\eta(x^N)$, where $x_1, \ldots, x_N$ are independent uniformly distributed random points from K, is equal to 0.211 for $N = 17 - 0.128$.

Table 1

Γ_n^1	0.227	0.151	0.127	0.103
Cubic lattice	0.193	0.157	0.130	0.104
LP_τ	0.241	0.179	0.142	0.115

1.5. Central terminal operation in the integration problem

In this section, we find the best guaranteed (after all informational computations) accuracy $\varepsilon_N(Z^N)$ and obtain the central terminal operation $\widetilde{\beta}_*$, see (4.3) and (4.4) of Chapter 1. Here, as before, $F = F_\rho$. Moreover, $S(f) = \int\limits_K f(x)dx$ and $B = \mathbb{R}$, see Section 1.1 of Chapter 1, i.e., the distance between $S(f_1)$ and $S(f_2)$ is $|S(f_1) - S(f_2)|$.

Theorem 1.4. *The central terminal operation is given by the formula*

$$\widetilde{\beta}_*(z^N)=\int_K \frac{1}{2}[\phi_{1N}(x)+\phi_{2N}(x)]dx.$$

Moreover,

$$\varepsilon_N(z^N)=\int_K \frac{1}{2}[\phi_{2N}(x)-\phi_{1N}(x)]dx.$$

Proof. Due to Lemmas 1.1 and 1.2, the functions ϕ_{1N} and ϕ_{2N} given by (1.3) provide sharp lower and upper envelopes for functions from $F(z^N)$. Therefore, the image of the set $F(z^N)$ under the mapping $S(f)=\int_K f(x)dx$ is the segment

$$S(F(z^N))=\left[\int_K \phi_{1n}9x)dx, \int_K \phi_{2N}(x)dx\right].$$

By definition (see Section 4.5 of Chapter 1), $\widetilde{\beta}_*(z^N)$ is the center of this segment and $\varepsilon_N(z^N)$ is its radius, i.e. half of its length. □

Remark 1. In general, the quadrature formula with optimal weights $p_i^0(x^N)$, $i=1,\dots,N$, does not coincide with the central terminal operation $\widetilde{\beta}_*(z^N)$. Indeed, let $\rho(u,v)=\max_{i=1,\dots,N}|u^i-v^i|$, $K=\{u=(u^1,\dots,u^n)\mid 0\leq u^i\leq 1,\ i=1,\dots,n\}$, $n=2$, $N=4$, $x_1=(1/4,1/4)$, $x_2=(3/4,1/4)$, $x_3=(3/4,3/4)$, $x^4=(1/4,3/4)$, $x^4=(x_1,x_2,x_3,x_4)$. Then $p_1^0(x_4)=\cdots=p_4^0(x_4)=1/4$. It is easily verified that, for $0\leq y\leq 1/2$,the situation $z^4=(x_4,y,0,0,0)$ is realizable (see Section 3.2 of Chapter 1), and the central terminal operation has the form $\widetilde{\beta}_*(z^4)=-5y^3/24+3y^2/16+y/4$, while $\sum_{i=1}^4 p_1^0(x^4)y_i=y/4$, where $(y_1,y_2,y_3,y_4)=(y,0,0,0)$.

On the other hand, in Section 3 of Chapter 1 we shall see that, for $n=1$ and arbitrary $x_1,\dots,x_N$, the quadrature formula with optimal weights coincides with the central terminal operation. □

Remark 2. With Theorem 1.4, we easily prove Theorem 1.3. Indeed, for any fixed $x_1,\dots,x_N\in K$

$$\min_{\widetilde{\beta}\in\widetilde{B}_N}\sup_{f\in F}\gamma(S(f),\widetilde{\beta}(x_1,\dots,x_N,f(x_1),\dots,f(x_N))$$

$$=\sup_{y^N\in l(x^N)}\varepsilon_N(z^N)\geq\varepsilon_N(x^N,0)=\int_K \min_{i=1,\dots,N}\rho(x,x_i)dx,$$

where the first equation is established in the proof of Theorem 4.3 of Chapter 1 and the inequality holds since $0\in l(x^N)$. Thus, no terminal operation

$\widetilde{\beta} \in B_N$ guarantees a result better than $\int\limits_K \min\limits_{i=1,\dots,N} \rho(x, x_i)dx$. On the other hand, Theorem 1.2 yields that this result is guaranteed by the quadrature formula with the weights $p_i^0(x^N)$, $i = 1, \dots, N$. Therefore, this quadrature formula is the optimal terminal operation for fixed $x_1, \dots, x_N$. □

Remark 3. The proof of Theorem 1.4 remains valid not only for $F = F_\rho$, but also for any functional class F, provided that ϕ_{1N} and ϕ_{2N} are sharp lower and upper envelopes for functions from the subset $F(z^N)$ of F. □

2. OPTIMAL QUADRATURES FOR FUNCTIONAL CLASSES DETERMINED BY MODULI OF CONTINUITY

In this section, we apply the results of Section 1 to functional classes traditionally treated in the theory of optimal quadratures. In some cases we succeed in obtaining explicit optimal or asymptotically optimal quadratures, in other cases we only estimate coefficients in the expression for the quadrature error.

The following argument is based on the expression for the error m_N of an optimal on F_ρ quadrature, obtained in Section 1 (see corollary of Theorem 1.2):

$$m_N = \inf_{x^N \in K^N} \int\limits_K \min_{i=1,\dots,N} \rho(x, x_i)dx. \tag{2.1}$$

(Recall that in this chapter all the functions $\rho(u, v)$ are assumed to be integrable with respect to u for any fixed v.)

2.1. Auxiliary statements

We introduce some notations necessary for the subsequent presentation and obtain some auxiliary results.

For any quasi-metric ρ, the set $\{u | \rho(u, v) \leq R\}$ is called a *ρ-ball of radius R with center v*.

Clearly, ρ-balls of radius $\sup\limits_{x \in K} \min\limits_{i=1,\dots,N} \rho(x, x_i)$ with centers $x_1, \dots, x_N$ cover K, while ρ-balls of any smaller radius with these centers do not. We call

$$R(x^N) \stackrel{def}{=} \sup_{x \in K} \min_{i=1,\dots,N} \rho(x, x_i) \tag{2.2}$$

the *radius of the covering determined by the centers $x_1, \dots, x_N$* (i.e., by the vector x^N). A covering of the minimal radius is called *optimal*.

Thus, obtaining an optimal covering amounts to computing

$$R_N \stackrel{def}{=} \inf_{x_1,\ldots,x_N \in K} \sup_{x \in K} \min_{i=1,\ldots,N} \rho(x, x_i), \tag{2.3}$$

which is called the *radius of optimal covering*, and finding the points $x_1, \ldots, x_N$ that deliver the infimum.

From here on in this section $\rho(u,v) = \|u - v\|$, where $\|\cdot\|$ is an arbitrary norm in n-dimensional space. Let $V(R)$ be the volume of the ball of radius R, i.e., of $\Omega(R) = \{x \mid \|x\| \leq R\}$, and let r_N be defined by

$$NV(r_N) = \mu(K), \tag{2.4}$$

where $\mu(K)$, as before, denotes the Lebesgue measure of K.

Lemma 2.1. *The following estimate is valid:*

$$N \int\limits_{\Omega(r_N)} \|x\| dx \leq m_N \leq N \int\limits_{\Omega(R_N)} \|x\| dx.$$

Proof. Compare the Lebesgue integral sums for the integrals $N \int\limits_{\Omega(r_N)} \|x\| dx$ and

$$\int\limits_K \min_{i=1,\ldots,N} \|x - x_i\| dx = \sum_{i=1}^N \int\limits_{K_i} \|x - x_i\| dx = \sum_{i=1}^N \int\limits_{K_i'} \|x\| dx,$$

where x_i's are arbitrary, $K_i = K_i(x^N)$, i.e., the sets K_i form a Voronoi-Dirichlet ρ-partition of K (see Section 1.4), and $K_i' = k_i - x_i$, $i = 1, \ldots, N$.

Consider any integer k, and let j_0 satisfy the condition

$$\frac{j_0 - 1}{k} r_N < \sup_{x \in K} \min_{i=1,\ldots,N} \|x - x_i\| = \max_{i=1,\ldots,N} \max_{x \in K_i'} \|x\| \leq \frac{j_0}{k} r_N.$$

Let

$$A_j = \{x \mid (j-1) r_N / k \leq \|x\| \leq j r_N / k\}.$$

Then

$$N \sum_{j=1}^N \frac{j}{k} r_N \mu(A_j) - \sum_{i=1}^N \sum_{j=1}^{j_0} \frac{j}{k} r_N \mu(A_j \cap K_i')$$

$$= \sum_{i=1}^N \left[\sum_{j=1}^k \frac{j}{k} r_N [\mu(A_j) - \mu(A_j \cap K_i')] - \sum_{j=k+1}^{j_0} \frac{j}{k} r_N \mu(A_j \cap K_i' \right]$$

$$\leq r_N \sum_{i=1}^N \left[\sum_{j=1}^k [\mu(A_j) - \mu(A_j \cap K_i')] - \sum_{j=k+1}^{j_0} \mu(A_j \cap K_i' \right]$$

$$= r_N \sum_{i=1}^N [V(r_N) - \mu(K_i')] = r_N \left[NV(r_N) - \sum_{i=1}^N \mu(K_i') \right] = 0,$$

since $NV(r_N)=\sum_{i=1}^{N}\mu(K_i')=\mu(K)$. Letting k tend to ∞ in the last inequality, we get

$$N\int_{\Omega(r_N)}\|x\|dx-\int_K \min_{i=1,\dots,N}\|x-x_i\|dx\leq 0.$$

Since $x_1,\dots,x_N$ are arbitrary, (2.1) yields now the left-hand inequality of the lemma.

To prove the right-hand inequality, fix any $\varepsilon\geq 0$ and $x_1,\dots,x_N\in K$ such that

$$\sup_{x\in K}\min_{i=1,\dots,N}\|x-x_i\|\leq R_N+\varepsilon.$$

Then

$$m_N\leq\int_K \min_{i=1,\dots,N}\|x-x_i\|dx$$

$$\leq\sum_{i=1}^{N}\int_{\{x\,|\,\|x-x_i\|\leq R_N+\varepsilon\}}\|x-x_i\|dx=N\int_{\Omega(R_N+\varepsilon)}\|x\|dx.$$

Since $\varepsilon\geq 0$ is arbitrary, this implies the right-hand inequality of the lemma. □

Lemma 2.2. *The following estimate is valid:*

$$\int_{\Omega(1)}\|x\|dx\frac{\mu(K)}{V(1)}r_N\leq m_N\leq\mu(K)R_N.$$

Proof. The right-hand inequality follows directly from the definitions of m_N and R_N. To establish the left-hand inequality, it suffices to observe that

$$\int_{\Omega(r)}\|x\|dx=r^{n+1}\int_{\Omega(1)}\|x\|dx,\quad V(r)=r^N V(1)$$

and to use the left-hand inequality of Lemma 2.1 and the definition of r_N. □

Lemma 2.3. *Let $A\subset\Omega(R)$, where A is a convex solid with a piecewise smooth boundary and A contains the origin. Let $\|x\|$ be one of the following three norms:*

$$\|x\|_0=\max|x^i|,$$

$$\|x\|_1=\sum_{i=1}^{N}|x^i|,\quad \|x\|_2=\sqrt{\sum_{i=1}^{N}(x^i)^2},$$

where $x=(x^1,\dots,x^n)$. *Then*

$$\int\limits_A \|x\|dx \le \frac{n}{n+1}R\mu(A) \quad \text{and} \quad \int\limits_{\Omega(R)} \|x\|dx = \frac{n}{n+1}RV(R).$$

Proof. Let $\|x\|=\|x\|_2$. Performing the spherical coordinate transformation

$$x^1 = r\cos\alpha_1, \quad x^i = r\sin\alpha_1 \dots \sin\alpha_{i-1}\cos\alpha_i, \; i=2,\dots,n-1,$$
$$x^n = r\sin\alpha_1 \dots \sin\alpha_{n-1}$$

with the Jacobian $r^{n-1}\sin^{n-2}\alpha_1 \sin^{n-3}\alpha_2 \dots \sin\alpha_{n-2}$ and taking into account that $\|x\|_2=r$, we obtain

$$\int\limits_A \|x\|dx = \int\limits_0^{2\pi} d\alpha_1 \dots \int\limits_0^{2\pi} d\alpha_{n-1} \int\limits_0^{r(\alpha_1,\dots,\alpha_{n-1})} r^n \sin^{n-2}\alpha_1 \dots \sin\alpha_{n-2}dr$$
$$= \frac{1}{n+1}\int\limits_0^{2\pi} d\alpha_1 \dots \int\limits_0^{2\pi} d\alpha_{n-2} \int\limits_0^{2\pi} [r(\alpha_1,\dots,\alpha_{n-1})]^{n+1}$$
$$\times \sin^{n-2}\alpha_1 \dots \sin\alpha_{n-2}d\alpha_{n-1}, \tag{2.5}$$

where the equation $r=r(\alpha_1,\dots,\alpha_{n-1})$ determines the boundary of A and the choice of the integration limits corresponds to the case where the origin is an interior point of A. Observing that

$$\mu(A) = \frac{1}{n}\int\limits_0^{2\pi} d\alpha_1 \dots \int\limits_0^{2\pi} d\alpha_{n-2} \int\limits_0^{2\pi} [r(\alpha_1,\dots,\alpha_{n-1})]^n$$
$$\times \sin^{n-2}\alpha_1 \dots \sin\alpha_{n-2}d\alpha_{n-1} \tag{2.6}$$

and using in (2.5) the estimate $[r(\alpha_1,\dots,\alpha_{n-1})]^{n+1} \le R[r(\alpha_1,\dots,\alpha_{n-1})]^n$, we obtain the claimed inequality. If $A=\Omega(R)$, then, obviously, $r(\alpha_1,\dots,\alpha_{n-1}) \equiv R$ and the claimed equation follows immediately from (2.5) and (2.6).

In the case where $\|x\|=\|x\|_1$, we replace the integral with the sum of 2^n integrals over the intersections of A and the 2^n orthants of $\mathbb{R}^N$. Then, in the orthant $x_1 \ge 0,\dots,x^n \ge 0$, for instance, we perform the coordinate transformation

$$x^1 = r(1-\beta_1), \quad x^i = r\beta_1 \dots \beta_{i-1}(1-\beta_i), \quad i=2,\dots,n-1,$$
$$x^n = r\beta_1 \dots \beta_{n-1}$$

with the Jacobian $r^{n-1}\beta_1^{n-2} \dots \beta_{n-2}$, $\|x\|_1 = r$.

In the case where $\|x\|=\|x\|_0$, the integral is replaced with the sum of $2n$ integrals over the intersections of A with the domains

$\{x \mid \|x\|_0 = x^i\}$, $\{x \mid \|x\|_0 = -x^i\}$, $i=1,\ldots,n$. Then, in the domain $\{x \mid \|x\|_0 = x^1\}$, for instance, we perform the coordinate transformation

$$x^1 = r, \quad x^i = r\gamma_{i-1}, \quad i=2,\ldots,n,$$

with the Jacobian r^{n-1}, $\|x\|_0 = r$.

Clearly, in the both cases $\|x\| = \|x\|_1$ and $\|x\| = \|x\|_0$ the proof of the lemma is now similar to the proof for $\|x\| = \|x\|_2$. □

Lemma 2.4. *Let $\|x\|$ be one of the three norms $\|x\|_i, i=0,1,2,$ and let K be a convex solid with a piecewise smooth boundary. Then*

$$\frac{n}{n+1} r_N \mu(K) \le m_N \le \frac{n}{n+1} R_N \mu(K).$$

Proof. The left-hand inequality follows directly from Lemmas 2.1 and 2.3. To prove the right-hand inequality, observe that

$$m_N \le \int_K \min_{i=1,\ldots,N} \|x - x_i^0\| dx$$

for any $x_1^0, \ldots, x_N^0$, including the case where these points are the centers of an optimal covering of K. For such x_i^0's we put $K_i^o = \{x \mid \|x - x_i^0\| = \min_{j=1,\ldots,N} \|x - x_j^0\|\}$ and observe that $K_i^0 - x_i^0 \subset \Omega(R_N)$, $i=1,\ldots,N$.

Applying Lemma 2.3, we obtain

$$\int_K \min_{i=1,\ldots,N} \|x - x_i^0\| dx = \sum_{i=1}^N \int_{K_i^0} \|x - x_i^0\|$$

$$= \sum_{i=1}^N \int_{K_i^0 - x_i^0} \|x\| dx \le \frac{n}{n+1} R_N \mu(K). \qquad \square$$

2.2. Basic results

We now proceed to the basic results of the section (Theorems 2.1-2.3). In these theorems $F = F_\rho$, where $\rho(u,v) = \|u - v\|$.

Theorem 2.1 (Sukharev, Timokhov and Fedorov [86]). *Let $K = \cup_{i=1}^N [\Omega(R_N) + x_i^0]$ and*

$$\mu[(\Omega(R_N) + x_i^0) \cap (\Omega(R_N) + x_j^0)] = 0 \quad \textit{if} \quad i \ne j.$$

Then the quadrature formula with the nodes x_i^0 and weights $p_i^0 = V(R_N) = \mu(K)/N$ is optimal on F. Moreover, $m_N = \int_{\Omega(R_N)} \|x\| dx$.

Proof. The part of the theorem concerning the nodes of the optimal quadrature and its guaranteed result follows immediately from Lemma 2.1 if we observe that, under the assumptions of the theorem, $r_N = R_N$. The part concerning the weights follows from Theorem 1.2. □

Theorem 2.1 leads directly to the following results.

Corollary 1. *Let*

$$K = \{x | 0 \le x^i \le 1, \quad i = 1, \dots, n\}, \quad \|x\| = \|x\|_0, \quad N = m^n.$$

Then the quadrature with the nodes

$$\left(\frac{j_1}{2m}, \dots, \frac{j_n}{2m}\right), \quad j_1, \dots, j_n \in \{1, 3, \dots, 2m-1\},$$

and weights $p_i^0 = 1/m^n, \quad i = 1, \dots, m^n$, *is optimal on* F.

Corollary 2. *Let*

$$n = 2, \quad K = \{x | |x^1| + |x^2| \le 1\}, \quad \|x\| = \|x\|_1, \quad N = m^2.$$

Then the quadrature with the nodes

$$\left(\frac{j_1 - j_2}{2m}, \frac{j_1 + j_2 - 2m}{2m}\right), \quad j_1, j_2 \in \{1, 3, \dots, 2m-1\},$$

and weights $p_i^0 = 2/N^2, \quad i = 1, \dots, m^2$, *is optimal on* F.

The optimal lattices described in Corollaries 1 and 2 are shown in Fig.2.

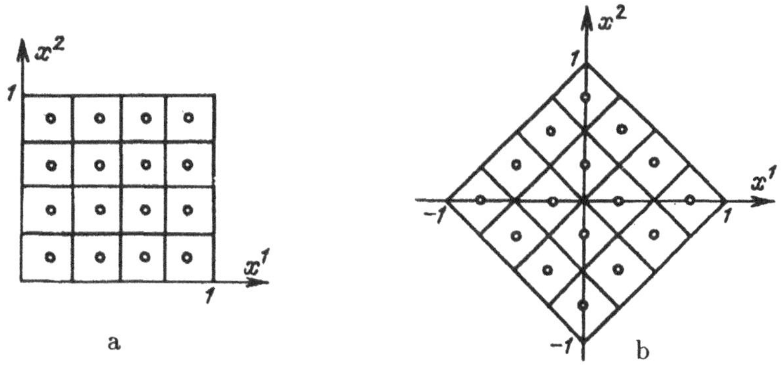

Fig. 2. The nodes of the optimal on F_ρ quadrature for $n = 2$, $m = 4$:
a) $\rho(u, v) = \|u - v\|_0$;
b) $\rho(u, v) = \|u - v\|_1$

In accordance with the general definitions from Section 4.2 of Chapter 1, we say that a sequence of quadratures with nodes $x_1(N), \dots, x_N(N)$ and weights $p_1(N), \dots, p_N(N)$ is *asymptotically optimal* on F as $N \to \infty$ iff

$$\sup_{f \in F} \left| \int_K f(x) dx - \sum_{i=1}^{N} p_i(N) f(x_i(N)) \right| \sim m_N \quad \text{as} \quad N \to \infty.$$

Theorem 2.2 (Sukharev [82a]) *Let K be a bounded set with a boundary of zero measure, and let $\|\cdot\|$ be a norm such that the balls obtained by translations of $\Omega(1)=\{x \mid \|x\|\leq 1\}$ cover the entire space, i.e., for some $x_1, x_2, \ldots$*

$$\cup_{i=1}^{\infty}(\Omega(1)+x_i)=\mathbb{R}^n, \quad \mu[(\Omega(1)+x_i)\cap(\Omega(1)+x_j)]=0, \quad i\neq j.$$

Let $R(N)$ be minimal among the numbers R such that the number of the balls $\Omega(R)+Rx_i$ whose intersections with K are of positive measure does not exceed N, and let $R(N)x_{i_j}$ $(j=1,\ldots,N_1;\quad N_1\leq N)$ be the centers of these balls.

Then the sequence of quadratures with the nodes $x_1(N),\ldots,x_N(N)$, where $x_j(N)=R(N)x_{i_j}$ for $j=1,\ldots,N_1$ and $x_j(N)$ are arbitrary points of K for $j=N_1+1,\ldots,N$, and weights

$$p_j(N)=\mu\{x\in K \mid \|x-x_j(N)\|=\min_{\nu=1,\ldots,N}\|x-x_\nu(N)\|\}, \quad j=1,\ldots,N,$$

is asymptotically optimal on F. Moreover,

$$m_N\sim\int_{\Omega(R(N))}\|x\|dx \quad as \quad N\to\infty.$$

Proof. Since the boundary of K is of zero measure, $NV(R(N))\to\mu(K)=NV(r_N)$ as $N\to\infty$, hence $R(N)\sim r_N$. On the other hand, $r_N\leq R_N\leq R(N)$, see (2.3), thus $r_N\sim R_N$, and, due to Lemma 2.1, $m_N\sim N\int_{\Omega(R(N))}\|x\|dx$. Theorem 1.2 yields that the result guaranteed by the quadrature with the nodes $x_j(N)$ and weights $p_j(N)$, $j=1,\ldots,N$, is equal to $\int_K \min_{i=1,\ldots,N}\|x-x_i(N)\|dx$. Clearly,

$$m_n\leq\int_K \min_{i=1,\ldots,N}\|x-x_i(N)\|dx\leq N\int_{\Omega(R(N))}\|x\|dx.$$

To complete the proof, it suffices to recall the already proved equivalence of the left- and right-hand sides of this inequality. □

This theorem makes it easy to construct asymptotically optimal quadratures for classes determined by norms such that the entire space can be covered with translations of the convex and symmetric with respect to the origin solid $\Omega(1)=\{x \mid \|x\|\leq 1\}$ (which is called the length indicatrix of the metric $\|u-v\|$). For $\|\cdot\|_0$ and $\|\cdot\|_1$, $n=2$, these quadratures were constructed by Babenko [76b] and Maung Cho Niun and Sharygin [71].

Since every convex solid symmetric with respect to the origin determines a norm, the question arises: which convex solids have this property (i.e., their translations cover the entire space without intersections of nonzero measure)? This question was treated in a number of papers. For instance, for $n=3$ all such solids (and, consequently, all such norms) in the case where the centers of the solids covering the space form a lattice were listed in Coxeter [62]. Convex polyhedrons with this property (parallelohedrons) for $n=5$ were treated by Ryshkov and Baranovskii [76].

Lemmas 2.2 and 2.4 lead to some results on optimal quadratures also in the cases where the class F of functions defined on some subset of n-dimensional coordinate space is described with the help of the norm $\|\cdot\|_1$ (for $n>2$) or $\|\cdot\|_2$. It is easily verified that in the both cases we cannot cover the space with translations of the corresponding convex solids $\Omega(1)$. We formulate these results below.

Theorem 2.3 (Sukharev [82a]). *Let K be a bounded set of non-zero measure with a boundary of zero measure. Then, for N large enough,*

$$a(n)N^{-1/n} \le m_N \le b(n)N^{-1/n},$$

where

$$a(n)=(\mu(K))^{(n+1)/n} \int_{\Omega(1)} \|x\|dx/(V(1))^{(n+1)/n},$$

$$b(n)=(\mu(K))^{(n+1)/n} \sqrt[n]{n\ln n+n\ln\ln n+5n}/(V(1))^{1/n}.$$

Moreover, if $K=\{x \mid 0\le x^i\le 1,\ i=1,\dots,n\}$, $\|x\|=\|x\|_1$, then

$$a(n)\sim b(n)\sim \frac{n}{2e} \quad as \quad n\to\infty,$$

and if $K=\{x \mid 0\le x^i\le 1,\ i=1,\dots,n\}$, $\|x\|=\|x\|_2$, then

$$a(n)\sim b(n)\sim \frac{\sqrt{n}}{\sqrt{2\pi e}} \quad as \quad n\to\infty.$$

Proof. Let $\|x\|$ be an arbitrary norm. According to Rogers [57], for $n\ge 3$, the least possible density θ_n (see, e.g., Rogers [64] for the definition) of a covering of n-dimensional coordinate space with translations of the convex solid $\Omega(1)$ satisfies the inequality

$$\theta_n \le n\ln n+n\ln\ln n+5n. \tag{2.7}$$

Due to Kolmogorov and Tikhomirov [59, Theorem 9]

$$\theta_n=\lim_{N\to\infty} NV(R_N)/\mu(K).$$

Taking into account that $V(r)=r^nV(1)$, we obtain for N large enough

$$R_N^n<(n\ln n+n\ln\ln n+5n)\mu(K)/(NV(1)).$$

It follows by the definition of r_N that $r_N^n=\mu(K)/(NV(1))$. Lemma 2.2 yields now that for N large enough

$$\frac{(\mu(K))^{(n+1)/n}\int_{\Omega(1)}\|x\|dx}{(V(1))^{(n+1)/n}}N^{-1/n}\le m_N$$

$$\le \frac{(\mu(K))^{(n+1)/n}\sqrt[n]{n\ln n+n\ln\ln n+5n}}{(V(1))^{1/n}}N^{-1/n}. \tag{2.8}$$

Using the expression for $\int\limits_{\Omega(1)} \|x\|dx$ from Lemma 2.3 and the well-known formulas

$$V(1)=\frac{2^n}{n!} \quad \text{for} \quad \|x\|=\|x\|_1, \quad V(1)=\frac{2\pi^{n/2}}{n\Gamma(n/2)} \quad \text{for} \quad \|x\|=\|x\|_2,$$

$$\sqrt[n]{\Gamma(n/2)} \sim \sqrt{n}/\sqrt{2e}, \quad \sqrt[n]{n!} \sim n/e \quad \text{as} \quad n \to \infty,$$

we prove the rest statements of the theorem. □

Remark 1. Let $\|x\|$ be one of the three norms $\|x\|_i$, $i=0,1,2$, and let K be a convex solid with a piecewise smooth boundary. Then the inequality analogous to (2.8) has the form

$$\frac{n}{n+1}(\mu(K))^{(n+1)/n}(V(1))^{-1/n}N^{-1/n} \leq m_N$$

$$\leq \frac{n}{n+1}(\mu(K))^{(n+1)/n}(V(1))^{-1/n}\sqrt[n]{\theta_n^*}N^{-1/n}.$$

It holds for N large enough. The left-hand side of this inequality coincides with that of (2.8), and the right-hand side is obtained just as in (2.8) with the only difference: we use Lemma 2.4 instead of Lemma 2.2 and some estimate $\theta < \theta_n^*$ instead of estimate (2.7). □

Remark 2. If $\|x\|=\|x\|_0$, then $\theta_n=1$. With the above remark, we get under the same assumptions

$$m_N \sim \frac{n}{2(n+1)}(\mu(K))^{(n+1)/n}N^{-1/n} \quad \text{as} \quad N \to \infty. \qquad \square$$

Remark 3. The essence of Theorem 2.3 does not consist in the assertion that $m_N \asymp N^{-1/n}$ as $n \to \infty$ (the order of the error was estimated in Bakhvalov [59] under more general assumptions). The main point is the formulas for $a(n)$ and $b(n)$. They enable us to obtain asymptotics with respect to the dimension of the space that contains the coefficient of $N^{-1/n}$ in the expression for the error m_N of optimal quadrature with N nodes. □

Remark 4. The proof of Theorem 2.3 shows that the error estimate $c(n)N^{-1/n}$ where $a(n) \leq c(n) \leq b(n)$ is attained for quadratures whose nodes determine optimal coverings. (The connection between lattices of nodes of optimal quadratures with optimal packings was revealed in Sobolev [74].) However, in general nodes of optimal quadratures coincide neither with the centers of optimal coverings nor with the centers of optimal packings. This can be easily recovered from the example

$$\|x\|=\|x\|_0, \quad K=\{x=(x^1,x^2) \mid 0 \leq x^1, x^2 \leq 1\}, \quad m^2 < N < (m+1)^2,$$

where m is integer. □

Remark 5. The proof of (2.7) given by Rogers [57], as well as the proofs of other estimates $\theta_n < \theta_n^*$ with the desired property $\sqrt[n]{\theta_n^*} \sim 1$ as

$n \to \infty$, is not constructive, i.e., does not lead to any method of constructing coverings for which the obtained bounds are attained. □

Remark 6. When selecting nodes of a quadrature for the class F determined by $\|x\| = \|x\|_2$ for n small, it sometimes makes sense to use the so-called basic Voronoi lattice Γ_1^n of the first type. This lattice determines the least dense lattice covering of the space with $n \leq 5$ (see Ryshkov and Baranovskii [76]) and enjoys other extremality properties, see Gametskii [63] and Ryshkov [67].

3. SEQUENTIALLY OPTIMAL AND ONE-STEP OPTIMAL INTEGRATION ALGORITHMS

Alongside studying quadrature formulas and nonadaptive algorithms for numerical integration, considerable attention has been paid to constructing adaptive integration algorithms (also called quadrature processes and algorithms with automatic step selection). Among the papers on the subject we can mention Bakhvalov [62, 64, 65, 66, 73], De Boor [71a, b], Dixon [74], Einarsson [74], Glinkin [81a, 83], Glinkin and Sukharev [85], Haber [75], Kaxaner [71], Korchanov [84, 86a, b], Lyness [69], Lyness and Kaganove [76], Malcolm and Simpson [75], Rice [74, 75, 76a, b, 83], Shapiro [84], Sukharev [79b, c, 84], Vasil'eva et al. [72]. In this section we deal with one-dimensional case. In Sections 5 and 6 we apply the obtained results to cumputing integrals of functions of several variables. From here to the end of the chapter we assume (unless otherwise stated explicitly) that $S(f) = \int_K f(x)dx$ and $B = \mathbb{R}$ (see Section 1.1 of Chapter 1), i.e., the distance between elements $S(f_1)$ and $S(f_2)$ is $|S(f_1) - S(f_2)|$.

3.1. Optimal error algorithms for classes of functions satisfying the Lipschitz condition

Consider the functional class

$$F = \{f \mid |f(u) - f(v)| \leq M|u - v|, \; u, v \in [a, b]\}, \tag{3.1}$$

and let informational computations be evaluations of the integrand. This class is a special case of the class F_ρ for $K = [a, b]$ and $\rho(u, v) = \|u - v\| = M|u - v|$. Theorem 2.1 yields immediately the following well-known result (see Bakhvalov [73], Berezin and Zhidkov [66]): a compound rectangle quadrature formula with equidistant nodes, i.e., the quadrature with the nodes and weights given by the vectors

$$x_0^N = \left(a + \frac{b-a}{2N}, a + 3\frac{b-a}{2N}, \ldots, a + (2N-1)\frac{b-a}{2N}\right),$$

$$p_0^N = \left(\frac{b-a}{N}, \ldots, \frac{b-a}{N}\right), \tag{3.2}$$

is optimal on the class (3.1). Moreover,

$$m_N = N \int\limits_{-(b-a)/(2N)}^{(b-a)/(2N)} M|x|dx = \frac{M(b-a)^2}{4N}. \tag{3.3}$$

With Theorem 1.4, we easily derive expressions for the central terminal operation and the best guaranteed (after all informational computations) accuracy. For the class under consideration we have

$$\phi_{1N}(x) = \max_{j=1,\ldots,N}\{y_j - M|x - x_j|\},$$

$$\phi_{2N}(x) = \min_{j=1,\ldots,N}\{y_j + M|x - x_j|\},$$

$$\begin{aligned}\widetilde{\beta}_*(z^N) &= \int_a^b \frac{1}{2}[\phi_{1N}(x) + \phi_{2N}(x)]dx \\ &= (x_1 - a)y_1 + \sum_{j=2}^{N}(x_j - x_{j-1})\frac{y_j + y_{j-1}}{2} + (b - x_N)y_N,\end{aligned} \tag{3.4}$$

$$\begin{aligned}\varepsilon_N(z^N) &= \int_a^b \frac{1}{2}[\phi_{2N}(x) - \phi_{1N}(x)]dx \\ &= \frac{M}{2}(x_1 - a)^2 + \frac{1}{4M}\sum_{j=2}^{N}[(x_j - x_{j-1})^2 M^2 \\ &\quad - (y_j - y_{j-1})^2] + \frac{M}{2}(b - x_N)^2,\end{aligned} \tag{3.5}$$

where we assume that $a \le x_1 < x_2 < \cdots < x_N \le b$.

Equations (3.4) and (3.5) are illustrated by Fig.3, where the slanted line segments have the angular coefficients $\pm M$. Quantity (3.4) is equal to the area of the figure bounded by the graph of the function $\frac{1}{2}(\phi_{1N} + \phi_{2N})$ (heavy line in Fig.3) and the x-axis. The best guaranteed accuracy (3.5) is equal to the area of the dashed figure.

Computing $\sup_{y^n} \varepsilon_N(z^N)$ and minimizing the obtained function with respect to x^N, we easily derive (3.3) and prove that the vector x_0^N given by (3.2) is optimal.

Identical transformations of the right-hand side of (3.4) give

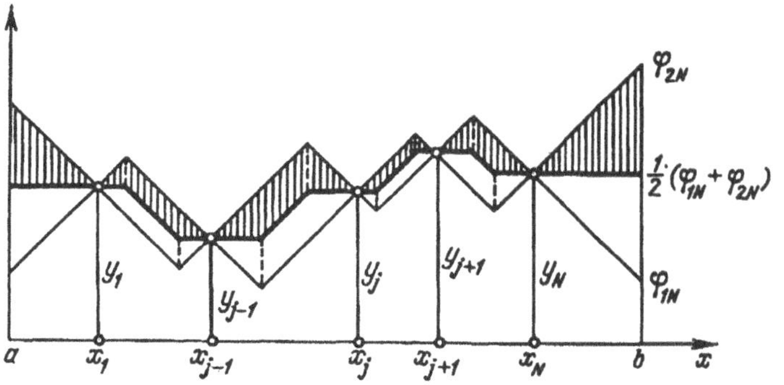

Fig. 3. The central terminal operation for integration and approximation of functions from the Lipschitz class F; the lower and upper envelopes ϕ_{1N} and ϕ_{2N} for the subclass $F(z^N)$

$$\widetilde{\beta}_*(z^N)=\left(\frac{x_1+x_2}{2}-a\right)y_1+\sum_{j=2}^{N-1}\left(\frac{x_j+x_{j+1}}{2}-\frac{x_{j-1}+x_j}{2}\right)y_j$$
$$+\left(b-\frac{x_{N-1}+x_N}{2}\right)y_N.$$

This equation shows that, for the class (3.1), the central terminal operation and the quadrature with optimal weights coincide. In the case being considered the quadrature with optimal weights is a compound rectangle quadrature (Fig.4).

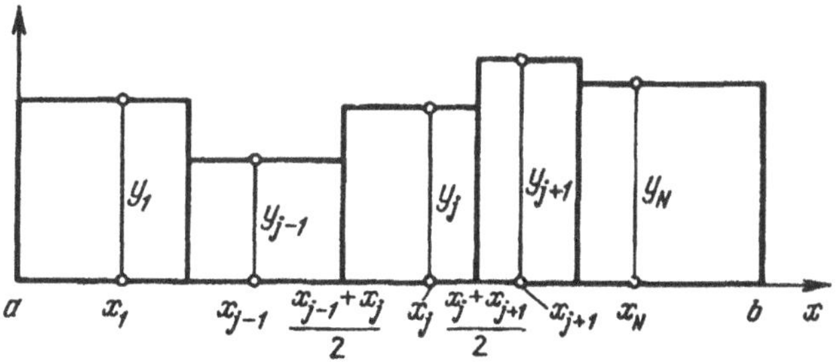

Fig. 4. The quadrature with optimal weights: a compound rectangle quadrature

Deriving sequentially optimal (by error) algorithms calls for obtaining optimal algorithms for subclasses of the class (3.1) that contain functions with fixed values at one or both endpoints of the segment of integration $[a,b]$. The solutions to these problems are given by the following two lemmas. In their formulations $x_0=a$, $x_{z+1}=b$, the number of informational computations is denoted by r, and F denotes the functional class (3.1). The classes $F(x_{r+1},y_{r+1})$, $F(x_0,y_0)$, and $F(x_0,x_{r+1},y_0,y_{r+1})$ contain functions with fixed values at the left, right and both endpoints of the segment of integration respectively.

Lemma 3.1. *The algorithm* $(x_R^r,\tilde{\beta}_*(x_R^r,x_{r+1},y^r,y_{r+1}))\in A^r$, *where*

$$x_R^r=\left(a+\frac{b-a}{2r+1},a+3\frac{b-a}{2r+1},\dots,a+(2r-1)\frac{b-a}{2r+1}\right)$$

and

$$\tilde{\beta}_*(x^r,x_{r+1},y^r,y_{r+1})=(x_1-a)y_1+\sum_{j=2}^{r+1}(x_j-x_{j-1})\frac{y_j+y_{j-1}}{2}$$

is the central terminal operation, is an optimal error algorithm both in the class A^r *of all nonadaptive algorithms and in the class* A^r *of all adaptive algorithms.*

The algorithm $(x_L^r,\tilde{\beta}_*(x_0,x_L^r,y_0,y^r))\in A^r$, *where*

$$x_L^r=\left(a+2\frac{b-a}{2r+1},a+4\frac{b-a}{2r+1},\dots,a+2r\frac{b-a}{2r+1}\right)$$

and

$$\tilde{\beta}_*(x_0,x_L^r,y_0,y^r)=\sum_{j=1}^{r}(x_j-x_{j-1})\frac{y_j+y_{j-1}}{2}+(b-x_r)y_r$$

is the central terminal operation, is an optimal error algorithm on $F(x_0,y_0)$ *in the both classes* A^r *and* $\tilde{A}^r$.

The best guaranteed result in the both cases is equal to $\frac{M}{4}\cdot\frac{(b-a)^2}{r+1/2}$.

Proof. Consider the functional class $F(x_{r+1},y_{r+1})$. Clearly, the subclass of all functions from $F(x_{r+1},y_{r+1})$ whose values at points $x_1,\dots,x_r$ are equal to $y_1,\dots,y_r$ respectively coincides with $F(z^{r+1})$. Since the central terminal operations enjoys the universal optimality property (see Theorem 4.1 of Chapter 1 and its corollary), it suffices to prove the optimality of x_R^r. Applying Theorem 4.3 of Chapter 1 to the class $F(x_{r+1},y_{r+1})$, we get

$$\begin{aligned}&\inf_{a\in A^r}\sup_{f\in F(x_{r+1},y_{r+1})}\varepsilon(\alpha,f)\\ &=\inf_{x^r\in X^r}\sup_{y^r\cap\{y^r|F(z^{r+1})\neq\emptyset\}}\inf_{\beta\in B}\sup_{f\in F(z^{r+1})}\gamma(S(f),\beta)\\ &=\inf_{x^r\in X^r}\sup_{y^r\in\{y^r|F(z^{r+1}\neq\emptyset\}}\varepsilon_{r+1}(z^{r+1}).\end{aligned}$$

According to (3.5),

$$\varepsilon_{r+1}(z^{r+1})=\frac{M}{2}(x_1-a)^2+\frac{1}{4M}\sum_{j=2}^{r+1}[(x_j-x_{j-1})^2M^2-(y_j-y_{j-1})^2].$$

It is easily seen that

$$\max_{y^r \in \{y^r | F(z^{r+1}) \neq \emptyset\}} \varepsilon_{r+1}(z^{r+1}) = \frac{M}{2}(x_1 - x_1)^2 + \frac{M}{4}\sum_{j=2}^{r+1}(x_j - x_{j-1})^2,$$

where the maximum is attained at $y^r = (y_{r+1}, \ldots, y_{r+1})$. Thus, to obtain an optimal algorithm and the best guaranteed result, we now have to solve the problem

$$\frac{M}{2}t_1^2 + \frac{M}{4}\sum_{j=2}^{r+1} t_j^2 \to \min,$$

$$t_1 + \cdots + t_{r+1} = b - a, \; t_1 \geq 0, \ldots, t_{r+1} \geq 0,$$

where $t_j = x_j - x_{j-1}$, $j = 1, \ldots, r+1$. Its solution shows that the algorithm $(x_R^r, \widetilde{\beta}_*(x_R^r, x_{r+1}, y^r, y_{r+1}))$ is optimal in A^r and the best guaranteed result (i.e., the minimum in the above problem of finding a conditional extremum) is equal to $\frac{M}{4} \cdot \frac{(b-a)^2}{r+1/2}$. Optimality of this algorithm in A^r follows from Theorem 5.2 of Chapter 1.

The statement of the lemma regarding the class $F(x_0, y_0)$ can be proved in the same way. □

Lemma 3.2. *The algorithm* $(x_C^r, \widetilde{\beta}_*(x_0, x_C^r, x_{r+1}, y_0, y_{r+1})) \in A^r$, *where*

$$x_C^r = \left(a + \frac{b-a}{r+1}, a + 2\frac{b-a}{r+1}, \ldots, a + r\frac{b-a}{r+1}\right)$$

and

$$\widetilde{\beta}_*(x_0, x^r, x_{r+1}, y_0, y^r, y_{r+1}) = \sum_{j=1}^{r+1}(x_j - x_{j-1})\frac{y_j + y_{j-1}}{2}$$

is the central terminal operation, is an optimal error algorithm on $F(x_0, x_{r+1}, y_0, y_{r+1})$ *in the both classes* A^r *and* $\widetilde{A}^r$. *The best guaranteed result is equal to*

$$v_r \stackrel{def}{=} \frac{1}{4M} \cdot \frac{(b-a)^2 M^2 - (y_{r+1} - y_0)^2}{r+1}.$$

Proof. Like in the proof of Lemma 3.1, we show that, in order to find the result guaranteed by the algorithm $(x^r, \widetilde{\beta}_*(x_0, x^r, x_{r+1}, y_0, y^r, y_{r+1}))$ on $F(x_0, x_{r+1}, y_0, y_{r+1})$, we have to maximize the function

$$\varepsilon_{r+2}(x_0, x^r, x_{r+1}, y_0, y^r, y_{r+1}) = \frac{1}{4M}\sum_{j=1}^{r+1}[(x_j - x_{j-1})^2 M^2 - (y_j - y_{j-1})^2].$$

with respect to $y^r \in \{y^r | F(x_0, x^r, x_{r+1}, y_0, y^r, y_{r+1}) \neq \emptyset\}$. It is easy to see that

$$\max_{y^r \in \{y^r | F(x_0,x^r,x_{r+1},y_0,y^r,y_{r+1}) \neq \emptyset\}} \varepsilon_{r+2}(x_0, x^r, x_{r+1}, y_0, y^r, y_{r+1})$$

$$\leq \max_{y^r \in \mathbb{R}^r} \frac{1}{4M} \sum_{j=1}^{r+1} [(x_j - x_{j-1})^2 M^2 - (y_j - y_{j-1})^2]$$

$$= \frac{M}{4} \sum_{j=1}^{r+1} (x_j - x_{j-1})^2 - \frac{1}{4M} \cdot \frac{(y_{r+1} - y_0)^2}{r+1}.$$

Substituting $x^r = x^r_C$ in this inequality, we show that the result guaranteed by the algorithm $(x^r_C, \tilde{\beta}_*(x_0, x^r_C, x_{r+1}, y_0, y^r, y_{r+1}))$ is not worse than v_r.

We now fix the central terminal operation $\tilde{\beta}_*(x_0, x^r, x_{r+1}, y_0, y^r, y_{r+1})$ and choose an arbitrary algorithm $\tilde{x}^r \in \tilde{X}^r$ (see (3.10), of Chapter 1). We show that this choice cannot guarantee a result better than v_r. Put $g(x) = y_0 + (y_{r+1} - y_0)(x - a)/(b - a)$. Let $x^r = (x_1, ..., x_r)$ be the vector obtained in accordance with (3.2), Chapter 1, when applying the algorithm $\tilde{x}^r$ to the function g. The result guaranteed by the algorithm $(\tilde{x}^r, \tilde{\beta}_*)$ is not better than

$$\varepsilon_{r+2}(x_0, x^r, x_{r+1}, y_0, g(x_1), \ldots, g(x_r), y_{r+1})$$

$$= \frac{1}{4M} \sum_{j=1}^{r+1} \left[(x_j - x_{j-1})^2 M^2 - \left(\frac{y_{r+1} - y_0}{b - a} \right)^2 (x_j - x_{j-1})^2 \right]$$

$$= \frac{1}{4M} \left[M^2 - \left(\frac{y_{r+1} - y_0}{b - a} \right)^2 \right] \sum_{j=1}^{r+1} (x_j - x_{j-i})^2$$

$$\geq \frac{1}{4M} \left[M^2 - \left(\frac{y_{r+1} - y_0}{b - a} \right)^2 \right] \frac{(b - a)^2}{r + 1} = v_r.$$

Thus, we have proved optimality of the algorithm $(x^r_C, \tilde{\beta}_*(x_0, x_{r+1}, y_0, y^r, y_{r+1})) \in A^r$ in $\tilde{A}^r$ and, therefore, in A^r. □

Note that, unlike the proof of Lemma 3.1, the proof of Lemma 3.2 cannot make use of Theorem 5.2 from Chapter 1 since the class $F(x_0, x_{r+1}, y_0, y_{r+1})$ is not symmetric for $y_0 \neq y_{r+1}$. However, Lemma 3.2 could be proved with the help of Theorem 5.1 from Chapter 1 (but not its corollary).

3.2. Sequentially optimal (by error) algorithm for the class of functions satisfying the Lipschitz condition

We now construct a sequentially optimal algorithm. For simplicity of notation, we put $K = [0, 1]$. According to the definition of a sequentially optimal algorithm, for the choice of x_1 at the first step we can make use of any optimal in $\widetilde{A}^N$ algorithm. In the case under consideration, due to (3.2)

$$x_1 \in \left\{ \frac{1}{2N}, \frac{3}{2N}, \ldots, \frac{2N-1}{2N} \right\}, \tag{3.6}$$

any of these points being acceptable as the point of the first informational computation.

Let i computations at points $x_1, \ldots, x_i$ have been performed, let $x_{i1}, \ldots, x_{ii}$ be the ordered permutation of the numbers $x_1, \ldots, x_i$, and let $y_{ij} = f(x_{ij})$, $j = 1, \ldots, i$.

Since, for the functional class being considered, the values the integrand f takes within any particular segment $[x_{i,j-1}, x_{ij}]$ $(j = 1, \ldots, i+1, x_{i0} \overset{def}{=} 0, x_{i,i+1} \overset{def}{=} 1)$ do not influence the sets of possible values of the integrand on the other segments (these sets are bounded by the lower and upper envelopes ϕ_{1i} and the ϕ_{2i}), choosing the next step x_{i+1} of a sequentially optimal algorithm amounts to finding the optimal distribution of the rest $N-i$ evaluation points among the segments $[x_{i,j-1}, x_{ij}]$, constructing the optimal nonadaptive (and, therefore, optimal adaptive) algorithms for the functional classes corresponding to these segments, and selecting any point of any of these algorithms as x_{i+1}.

Assume that the rest $N-i$ informational evaluations are distributed in such a way that on the segment $[x_{i,j-1}, x_{i,j}]$ n_j evaluations are to be performed, $j = 1, \ldots, i+1$, and for each segment an optimal algorithm is used. The best guaranteed results for the classes of functions with a fixed value at one endpoint of the segment can be computed by Lemma 3.1 and are equal to $\frac{M}{4} \cdot \frac{x_{i1}^2}{n_1 + 1/2}$ and $\frac{M}{4} \cdot \frac{(1-x_{ii})^2}{n_{i+1} + 1/2}$ for the classes corresponding to the segments $[x_{i0}, x_{i1}]$ and $[x_{ii}, x_{i,i+1}]$ respectively. The best guaranteed results for the classes of functions with fixed values at the both endpoints of the segment can be computed by Lemma 3.2 and are equal to $\frac{1}{4M} \cdot \frac{(x_{ij} - x_{i,j-1})^2 M^2 - (y_{ij} - y_{i,j-1})^2}{n_j + 1}$, $j = 2, \ldots, i$. The guaranteed result in the situation $\widetilde{z}^i$ for fixed $n_1, \ldots, n_{i+1}$ is obviously equal to the sum of the best guaranteed results for the classes corresponding to the segments $[x_{i,j-1}, x_{ij}]$, $j = 1, \ldots, i+1$, i.e., it is equal to

$$\frac{M}{4} \cdot \frac{x_{i1}^2}{n_1 + 1/2} + \frac{1}{4M} \sum_{j=2}^{i} \frac{(x_{ij} - x_{i,j-1})^2 M^2 - (y_{ij} - y_{i,j-1})^2}{n_j + 1} + \frac{M}{4} \cdot \frac{(1-x_{ii})^2}{n_{i+1} + 1/2}.$$

To distribute the rest $N-i$ evaluations in the optimal way, we have to find a set $n_1^i, \ldots, n_{i+1}^i$ giving the result $\varepsilon_N(z^i)$ defined by formula (6.7) of

Chapter 1, i.e., delivering the minimum in the following formula:

$$\varepsilon_N(z^i) = \min_{\substack{n_1,\dots,n_{i+1}\in\{0,1,2,\dots\} \\ \sum_{j=1}^{i+1} n_j = N-i}} \left[\frac{M}{4}\cdot\frac{x_{i1}^2}{n_1+1/2} + \frac{1}{4M}\sum_{j=2}^{i}\frac{(x_{ij}-x_{i,j-1})^2M^2-(y_{ij}-y_{i,j-1})^2}{n_j+1} + \frac{M}{4}\cdot\frac{(1-x_{ii})^2}{n_{i+1}+1/2}\right].$$

Having found the solution $n_1^i,\dots,n_{i+1}^i$, we can set x_{i+1} equal to any point from the lattice obtained as a union of the optimal sets given by Lemmas 3.1 and 3.2, i.e., of the sets x_R^r for $r=n_1^i$, x_C^r for $r=n_j^i$, $j=2,\dots,i$, and x_L^r for $r=n_{i+1}^i$, corresponding to the segments $[x_{i,j-1},x_{ij}]$, $j=1,\dots,i+1$:

$$\begin{aligned} x_{i+1} \in & \left\{ (2k_1-1)\frac{x_{i1}}{2n_1^i+1} \;\middle|\; k_1=1,\dots,n_1^i \right\} \\ & \cup \left(\bigcup_{j=2}^{i} \left\{ x_{i,j-1}+k_j\frac{x_{ij}-x_{i,j-1}}{n_j^i+1} \;\middle|\; k_j=1,\dots,n_j^i \right\} \right) \\ & \cup \left\{ x_{ii}+2k_{i+1}\frac{1-x_{ii}}{2n_{i+1}^i+1} \;\middle|\; k_{i+1}=1,\dots,n_{i+1}^i \right\}, \\ & \qquad\qquad i=1,\dots,N-1 \end{aligned} \tag{3.8}$$

(if $n_j^i=0$, the corresponding set is considered to be empty).

Lemmas 3.1 and 3.2 together with the above considerations give a complete proof of the following theorem:

Theorem 3.1 (Sukharev [79b]). *The algorithm* (3.6), (3.8) *is sequentially optimal (by error) on the class* (3.1) *with* $[a,b]=[0,1]$. □

Remark. By the definition of sequentially optimal algorithms, they make the best use of the accumulated information about the function f in every situation. That is, they most severely punish "nature" responsible for the "choice" of f for every commited mistakes, i.e., for every choice of a variant which is not worst for "the computer". However, sequentially optimal algorithms (3.6) and (3.8) differ in what opportunities they give to "nature" for committing mistakes the algorithm can benefit from. For instance, it is easy to see that the algorithm $x_1=1/(2N)$, $x_2=3/(2N)$, ..., $x_N=(2N-1)/(2N)$ is sequentially optimal by error. At the same time, this algorithm gives "nature" absolutely no chance to commit a mistake. That is why it is important to make a good use of the freedom given by formulas (3.6) and (3.8). For example, we can choose the next point randomly from the set of possible points (3.6), (3.8) with equal probability. Another way is to choose the point from the set (3.6), (3.8) which is closest to the point the one-step optimal algorithm a wiuld suggest. The points x_1, x_{i+1}, $i\geq 1$, suggested by the one-step optimal algorithm will be constructed in Section 4.

3.3. Solution of auxiliary integer optimization problems

To apply a sequentially optimal algorithm, we have to solve a nonlinear integer optimization problem of finding $\varepsilon_N(z^i)$ at every step. Such problems have the form

$$\sum_{j=1}^{m} \frac{A_j}{n_j+\alpha_j} \to \min_{n_1,\dots,n_m}, \quad n_1,\dots,n_m \in \{0,1,2,\dots\}, \quad \sum_{j=1}^{m} n_j = p, \tag{3.9}$$

where $A_j > 0$ and $\alpha_j = 1/2$ or $\alpha_j = 1$. The algorithm for solving such problems developed by Gross [56] (see also Saaty [70, Chapter 4] and Sukharev, Timokhov and Fedorov [86, Chapter 7]) makes use of convexity of the functions $\phi_j(n_j) = A_j/(n_j+\alpha_j)$ and is aimed at solving problems of the form

$$\sum_{j=1}^{m} \phi_j(n_j) \to \min, \quad \sum_{j=1}^{m} n_j = p,$$
$$n_j \in \{0,1,\dots,p\}, \quad j=1,\dots,m, \tag{3.10}$$

where for all $k=1,\dots,p-1$, $j=1,\dots,m$,

$$\phi_j(k) - \phi_j(k-1) \le \phi_j(k+1) - \phi_j(k). \tag{3.11}$$

Obviously, the functions $\phi_1,\dots,\phi_m$ convex on $[0,p]$ satisfy conditions (3.11). These conditions are analogous to the definition of convexity for functions of an integer argument. Therefore, the problem (3.9) is a special case of the problem (3.10). Note that problems of integer minimization of a sum of convex functions were considered, besides Gross [56], also by Mjelde [83], Shih [74], Veinott [66], etc.

To obtain an algorithm for solving the problem (3.10), we have to carry out some simple transformations. Since

$$\phi_j(n_j) = (\phi_j(n_j) - \phi_j(n_j-1)) + (\phi_j(n_j-1)$$
$$- \phi_j(n_j-2) + \dots + (\phi_j(1) - \phi_j(0)) + \phi_j(0),$$

we can write

$$\sum_{j=1}^{n} \phi_j(n_j) = \sum_{j\in\{j|1\le j\le m, n_j>0\}} \sum_{k=1}^{n_j} a_{jk} + C, \tag{3.12}$$

where

$$a_{jk} = \phi_j(k) - \phi_j(k-1), \quad j=1,\dots,m, \quad k=1,\dots,p, \quad C = \sum_{j=1}^{m} \phi_j(0).$$

Thus, the original problem (3.10) is reduced to the problem of minimizing a double sum containing p elements of the matrix (a_{jk}). Taking into account inequalities (3.11), which can be rewritten as

$$a_{jk} \le a_{j2} \le \cdots \le a_{jp}, \quad j=1,\ldots,m,$$

we see that, to solve (3.10), it is sufficient to select p minimal elements from the matrix $\begin{pmatrix} a_{11} & \ldots & a_{1p} \\ \vdots & \ddots & \vdots \\ a_{m1} & \ldots & a_{mp} \end{pmatrix}$. If we have chosen n_j elements from the jth row, $j=1,\ldots,m$, then $(n_1,\ldots,n_m)$ will solve the problem (3.10). Actually, the above argument proves the following lemma:

Lemma 3.3 (Gross's criterion, see Gross [56]). *The condition*

$$\max_{j\in\{j|1\le j\le m, n_j>0\}} [\phi_j(n_j)-\phi_j(n_j-1)] \le \min_{j=1,\ldots,m} [\phi_j(n_j+1)-\phi_j(n_j)]$$

is necessary and sufficient for a vector $(n_1,\ldots,n_m)$ *satisfying the restrictions of the problem* (3.10), (3.11) *to be its solution.*

Proof. The above condition combined with inequalities (3.11) shows that the maximum of the elements selected from the matrix $(a_{jk})=(\phi_j(k)-\phi_j(k-1))$ does not exceed the minimum of the rest elements of the matrix, so we have really selected p minimal elements. □

Now we can formulate the following algorithm:

Algorithm 1.
Step 0: $n_j:=0$, $j=1,\ldots,m$.
Step s $(1\le s\le p)$: $\mu:=\arg\min_{j=1,\ldots,m}(\phi_j(n_j+1)-\phi_j(n_j))$, $n_\mu:=n_\mu+1$. □

Taking into account (3.11), we see that, having carried out Step s for $s=1,\ldots,p$, we will obtain p minimal elements from the sum (3.12) and thus solve the problem (3.10).

If we have a good initial approximation $n_1^0,\ldots,n_m^0$ such that $\sum_{j=1}^{m} n_j^0=p$, we may reasonably use a slightly different algorithm:

Algorithm 2.
Step 0: $n_j:=n_j^0$, $j=1,\ldots,m$.
Step 1: $\mu:=\arg\min_{j=1,\ldots,m}(\phi_j(n_j+1)-\phi_j(n_j))$,
$\nu:=\arg\max_{j\in\{j|1\le j\le m,n_j>0\}}(\phi_j(n_j)-\phi_j(n_j-1))$;
if $\phi_\mu(n_\mu+1)-\phi_\mu(n_\mu)\ge\phi_\nu(n_\nu)-\phi_\nu(n_\nu-1)$, then stop, else execute the operators $n_\mu:=n_\mu+1$, $n_\nu:=n_\nu+1$ and repeat Step 1. □

Clearly, Algorithm 2 also selects p minimal elements from (a_{jk}), i.e., solves the problem (3.10). As an initial approximation for the problem (3.9), we can use either the solution of the integer problem from the previous step or rounded solution of the problem

$$\sum_{j=1}^{m} \frac{A_j}{n_j+\alpha_j} \longrightarrow \min_{n_1,\dots,n_m:\sum_{j=1}^{m} n_j=p}, \tag{3.13}$$

which can easily be obtained in an explicit form:

$$n_j^* = \frac{\sqrt{A_j}(p+\alpha)}{A} - \alpha_j, \quad j=1,\dots,m,$$

where

$$\alpha = \sum_{j=1}^{m} \alpha_j, \quad A = \sum_{j=1}^{m} \sqrt{A_j}.$$

The case $A=0$ means that we have already found the exact value of the integral $\int_0^1 f(x)dx$ and there is no need of solving (3.13).

It is easy to see that if $i=1$, i.e., if we have to select x_2, the problem is trivial and

$$x_2 \in \left\{ \frac{1}{2N}, \frac{3}{2N}, \dots, \frac{2N-1}{2N} \right\} \setminus \{x_1\}.$$

3.4. One-step optimal algorithm for the Lipschitz functional class

The obtained result also makes it easy to construct an algorithm with the property of one-step optimality. To select the point x_1 in the optimal way, it is enough to solve the problem of minimizing the result (3.5) guaranteed after the first step:

$$\varepsilon_1(z^1) = \frac{M}{2}x_1^2 + \frac{M}{2}(1-x_1)^2 \to \min, \quad x_1 \in [0,1].$$

This gives

$$x_1 = 1/2. \tag{3.14}$$

To find x_{i+1} for $i \geq 1$, we have to solve the problem (3.7) for $N-i=1$. In other words, we have to solve the problem (3.9) for $m=i+1$, $p=1$, $A_1 = Mx_{i1}^2/4$, $\alpha_1 = 1/2$, $A_{i+1} = M(1-x_{ii})^2/4$, $\alpha_{i+1} = 1/2$, $A_j = [(x_{ij} - x_{i,j-1})^2 M^2 - (y_{ij} - y_{i,j-1})^2]/(4M)$, $\alpha_j = 1$, $j=2,\dots,i$, or the problem (3.10) for $\phi_j(n_j) = A_j/(n_j+\alpha_j)$. Using Algorithm 1 for solving this problem, we get, in accordance with (3.8),

$$x_{i+1} = \begin{cases} \dfrac{x_{i1}}{3} & \text{if } \frac{A_1}{1+\alpha_1} - \frac{A_1}{\alpha_1} = m, \\ \dfrac{x_{i,k-1}+x_{ik}}{2} & \text{if } \frac{A_k}{1+\alpha_k} - \frac{A_k}{\alpha_k} = m, \ k \in \{2,\dots,i\}, \\ x_{ii} + \dfrac{2}{3}(1-x_{ii}) & \text{if } \frac{A_{i+1}}{1+\alpha_{i+1}} - \frac{A_{i+1}}{\alpha_{i+1}} = m, \end{cases} \tag{3.15}$$

where $m = \min_{j=1,\dots,i+1} \{A_j/(1+\alpha_j) - A_j/\alpha_j\}$. If (3.15) does not determine a unique point x_{i+1}, we can choose any possible point. This gives us a one-step optimal algorithm.

Theorem 3.2. *The algorithm* (3.14), (3.15) *is one-step optimal on the class* (3.1) *with* $[a,b]=[0,1]$. □

3.5. Sequentially optimal (counting informational computations) algorithm for the Lipschitz functional class

Let the required accuracy ε of the solution be fixed. We briefly describe the process of constructing an algorithm sequentially optimal (counting informational computations) on the class (3.1) for $[a,b]=[0,1]$.

Due to Theorem 5.2 of Chapter 1, for this functional class the best guaranteed result ε_N in the class of all adaptive algorithms coinsides with the best guaranteed result m_N in the class of all nonadaptive algorithms, and, by (3.3), $m_N = M/(4N)$. Hence, we have

$$N_\varepsilon = \min\{N \mid \varepsilon_N \le \varepsilon\} = \min\left\{N \;\middle|\; \frac{M}{4N} \le \varepsilon\right\} = \left\lceil \frac{M}{4\varepsilon} \right\rceil. \tag{3.16}$$

In accordance with the definitions of optimal (counting informational computations) algorithm (see Section 4.13 of Chapter 1) and sequentially optimal (counting informational computations) algorithm (see Section 6.4 of Chapter 1), the choice of x_1 in a sequentially optimal algorithm is given by (3.6) for $N = N_\varepsilon$.

If after i steps ($i \ge 1$) in a situation z^i the stopping criterion (4.20) from Chapter 1 is not satisfied, we find $N_\varepsilon(z^i)$ from (3.7) and then select the point x_{i+1} in accordance with (3.8). The solution of the problem (3.7) for $N = N_\varepsilon(z^i)+i$ will be $n_1^i, \dots, n_{i+1}^i$. This completes construction of the algorithm.

Let the accuracy ε still be fixed. We now extract from the set of all sequentially optimal (counting informational computations) algorithms the algorithm that corresponds to trivial problems of the form (3.9) and hence is easy to program.

This algorithm is determined by the following condition. At every step we choose the leftmost point from the set of all points admissible for a sequentially optimal (counting informational computations) algorithm.

Thus (see (3.16)),

$$x_1 = \frac{1}{2N_\varepsilon} = \frac{1}{2\lceil M/(4\varepsilon)\rceil}. \tag{3.17}$$

In order to find x_2, we have to solve the following problem of the form (3.7):

$$\frac{M}{4}\cdot\frac{x_1^2}{n_1+1/2}+\frac{M}{4}\cdot\frac{(1-x_1)^2}{n_2+1/2}\to\min,$$

$$n_1,n_2\in\{0,1,2,\dots\},\ n_1+n_2=N_\varepsilon(z^1).$$

We show that $(0,N_\varepsilon(z^1))$ is the solution of the problem. Due to Lemma 3.3, it amounts to verification of the inequality

$$\frac{\frac{M}{4}x_1^2}{3/2}-\frac{\frac{M}{4}x_1^2}{1/2}\geq\frac{\frac{M}{4}(1-x_1)^2}{N_\varepsilon(z^1)+1/2}-\frac{\frac{M}{4}(1-x_1)^2}{N_\varepsilon(z^1)-1/2}.$$

This inequality follows from (3.17) and (6.12) of Chapter 1. So, in accordance with (3.8) and the condition determining the algorithm, we get

$$x_2=x_1+2\frac{1-x_i}{2N_\varepsilon(z^1)+1}. \tag{3.18}$$

(It can be checked that $N_\varepsilon(z^1)=N_\varepsilon-1$, and so $x_2=3/(2N_\varepsilon)$.)

Assume that $x_1<x_2<\dots<x_{i-1}$ and the solution to the following problem of the type (3.9)

$$\frac{M}{4}\cdot\frac{x_1^2}{n_1+1/2}+\frac{1}{4M}\sum_{j=2}^{i-1}\frac{(x_j-x_{j-1})^2M^2-(y_j-y_{j-1})^2}{n_j+1}$$

$$+\frac{M}{4}\cdot\frac{(1-x_{i-1})^2}{n_i+1/2}\to\min_{n1,\dots,n_i},$$

$$n_1,\dots,n_i\in\{0,1,2,\dots\},\quad \sum j=1^in_j=N_\varepsilon(z^{i-1}),$$

has the form $(0,\dots,0,N_\varepsilon(z^{i-1}))$. In this case, by the definition of the algorithm we have

$$x_i=x_{i-1}+2\frac{1-x_{i-1}}{2N_\varepsilon(z^{i-1})+1}, \tag{3.19}$$

and, due to Lemma 3.3,

$$\frac{\frac{M}{4}x_1^2}{3/2}-\frac{\frac{M}{4}x_1^2}{1/2}\geq\frac{\frac{M}{4}(1-x_{i-1})^2}{N_\varepsilon(z^{i-1})+1/2}-\frac{\frac{M}{4}(1-x_{i-1})^2}{N_\varepsilon(z^{i-1})-1/2}, \tag{3.20}$$

$$\frac{\frac{1}{4M}[(x_j-x_{j-1})^2M^2-(y_j-y_{j-1})^2]}{2}-\frac{\frac{1}{4M}[(x_j-x_{j-1})^2M^2-(y_j-y_{j-1})^2]}{1}$$

$$\geq\frac{\frac{M}{4}(1-x_{i-1})^2}{N_\varepsilon(z^{i-1})+1/2}-\frac{\frac{M}{4}(1-x_{i-1})^2}{N_\varepsilon(z^{i-1})-1/2},$$

$$j=2,\dots,i-1. \tag{3.21}$$

We now prove that the solution of the problem

$$\frac{M}{4}\cdot\frac{x_1^2}{n_1+1/2}+\frac{1}{4M}\sum_{j=2}^{i}\frac{(x_j-x_{j-1})^2M^2-(y_j-y_{j-1})^2}{n_j+1}$$

$$+\frac{M}{4}\cdot\frac{(1-x_i)^2}{n_{i+1}+1/2}\to\min_{n1,\dots,n_i}, \tag{3.22}$$

$$n_1,\dots,n_{i+1}\in\{0,1,2,\dots\},\quad \sum j=1^{i+1}n_j=N_\varepsilon(z^i),$$

has the form $(0,\dots,0,N_\varepsilon(z^{i-1}))$. It is natural to assume that $N_\varepsilon(z^i)\geq 1$ (otherwise the required accuracy ε has already been achieved). Due to Lemma 3.3, the proof amounts to verification of the following inequalities:

$$\frac{\frac{M}{4}x_1^2}{3/2}-\frac{\frac{M}{4}x_1^2}{1/2}\geq\frac{\frac{M}{4}(1-x_i)^2}{N_\varepsilon(z^i)+1/2}-\frac{\frac{M}{4}(1-x_i)^2}{N_\varepsilon(z^i)-1/2},\tag{3.23}$$

$$\frac{\frac{1}{4M}[(x_j-x_{j-1})^2M^2-(y_j-y_{j-1})^2]}{2}-\frac{\frac{1}{4M}[(x_j-x_{j-1})^2M^2-(y_j-y_{j-1})^2]}{1}$$

$$\geq\frac{\frac{M}{4}(1-x_i)^2}{N_\varepsilon(z^i)+1/2}-\frac{\frac{M}{4}(1-x_i)^2}{N_\varepsilon(z^i)-1/2},\quad j=2,\dots,i.\tag{3.24}$$

Rewrite (3.19) in the following form:

$$\frac{M}{4}(1-x_i)^2=\frac{M}{4}(1-x_{i-1})^2\left(\frac{N_\varepsilon(z^{i-1})-1/2}{N_\varepsilon(z^{i-1})+1/2}\right)^2.$$

From here, taking into account that $N_\varepsilon(z^{i-1})\geq N_\varepsilon(z^i)+1\geq 2$ (see (6.12), Chapter 1), we get

$$\frac{\frac{M}{4}(1-x_i)^2}{N_\varepsilon(z^i)+1/2}-\frac{\frac{M}{4}(1-x_i)^2}{N_\varepsilon(z^i)-1/2}\leq\frac{-\frac{M}{4}(1-x_{i-1})^2\left(\frac{N_\varepsilon(z^{i-1})-1/2}{N_\varepsilon(z^{i-1})+1/2}\right)^2}{(N_\varepsilon(z^{i-1})-1/2)(N_\varepsilon(z^{i-1})-3/2)}.$$

$$\leq\frac{-\frac{M}{4}(1-x_{i-1})^2}{(N_\varepsilon(z^{i-1})-1/2)(N_\varepsilon(z^{i-1})+1/2)}=\frac{\frac{M}{4}(1-x_{i-1}^2}{N_\varepsilon(z^{i-1})+1/2}-\frac{\frac{M}{4}(1-x_{i-1}^2}{N_\varepsilon(z^{i-1})-1/2}\tag{3.25}$$

(the second inequality in this chain can be verified directly). Now inequality (3.24) for $j=2,\dots,i-1$ and inequality (3.23) follow from (3.21) and (3.20) respectively. Thus, it suffices to check (3.24) for $j=i$.

We have

$$\frac{\frac{M}{4}(1-x_i)^2}{N_\varepsilon(z^i)+1/2}-\frac{\frac{M}{4}(1-x_i)^2}{N_\varepsilon(z^i)-1/2}\leq\frac{-\frac{M}{4}(1-x_{i-1})^2(N_\varepsilon(z^{i-1})-1/2)}{(N_\varepsilon(z^{i-1})-1/2)(N_\varepsilon(z^{i-1})-3/2)}$$

$$\leq\frac{-\frac{M}{4}(1-x_{i-1})^2}{2(N_\varepsilon(z^{i-1})+1/2)^2}=-\frac{M}{8}(x_i-x_{i-1})^2$$

$$\leq-\frac{1}{8M}[(x_i-x_{i-1})^2M^2-(y_i-y_{i-1})^2]$$

$$=\frac{\frac{1}{4M}[(x_i-x_{i-1})^2M^2-(y_i-y_{i-1})^2]}{2}-\frac{\frac{1}{4M}[(x_i-x_{i-1})^2M^2-(y_i-y_{i-1})^2]}{1},$$

where the first inequality coincides with the first inequality in (3.25), the second inequality is verified directly, and the third inequality follows from (3.19). This establishes validity of inequalities (3.23) and (3.24), and thus proves by induction that for any i such that $N_\varepsilon(z^i)\geq 1$ the solution of the problem (3.22) has the form $(0,\dots,0,N_\varepsilon(z^i))$. Hence, in accordance with (3.8) and the definition of the algorithm, we get

$$x_{i+1}=x_i+2\frac{1-x_i}{2N_\varepsilon(z^i)+1}. \tag{3.26}$$

Due to (6.9), Chapter 1, the number $N_\varepsilon(z^i)$ is defined by the formula

$$N_\varepsilon(z^i)=\min\left\{N \;\middle|\; \frac{Mx_1^2}{2}+\frac{1}{4M}\sum_{j=2}^{i}[(x_j-x_{j-1})^2M^2 - (y_j-y_{j-1})^2]+\frac{M}{4}\cdot\frac{(1-x_i)^2}{N+1/2}\le\varepsilon\right\}.$$

It is easily seen that

$$N_\varepsilon(z^i)=\left\lceil\frac{\frac{M}{4}(1-x_i)^2}{\varepsilon-\frac{Mx_1^2}{2}-\frac{1}{4M}\sum_{j=2}^{i}[(x_j-x_{j-1})^2M^2-(y_j-y_{j-1})^2]}\right\rceil-\frac{1}{2}. \tag{3.27}$$

Thus, we have given a complete proof of the following theorem:

Theorem 3.3. *The algorithm* (3.17), (3.26), (3.27) *is sequentially optimal (counting informational computations) on the class* (3.1) *for* $[a,b]=[0,1]$. □

As we have already mentioned, the constructed algorithm is easy to program and has low combinatory complexity, i.e., does not require much computer time for its "inner needs". At the same time, if we use this algorithm, "nature" has less chance of "making mistakes" compared to the case where we use other sequentially optimal algorithms (in this connection, see the remark after Theorem 3.1).

3.6. Optimal and sequentially optimal algorithms for the class of monotonic functions

The methods we have described also enable us to construct sequentially optimal algorithms for functional classes whose nature is completely different from that of the class (3.1).

Consider the class of the functions monotonic on $[a,b]$ and having fixed values at the endpoints of the segment:

$$F=\{f \mid f \text{ is nondecreasing on } [a,b],\ f(a)=y_0,\ f(b)=y_{N+1}\}. \tag{3.28}$$

Let informational computations be evaluations of the function being integrated. Note that the class (3.28), in addition to containing nonsmooth functions, like the class (3.1), also contains discontinuous functions. This class has been studied by Glinkin [81a] and Kiefer [57].

Suppose we have computed values of a function f from the class (3.28) at points $x_1, x_2, \dots, x_N$, $x_1 < x_2 < \dots < x_N$, and these values are $y_1, y_2, \dots, y_N$ respectively. It is obvious that the functions

$$\phi_{1N}(x) = \begin{cases} y_{j-1} & \text{for } x_{j-1} \le x < x_j, \\ y_{N+1} & \text{for } x = x_{N+1}, \end{cases}$$

$$\phi_{2N}(x) = \begin{cases} y_0 & \text{for } x = x_0, \\ y_j & \text{for } x_{j-1} < x \le x_j, \end{cases}$$

where $x_0 = a$, $x_{N+1} = b$, and $j = 1, \dots, N+1$, are sharp lower and upper envelopes respectively for functions $f \in F(z^N) = \{f \in F \mid f(x_j) = y_j,\ j = 1, \dots, N\}$, see Fig.5. The central terminal operation is easily expressed through the lower and upper envelopes (see Remark 3 after Theorem 1.4):

$$\widetilde{\beta}_*(z^N) = \int_a^b \frac{1}{2}(\phi_{1N}(x) + \phi_{2N}(x))dx = \sum_{j=1}^{N+1} (x_j - x_{j-1}) \frac{y_j + y_{j-1}}{2}. \qquad (3.29)$$

The value of $\widetilde{\beta}_*(z^N)$ is equal to the area of the figure consisting of $N+1$ trapezoids and lying between the heavy line (Fig.5) and x-axis. (Apparently, this area is equal to the area between x-axis and the graph of the function $\frac{1}{2}(\phi_{1N} + \phi_{2N})$.) Thus, the central terminal operation coincides with the compound trapezoid quadrature formula with the nodes $x_0 = a$, $x_1, \dots, x_N$, $x_{N+1} = b$. The guaranteed accuracy in a situation z^N is given by the following formula (see Remark 3 after Theorem 1.4):

$$\varepsilon_N(z^N) = \int_a^b \frac{1}{2}(\phi_{2N}(x) - \phi_{1N}(x))dx = \sum_{j=1}^{N+1} (x_j - x_{j-1}) \frac{y_j - y_{j-1}}{2}. \qquad (3.30)$$

This is equal to the area of the shaded figure in Fig.5.

Theorem 3.4. *The algorithm $\alpha_0(x_0^N, \widetilde{\beta}_*)$ for $x_0^N = (x_1^0, \dots, x_N^0)$, $x_j^0 = a + j(b-a)/(N+1)$, $j = 1, \dots, N$ (i.e., the compound trapezoid quadrature formula with equidistant nodes $x_0^0 = a$, $x_1^0, \dots, x_n^0$, $x_{N+1}^0 = b$) is an optimal error algorithm both in the class of all quadrature formulas with n nodes, in the class A^N of all nonadaptive algorithms, and in the class $\widetilde{A}^N$ of all adaptive algorithms. The best guaranteed result is equal to* $\dfrac{(y_{N+1} - y_0)(b-a)}{2(N+1)}$.

Proof. For any adaptive algorithm $\alpha = (\widetilde{x}^N, \widetilde{\beta}) \in \widetilde{A}^N$ (see (3.11), Chapter 1) and any realizable for the algorithm α situation z^N (see Section 3.2, Chapter 1), we have

$$\begin{aligned} \sup_{j \in F} \varepsilon(\alpha, f) &\ge \sup_{f \in F(z^N)} \varepsilon(\alpha, f) = \sup_{f \in F(z^N)} |S(f) - \widetilde{\beta}(z^N)| \\ &\ge \inf_{\beta \in \mathbb{R}} \sup_{f \in F(z^N)} |S(f) - \beta| = \frac{1}{2}\left(\sup_{f \in F(z^N)} S(f) - \inf_{f \in F(z^N)} S(f) \right) \\ &= \frac{1}{2}(S(\phi_{2N}) - S(\phi_{1N})), \end{aligned} \qquad (3.31)$$

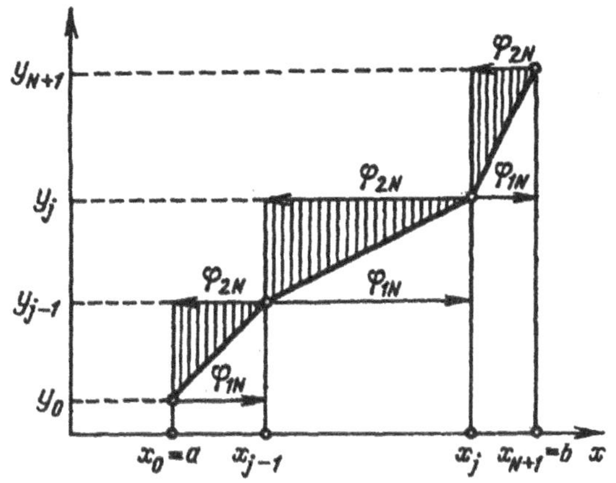

Fig. 5. The central terminal operation for integration problem on the class F of monotonic functions; sharp lower envelope ϕ_{1N} and sharp upper envelope ϕ_{2N} for a subclass $F(z^N)$

where $S(f)=\int_a^b f(x)dx$. Let $x_1,\ldots,x_N$ be the points we obtain when applying the algorithm α to the function $g(x)=y_0+(y_{N+1}-y_0)(x-a)/(b-a)$. For our notation not to grow too complex, assume that $x_1\le x_2\le\cdots\le x_N$. Put $y_j=g(x_j)$, $j=1,\ldots,N$. Then

$$\begin{aligned} S(\phi_{2N})-S(\phi_{1N}) &= \sum_{j=1}^{N+1}(y_j-y_{j-1}) \\ &= \sum_{j=1}^{N+1}(g(x_j)-g(x_{j-1}))(x_j-x_{j-1}) \\ &= \frac{y_{N+1}-y_0}{b-a}\sum_{j=1}^{N+1}(x_j-x_{j-1})^2 \\ &\ge \frac{(y_{N+1}-y_0)(b-a)}{N+1}. \end{aligned}$$

To verify the inequality in this formula, it suffices to find the following conditional extremum with the help of Lagrange's method of undetermined multipliers:

$$\sum_{j=1}^{N+1}(x_j-x_{j-1})^2\to\min,\quad \sum_{j=1}^{N+1}(x_j-x_{j-1})=b-a.$$

From here and from (3.31) we get for any $\alpha\in\widetilde{A}^N$

$$\sup_{f\in F}\varepsilon(\alpha,f)\ge\frac{(y_{N+1}-y_0)(b-a)}{2(N+1)}. \tag{3.32}$$

Thus we have shown that no algorithm $\alpha \in \widetilde{A}^N$ can guarantee a better accuracy than $\dfrac{(y_{N+1}-y_0)(b-a)}{2(N+1)}$.

Now it suffices to show that the algorithm $\alpha_0 = (x_0^N, \widetilde{\beta}_*)$ does guarantee this accuracy. We have

$$\begin{aligned}\sup_{f\in F}(\alpha_0, f) &= \sup_{y^N} \sup_{f\in F(x_0^N, y^N)} \varepsilon(\alpha_0, f)\\ &= \sup_{y^N} \sup_{f\in F(x_0^N, y^N)} |S(f) - \widetilde{\beta}_*(x_0^N, y^N)| = \sup_{y^N} \varepsilon_N(x_0^N, y^N)\\ &= \sup_{y^N} \sum_{j=1}^{N+1} (x_j^0 - x_{j-1}^0)\frac{y_j - y_{j-1}}{2} = \frac{(y_{N+1}-y_0)(b-a)}{2(N+1)}. \end{aligned} \tag{3.33}$$

Here the supremum with respect to y^N is taken over the set $\{y^N \mid F(x_0^N, y^N) \neq \emptyset\}$, the first equation holds since $F = \bigcup_{y^N} F(x_0^N, y^N)$, the second equation holds by the definitions of the criterion $\varepsilon(\alpha_0, f)$ and the algorithm α_0, the third equation is based on (4.3) and (4.4), Chapter 1, the fourth one follows from (3.30), and the fifth one makes use of the definition of the points $x_0^0, x_1^0, \ldots, x_{N+1}^0$. Formulas (3.32) and (3.33) imply that the algorithm α_0 is optimal in the class $\widetilde{A}^N$ of all adaptive algorithms, hence also in the class of all quadrature formulas with N nodes and in the class A^N of all nonadaptive algorithms. The best guaranteed result in each of the classes is equal to the result $\dfrac{(y_{N+1}-y_0)(b-a)}{2(N+1)}$ guaranteed by the optimal algorithm α_0. □

Note that we could give another proof of this theorem based on Theorems 4.2 and 5.1 of Chapter 1.

A sequentially optimal algorithm for the class (3.28) can be constructed according to the same scheme we used for the class (3.1) of functions satisfying the Lipschitz condition. At the first step we select any of the N equidistant points determining the optimal algorithm in accordance with Theorem 3.4:

$$x_1 \in \left\{ a + \frac{b-a}{N+1}, a + 2\frac{b-a}{N+1}, \ldots, a + N\frac{b-a}{N+1} \right\}. \tag{3.34}$$

Suppose we have carried out i informational computations at points $x_1, \ldots, x_i$. Let $x_{i1}, \ldots, x_{ii}$ be the permutation of the points $x_1, \ldots, x_i$ in the increasing order, and let $y_{ij} = f(x_{ij})$, $j = 1, \ldots, i$, $x_{i0} = a$, $x_{i,i+1} = b$, $y_{i0} = y_0$, $y_{i,i+1} = y_{N+1}$. Suppose the remaining $N-i$ informational computations are allocated in such a way that n_j computations are to be performed within $[x_{i,j-1}, x_{ij}]$, $j = 1, \ldots, i+1$, and for each of these segments the optimal algorithm is applied. Due to Theorem 3.4, if the number n_j of informational computations is fixed, the best guaranteed result for the functional class corresponding to the segment $[x_{i,j-1}, x_{ij}]$ with fixed values at the both endpoints is equal to $\dfrac{(y_{ij}-y_{i,j-1})(x_{ij}-x_{i,j-1})}{2(n_j+1)}$, $j=1,\dots,i+1$. The guaranteed result in the situation z^i for given $n_1,\dots,n_{i+1}$ is obviously equal to $\sum_{j=1}^{i+1}\dfrac{(y_{ij}-y_{i,j-1})(x_{ij}-x_{i,j-1})}{2(n_j+1)}$. In order to select the point x_{i+1}, we have to solve a problem of the type (3.9) – to find

$$\varepsilon_N = \min_{\substack{n_1,\dots,n_{i+1}\in\{0,1,2,\dots\}\\ \sum_{j=1}^{i+1} n_0 = N-i}} \sum_{j=1}^{i+1}\frac{(y_{ij}-y_{i,j-1})(x_{ij}-x_{i,j-1})}{2(n_j+1)}.$$

This problem can be solved with the help of the algorithms described in Section 3.3. After we have found the solution $n_1^i,\dots,n_{i+1}^i$, any point from the unification of optimal lattices specified by Theorem 3.4 for the functional classes corresponding to the segments $[x_{i,j-1}, x_{ij}]$, $j=1,\dots,i+1$, can be selected as the point of the $(i+1)$st informational computation:

$$x_{i+1}\in\bigcup_{j=1}^{i+1}\left\{x_{i,j-1}+k_j\frac{x_{ij}-x_{i,j-1}}{n_j^i+1}\;\middle|\; k_j=1,\dots,n_j^i\right\},$$

$$i=1,\dots,N-1 \tag{3.35}$$

(if $n_j^i=0$, the corresponding set is considered to be empty). Thus we have proved the following theorem:

Theorem 3.5. *The algorithm* (3.34), (3.35) *is sequentially optimal (by error) on the functional class* (3.28). □

A one-step optimal algorithm can be constructed according to the same scheme we used for the functional class (3.1). We represent this algorithm in the following form:

$$x_1=\frac{a+b}{2},\quad x_{i+1}=\frac{x_{i,k-1}+x_{ik}}{2},\quad i\geq 1, \tag{3.36}$$

where k is determined by

$$(y_{ik}-y_{i,k-1})(x_{ik}-x_{i,k-1})=\max_{j=1,\dots,i+1}\{(y_{ij}-y_{i,j-1})(x_{ij}-x_{i,j-1})\}.$$

Algorithms sequentially optimal counting informational computations are also easy to construct.

3.7. Requirements to the computation model ensuring possibility of constructing sequentially optimal algorithms according to a universal scheme

We start with mentioning another two functional classes such that sequentially optimal algorithms for them can be constructed according to the same scheme as for the classes (3.1) and (3.28).

Glinkin [81a] used this scheme for constructing a sequentially optimal algorithm for the class of all functions nondecreasing on $[a,b]$ and satisfying on this segment the Lipschitz condition with a constant M. Molchanova [86] has constructed a sequentially optimal algorithm for the class of functions satisfying the following condition:

$$-M_1(u-v)\leq f(u)-f(v)\leq M_2(u-v),$$

where $u,v\in[a,b]$, $u<v$, $M_1>0$, $M_2>0$.

All these functional classes together with the information about functions implied by our computation model have something in common. First of all, they consist of functions of one variable. Informational computations are evaluations of the function (and, maybe, its derivatives) at some points within its domain. The crucial characteristic of the computation model that enables us to reduce the problem of constructing a sequentially optimal algorithm to a set of problems of constructing optimal algorithms is the following.

For any realizable situation z^i, any $f_1,f_2\in F(z^i)$, and any $x_{i,j-1},x_{ij}$ (where $x_{i1},\ldots,x_{ii}$ is the ordered permutation of $x_1,\ldots,x_i$), we have

$$g(x)\stackrel{def}{=}\begin{cases} f_1(x), & x\in[x_{i,j-1},x_{ij}] \\ f_2(x), & x\in K\setminus[x_{i,j-1},x_{ij}] \end{cases}\in F(z^i). \tag{3.37}$$

In other words, we can vary a function from $F(z^i)$ within the limits of $F(z^i)$ independently on every segment $[x_{i,j-1},x_{ij}]$.

This property allows us to construct, for a given resource allocation, optimal algorithms for the functional classes corresponding to all the segments $[x_{i,j-1},x_{ij}]$, and then to solve the problem of optimal resource allocation and thus obtain a sequentially optimal algorithm.

Condition (3.37) holds for many computation models; the above list of examples could be extended. For instance, this condition is satisfied if $F=W^r_\infty([a,b],M)$ (see Section 2.4 of Chapter 1) and the computational model specifies Hermitian information of order r, i.e., informational computation at x is evaluation of the operator

$$x(f)=(f(x),f'(x),\ldots,f^{(r-1)}(x)). \tag{3.38}$$

Thus, the problem of constructing sequentially optimal algorithms can be reduced to a set of problems of constructing optimal algorithms for functional classes $W^r_\infty([a,b],M)$ with given information (3.38) at one or both endpoints of the integration segment.

4. NUMERICAL TESTS

In this section, based on Glinkin and Sukharev [80, 85], we dwell on some special characteristics of computer implementation of the algorithms from Section 3. We focus on numerical tests of the algorithms and analysis of their results. Numerical tests are so important because we often have to apply an algorithm optimal in a certain sense on some functional class F to a function $f \in F$ about which we have more *a priori* information than the mere fact that it belongs to F. Say, we may know that f is monotonic and has given values at the endpoints of the segment, i.e., belongs to the class (3.28), and, in addition, f is piecewise smooth on this segment. Theoretical investigation of additional opportunities this extra information gives is often very difficult, as well as theoretical assessment of performance of algorithms constructed for the class F on the subclass of F determined by the extra information. This makes it natural to use numerical tests. Moreover, we can use some heuristic considerations to modify an algorithm obtained within a certain computation model in order to apply it in a situation that does not fit into this model perfectly. In this case, numerical tests will also be necessary for studying performance of the modified algorithm.

4.1. Concrete versions of integration algorithms

Here we list concrete versions of the algorithms from Section 3 whose performance we are going to compare.

For all the sequentially optimal algorithms (SOA) considered here we use the version that at every step selects, among all the admissible points, the point closest to that prescribed by the one-step optimal algorithm (OSOA). We also consider OSOA and so-called "approximate" sequentially optimal algorithm (ASOA). When ASOA selects the next point, the problem (3.9) is replaced by the problem (3.13) and its solution is rounded off. We compare all these algorithms with the optimal nonadaptive algorithm. For all the functional classes discussed in Section 3, it is also optimal among all adaptive algorithms, and so in what follows we just call it "optimal algorithm" (OA).

All these algorithms are applied to two functional classes: (3.1) and (3.28), denoted in this section by F_1 and F_2 respectively.

For every algorithm, two versions have been programmed: the one with fixed number of informational computations and the one with fixed accuracy. When integrating functions from F_1, the user may not know (or not know exactly) the constant M. In the modified versions of the algorithms (MSOA, MOSOA, MASOA) this constant is estimated adaptively in the process of computations using the method described below. Certainly, the modifications do not enjoy the optimality properties of the original algorithms.

4.2. Methods for estimating the Lipschitz constant

We describe the method for estimating the constant M we use in the modified versions of the algorithms (this method was presented in Strongin [78]) and also some other methods.

After i computations at points $x_1, \ldots, x_i$ $(i \geq 2)$ have been performed, the current value of M in the modified versions of integration algorithms mentioned in Section 4.1 is set equal to

$$M_i = \sigma R_i, \tag{4.1}$$

where

$$R_i = \max_{\substack{1 \leq j,\ k \leq i \\ j \neq k}} \left| \frac{f(x_j) - f(x_k)}{x_j - x_k} \right|$$

is the maximum of absolute values of the first-order divided differences and σ is some "caution coefficient". Let $x_{i1}, \ldots, x_{ii}$ be the ordered permutation of $x_1, \ldots, x_i$, and let $x_{i,j-1} < x_{i+1} < x_{ij}$. Then

$$M_{i+1} = \max \left\{ M_i, \sigma \left| \frac{f(x_{i+1}) - f(x_{i,j-1})}{x_{i+1} - x_{i,j-1}} \right|, \sigma \left| \frac{f(x_{i+1}) - f(x_{ij})}{x_{i+1} - x_{ij}} \right| \right\}, \tag{4.2}$$

and so computing estimates of the Lipschitz constant does not require much machine time. This is the way we estimate the constant M in the modified algorithms.

The method (4.1) of estimating the Lipschitz constant can also be applied in many dimensions; all we have to do is to replace $|x_j - x_k|$ with $r(x_j, x_k)$ in the formula for R_i, where r is the distance. But then we cannot use (4.2) any longer, and at the $(i+1)$st step we have to compute i first-order divided differences instead of two. The method (4.1) can be improved by adding a refinement procedure after every s steps, i.e., at the $(i+1)$st step for $i = s, 2s, 3s, \ldots$. The procedure works in the following way. After i steps we choose some divided difference $(f(x_j) - f(x_k))/r(x_j, x_k)$, partition $[x_j, x_k]$ into several equal parts with points $u_1, \ldots, u_m$, compute $f(u_1), \ldots, f(u_m)$, and put the estimate of M equal to

$$M_i = \sigma \max \left\{ R_i, \frac{|f(u_1) - f(x_j)|}{r(u_1, x_j)}, \frac{|f(u_2) - f(u_1)|}{r(u_2, u_1)}, \ldots \right.$$
$$\left. \frac{|f(u_m) - f(u_{m-1})|}{r(u_m, u_{m-1})}, \frac{|f(x_k) - f(u_m)|}{r(x_k, u_m)} \right\}.$$

It makes sense to choose j and k in such a way that $|f(x_j) - f(x_k)|/r(x_j, x_k)$ is the maximum of absolute values of the divided differences to which we have not applied the refinement procedure yet.

There also exist other heuristic methods for estimating the constant M, applicable to both one-dimensional and multidimensional cases. Lbov and Grunov [76] suggested that, alongside R_i, average increment of the divided difference over i steps $\Delta R_i = (R_i - R_2)/(i-2)$ be computed and the estimate of M after i steps (for fixed N) be set equal to $M_i = R_i + (N-i)\Delta R_i$.

It is also possible to compute the geometric mean δR_i of the increments $R_j - R_{j-1}$, $j=3,\dots,i$, instead of their arithmetic mean ΔR_i, and put the estimate of M after i steps equal to $M_i = \tau_i R_i (\delta r_i)^{N-i}$, where τ_i is some correction coefficient.

Another heuristic trick works as follows. After i steps we construct an approximating function $R^{(i)}$ such that $R^{(i)}(j) \approx R_j$, $j=2,\dots,i$, and after i steps either $\sigma R^{(i)}(N)$ or (in the case of appropriate choice of $R^{(i)}$) $\lim_{t\to\infty} R^{(i)}(t)$ serves as an estimate of M. For instance, we can choose $R^{(i)}(t) = a_i/t + b_i$ and compute the coefficients $a_i < 0$, b_i with the help of the least squares method using either the last m values $R_{i-m+1}, \dots, R_i$ (where m is some fixed number) or the values $R_{\lfloor i/m \rfloor}, R_{\lfloor 2i/m \rfloor}, \dots, R_i$. In this case we take b_i as an estimate of M.

One can also use various modifications and combinations of the above methods. Note that it may prove useful to apply such techniques only at the first stage of the process of computations, say, only at the first N_1 steps, without changing the obtained estimate at the subsequent steps.

Finally, it is also possible to use *a priopi* estimates of M. For example, we can estimate the norm of gradient of f before any informational computations.

All these methods of estimating the Lipschitz constant can be applied not only to integration problems but also to other problems of numerical analysis.

4.3. Methods for testing computational algorithms

There exist different methods for testing computational algorithms. We can use sets of test functions either constructed on the basis of some real-life problems (see De Boor [71a], Dixon [74], Einarsson [74], Kahaner [71]) or consisting of functions traditionally considered to be "difficult" for integration (see De Boor [71a]).

Our approach is close to that developed in Alperovich, Batishchev and Strongin [73] and Batishchev [75], where it was suggested that algorithms be tested on random samples of functions from some family and the results be averaged over the sample. Though the families of test functions we consider are not directly connected to any real-life problems, we take into account characteristic features of the problems that are frequently encountered in real life.

Testing algorithms has several aims.

The first one is to test performance of the algorithms on representative families of test functions.

The second one is to test experimentally performance of the algorithms on classes of functions (test families) for which we can not carry out theoretical investigation.

The third one is to check experimentally if the concept of sequential optimality meets the requirements to organization of real-life computational processes. Here we compare results of SOA, ASOA, OSOA, and OA.

Now we describe a random technique of constructing continuous piecewise smooth (with m break points) functions that comprise test families. Fix a set of smooth functions $f_1, \ldots, f_p$ defined on segments $l_1, \ldots, l_p$ respectively. In what follows we call them ***base functions***. The process of constructing each test function consists of three stages.

I. Within the segment of integration $[a,b]=[0,1]$, we select m uniformly distributed random points $t_1, \ldots, t_m$ and arrange them in the increasing order: $0=t_0<t_1<\cdots<t_m<t_{m+1}=1$.

II. For each $i=0,1,\ldots,m$, we select a number k_i of the base function among equiprobable numbers $1,\ldots,p$, and then a random point ξ_i within the segment l_{k_i}.

III. The test function is defined in the following way:

$$f(x)=d_i+f_{k_i}(\xi_i+x-t_i)-f_{k_i}(\xi_i), \quad x\in[t_i,t_{i+1}],$$

where i takes the values $0,1,\ldots,m$ successively, $d_0=0$, and $d_i=f(t_i)$ for $i\geq 1$. Random choice of $\xi_i \in l_{k_i}$ is performed in such a way that the point $\xi_i+t_{i+1}-t_i$ lies within the segment l_{k_i}.

When constructing test functions that belong to the functional class F_1, we selected three base functions equal to sums of the first four terms of the Fourier series with random coefficients:

$$\sum_{i=1}^{4}(a_k \cos kx + b_k \sin kx),$$

where a_k and b_k are independent realizations of uniformly distributed on $[0,1]$ random variables.

When constructing test functions that belong to the functional class F_2, we selected five monotonically increasing base functions: e^x on $[-5,2]$, $\ln x$ on $[0.3,5]$, x^2 on $[0,2]$, $-1/x$ on $[0.5,5]$, and $\sin x$ on $[-\pi/2,\pi/2]$.

4.4. Results of numerical tests

Denote the estimate accuracy (the accuracy $\varepsilon_N(z^N)$ guaranteed after N informational computations and given by (4.15), Chapter 1) by $\varepsilon_{\mathrm{OA}}$, $\varepsilon_{\mathrm{SOA}}$, etc., depending on what algorithm is applied.

Table 2

Number of parameter	m	N = 7	15	31	51
1		.88	.87	.84	.80
2		80	90	100	98
3	3	98	86	76	68
4		99	.98	.97	.97
5		50	70	94	90
1		.84	.76	.70	.63
2		92	100	100	100
3	8	80	70	66	68
4		1.01	.99	.98	.98
5		34	60	74	84
1		.85	.70	.63	.62
2		92	100	100	100
3	20	88	64	62	52
4		1.02	1.01	1.00	.99
5		34	38	46	74

Table 2 shows the results of numerical tests for the class F_2. Every result is an average of 50 random tests, i.e., average of the results of applying the algorithms being compared to 50 functions from the test family generated in the random way that was described in Section 4.3.

The first parameter in the table is the ratio $\varepsilon_{\mathrm{SOA}}/\varepsilon_{\mathrm{OA}}$; the second one is the percentage of tests in the series where SOA gives a better accuracy than OA; the third one is the percentage of tests where the estimate accuracy of SOA is not worse than that of ASOA; the fourth one is the ratio $\varepsilon_{\mathrm{SOA}}/\varepsilon_{\mathrm{OSOA}}$; the fifth one is the percentage of tests where $\varepsilon_{\mathrm{SOA}} < \varepsilon_{\mathrm{OSOA}}$.

Note that the actual accuracy of integration turns out to be approximately two orders better than the estimate accuracy.

In testing algorithms that guarantee a prescribed accuracy, the same tendencies show as for a fixed N. The most apparent effect is reduction of the number of nodes required to achieve a fixed ε for SOA and ASOA compared to OA. Table 3 contains parameters characterizing performance of SOA and OA for different values of ε, both algorithms being applied to two piecewise smooth functions (one from F_1 and one from F_2) generated

Table 3

ε	Functional class					
	F_1			F_2		
0.02	13;	10;	0.77	18;	14;	0.82
0.01	25;	16;	0.62	36;	26;	0.74
0.005	50;	31;	0.61	72;	51;	0.73
2.5×10^{-4}	100;	59;	0.59	144;	97;	0.69
1.25×10^{-4}	200;	114;	0.57	288;	188;	0.67
6×10^{-4}	416;	235;	0.56	584;	379;	0.65

by the random method that was described in Section 4.3. We have chosen $m=3$ for the function from F_1 and $m=20$ for the function from F_2.

In every column of Table 3, the first number is the number of informational computations (nodes) guaranteeing the prescribed accuracy for OA; the second number is that of informational computations required for ASOA to achieve the same accuracy; the third one is the ratio of the second and the first numbers.

More results of numerical tests can be found in Glinkin and Sukharev [85].

4.5. Conclusions

The results of numerical tests part of which we presented and commented on in Section 4.4 enable us to make the following general conclusions.

I. All the algorithms under investigation have proved rather efficient and reliable methods of numerical integration. An important characteristic of the nonmodified algorithms is their ability to give a guaranteed estimate of the accuracy of integration or to compute the integral with a given accuracy, provided that the information we have about the function to be integrated ensures that it belongs to one of the functional classes F_1, F_2 and thus allows us to choose an appropriate integration algorithm.

II. For the algorithms with a fixed number N of informational computations (nodes), as N grows, SOA gains a growing advantage in accuracy over OA and OSOA (see Table 2). Additional tests (not reflected in the table) have shown that, if the number of nodes is large enough, this advantage exists in 100 percent cases. The same tendency is quite clear for the algorithms with a prescribed accurace ε. As ε grows smaller, the advantage ASOA has over OA in the number of nodes becomes more and more apparent (see Table 3).

III. When N is not too lagre, SOA has an obvious advantage over ASOA (see Table 2). As N grows, the advantage gradually goes to zero. This is quite explicable since approximation of the solution of the integer problem (3.9) by the solution of its continuous analogue (3.13) becomes better and better.

As we see it, the conclusions II and III demonstrate on the material of numerical integration that the concept of sequential optimality reflects specific features of real-life computational processes correctly and quite fully. Thus, approximation of the problem (3.9) we solve in the process of constructing SOA, or, more precisely, replacement of the problem (3.9) by its continuous analogue (3.13), leads to stable growth of the integration error. The advantage of SOA over OSOA is also very demonstrative, especially if we recall that our version of SOA at every step chooses from the set of all possible points the point closest to that prescribed by OSOA.

For the modified algorithms, we see the same main tendencies as for the modified ones, but not so pronounced. The estimate accuracy of the modified algorithms becomes worse as the "caution coefficient" σ grows since this leads to the estimate of the constant M being set too high. However, setting σ small is risky: in our tests, for small numbers of nodes $(7 \div 15)$, the actual accuracy of MSOA for $\sigma = 1.1$ turned out to be worse than the estimate accuracy in about 10 persent cases, while for $\sigma = 1.5$ - only in 3 persent cases.

In conclusion, we stress that SOA may prove most useful for integrating functions difficult to compute, since the combinatory complexity ("inner needs") of SOA is considerably larger than that of OA. Besides, if the number N of informational computations is large enough, it makes sense to use ASOA instead of SOA since the latter has a larger combinatory complexity.

5. OPTIMAL COMPUTATION OF ITERATED INTEGRALS

In this section, based on Sukharev [79c], we consider computing iterated integrals of functions of two variables. However, the results obtained can easily be extended to an arbitrary number of variables.

5.1. Problem of iterated integration

Consider the class F' of functions depending on two variables and satisfying the Lipschitz condition with respect to each of them:

$$F = \{g \mid |g(x,u) - g(x',u)| \le M|x - x'|,$$
$$|g(x,u) - g(x,u')| \le P|u - u'|,\ x, u, x', u' \in [0,1]\}.$$

Due to Lemma 2.4 of Chapter 1, this functional class can be defined with the help of some quasi-metric. If $g \in F'$, then the function $f(x) = \int_0^1 g(x,u)du$ belongs to the class

$$F = \{f \mid |f(x) - f(x')| \leq M|x - x'|, \ x, x' \in [0,1]\}$$

since

$$|f(x) - f(x')| \leq \int_0^1 |g(x,u) - g(x',u)|du' \leq M|x - x'|.$$

If all the information we have about g is $g \in F'$, then all the information we have about f is $f \in F$. Indeed, any function from F can be represented by an integral of some function from F'.

We wish to construct optimal algorithms for computing double integral $\iint_{[0,1]^2} g(x,u)dxdu$ regarded as an iterated integral, that is, we shall make use of the formula

$$\iint_{[0,1]^2} g(x,u)dxdu = \int_0^1 f(x)dx.$$

An important feature of the problem is that, in general, precise evaluation of the function f is impossible. Optimal algorithms for computing values of $f(x) = \int_0^1 g(x,u)du$ with any prescribed accuracy were constructed in Sections 2 and 3.

Note that the results of this section are applicable not only to $f(x) = \int_0^1 g(x,y)du$, but also to some other cases where we have to compute an integral of a function f which can be evaluated with any prescribed accuracy, for instance, $f(x) = \max_u g(u,x)$ (optimal evaluation of such functions is discussed in Chapter 4).

The results of this section can be directly extended to n-dimensional case. Having constructed an algorithm for integrating functions of two variables, we can integrate functions of three variables by the formula

$$\iiint_{[0,1]^3} h(x,u,t)dxdudt = \int_0^1 f(x)dx,$$

where

$$f(x) = \iint_{[0,1]^2} h(x,u,t)dudt,$$

and so on.

5.2 Optimal nonadaptive algorithm for iterated integration

Constructing an optimal nonadaptive algorithm for iterated integration can be reduced to the problem of optimal integration of a function that is evaluated only approximately.

Suppose we compute values of the function f at ν points $x_1, \dots, x_\nu$, $0 \le x_1 \le x_2 < \dots < x_\nu \le 1$; let $y_1, \dots, y_\nu$ be the obtained approximate values and $\delta_1, \dots, \delta_\nu$ be the errors of computation:

$$|f(x_i) - y_i| \le \delta_i, \quad i = 1, \dots, \nu \tag{5.1}$$

(in this case we can speak about computations with indeterminate errors). Using (5.1) and the Lipschitz condition, we get for any $x \in [0, 1]$

$$y_i - \delta_i - M|x - x_i| \le f(x) \le y_i + \delta_i + M|x - x_i|, \quad i = 1, \dots, \nu,$$

which is equivalent to

$$\phi_{1\nu}(x) \stackrel{def}{=} \max_{i=1,\dots,\nu} \{y_i - \delta_i - M|x - x_i|\} \le f(x)$$

$$\le \min_{i=1,\dots,\nu} \{y_i + \delta_i + M|x - x_i|\} \stackrel{def}{=} \phi_{2\nu}(x), \quad x \in [0, 1].$$

It is easy to verify that $\phi_{1\nu}$ and $\phi_{2\nu}$ are sharp lower and upper envelopes[1)] respectively in the situation $w^\nu \stackrel{def}{=} (x^\nu, \delta^\nu, y^\nu)$, where $x^\nu = (x_1, \dots, x_\nu)$, $y^\nu = (y_1, \dots, y_\nu)$, and $\delta^\nu = (\delta_1, \dots, \delta_\nu)$, i.e., for functions from F satisfying (5.1).

Similarly to the case of exact evaluations (see (3.4)), the function

$$\widetilde{\beta}_*(w^\nu) \stackrel{def}{=} \int_0^1 \frac{1}{2}[\phi_{1\nu}(x) + \phi_{2\nu}(x)]dx$$

$$= x_1 y_1 + \sum_{i=2}^{\nu} (x_i - x_{i-1}) \frac{y_i + y_{i+1}}{2} + (1 - x_\nu) y_\nu \tag{5.2}$$

determines the central terminal operation. We will take (5.2) as an approximate value of the integral in the situation w^ν. The worst-case error in the situation w^ν is

$$\varepsilon_\nu(w^\nu) \stackrel{def}{=} \int_0^1 [\phi_{2\nu}(x) - \phi_{1\nu}(x)]dx$$

$$= \frac{1}{2}\left\{ M z_1^2 + 2\delta_1 z_1 + \sum_{i=2}^{\nu} \left[\frac{M z_i^2}{2} - \frac{(y_i - y_{i-1})^2}{2M} \right.\right.$$

$$\left.\left. - \frac{(\delta_i - \delta_{i-1})^2}{2M} + (\delta_i + \delta_{i-1}) z_i \right] + M z_{\nu+1}^2 + 2\delta_n u z_{nu+1} \right\},$$

1) We hope that using the same notation for sharp lower and upper envelopes in the both cases where evaluations of the function are exact and approximate will not lead to any confusion.

where

$$z_1 = x_1; \quad z_i = x_i - x_{i-1}, \quad i = 2, \dots, \nu; \quad z_{\nu+1} = 1 - x_\nu.$$

Let N be the total number of informational computations, i.e., the number of evaluations of g, and let n_i be the number of values of g used for computing the approximate value y_i of f at the point x_i, $i = 1, \dots, \nu$. Due to (3.3), we have $\delta_i = P/(4n_i)$. Assume that N and ν are fixed and $n_1, \dots, n_\nu, x_1, \dots, x_\nu$ are variables to be determined. Replacing $n_1, \dots, n_\nu$, $x_1, \dots, x_\nu$, by $\delta_1, \dots, \delta_\nu$, $z_1, \dots, z_{\nu+1}$, we can rewrite restrictions on the variables in the following form:

$$\sum_{i=1}^{\nu} 1/\delta_i = c, \quad \delta_i > 0, \quad i = 1, \dots, \nu,$$

$$\sum_{i=1}^{\nu+1} z_i = 1, \quad z_i \geq 0, \quad z_{\nu+1} \geq 0, \quad |\delta_i - \delta_{i-1}| < M z_i, \quad i = 2, \dots, \nu. \tag{5.3}$$

Here in the first equation $c \stackrel{def}{=} 4N/P$, i.e., the first restriction is just another form of the condition $\sum_{i=1}^{\nu} n_i = N$. The second series of restrictions prohibits useless computations and ensures that every informational computation gives some new information. For instance, if we have $\delta_i - \delta_{i-1} \geq M z_i$ for some $i \in \{2, \dots, \nu\}$, then it is easy to see that in the worst case $y_{i-1} = y_i$ evaluating F at x_i gives no new information.

Note that in (5.3) and in the following construction we do not take into account that $n_1, \dots, n_\nu$ are integer.

It follows from (5.1) and $f \in F$ that y^ν must satisfy the resrictions

$$|y_i - y_{i-1}| \leq M z_i + \delta_{i-1} + \delta_i, \quad i = 2, \dots, \nu. \tag{5.4}$$

The following definition is a natural extension of the definition of optimality from Section 4.1 (see also Theorem 4.3 of Chapter 1) to the case of approximate computations.

A vector (x_0^ν, δ_0^ν) (together with the terminal operation (5.2)) is called an *optimal (by error) on the class F' nonadaptive algorithm of iterated integration (for fixed N and ν)*, iff

$$\sup_y \varepsilon_\nu(x_0^\nu, \delta_0^\nu, y^\nu) = \min_{x, \delta} \sup_y \varepsilon_\nu(x^\nu, \delta^\nu, y^\nu),$$

where x^ν, δ^ν, y^ν satisfy (5.3) and (5.4).

Later we will abandon the assumption that ν is fixed and find its optimal value ν_0.

Clearly,

$$\sup_y \varepsilon_\nu(x_0^\nu, \delta_0^\nu, y^\nu) = \varepsilon_\nu(x^\nu, \delta^\nu, (0, \dots, 0)) \stackrel{def}{=} \omega(x^\nu, \delta^\nu)$$

$$= \frac{1}{2}\left\{ M z_1^2 + 2\delta_1 z_1 + \sum_{i=2}^{\nu} \left[\frac{M z_i^2}{2} - \frac{(\delta_i - \delta_{i-1})^2}{2M} + (\delta_i + \delta_{i-1}) z_i \right] + M z_{\nu+1}^2 + 2\delta_\nu z_{\nu+1} \right\},$$

and so the problem amounts to finding $\min_{x\ ,\delta} \omega(x^\nu, \delta^\nu)$ and also x_0^ν and δ_0^ν at which this minimum is attained.

Temporarily, we disregard the restrictions

$$z_1 \geq 0, \quad z_{\nu+1} \geq 0, \quad |\delta_i - \delta_{i-1}| < M z_i, \quad i=2,\ldots,\nu.$$

It is easy to see that

$$\begin{aligned}\min_x \omega(x^\nu, \delta^\nu) = \min_{(z_1,\ldots,z\ {}_{+1})\in\ {}_{+1}} \frac{1}{2}\Big\{ & M z_1^2 + 2\delta_1 z_1 \\ & + \sum_{i=2}^{\nu}\left[\frac{M z_i^2}{2} - \frac{(\delta_i-\delta_{i-1})^2}{2M} + (\delta_i+\delta_{i-1})z_i\right] \\ & + M z_{\nu+1}^2 + 2\delta_\nu z_{\nu+1}\Big\} \\ = \frac{1}{4\nu M}\bigg[\Big(M + 2\sum_{i=1}^{\nu}\delta_i\Big)^2 & - 4\nu\sum_{i=1}^{\nu}\delta_i^2\bigg],\end{aligned}$$

where

$$S_{\nu+1} = \left\{(z_1,\ldots,z_{\nu+1}) \;\middle|\; \sum_{i=1}^{\nu} z_i = 1\right\},$$

and the minimum is attained at

$$z_1^0 = \frac{1}{2\nu} + \frac{\delta-\delta_1}{M}, \quad z_i^0 = \frac{1}{\nu} + \frac{2\delta-\delta_i-\delta_{i-1}}{M}, \quad i=2,\ldots,\nu,$$

$$z_{\nu+1}^0 = \frac{1}{2\nu} + \frac{\delta-\delta_\nu}{M}; \quad \delta \overset{def}{=} \sum_{i=1}^{\nu}\frac{\delta_i}{\nu}.$$

Hence, $\min_x \omega(x^\nu, \delta^\nu)$ is attained at $x_0^\nu = (x_1^0, \ldots, x_\nu^0)$, where $x_i^0 = \sum_{j=1}^{i} z_j^0$, $i = 1,\ldots,\nu$. We have

$$\min_{x\ ,\delta} \omega(x^\nu, \delta^\nu) = \frac{1}{4\nu M}\min_{\delta\geq\nu/}\ \min_{\delta_1,\ldots,\delta\ >0}\left[(M+2\nu\delta)^2 - 4\nu\sum_{i=1}^{\nu}\delta_i^2\right], \tag{5.5}$$

where

$$\sum_{i=1}^{\nu}\frac{1}{\delta_i} = c, \quad \sum_{i=1}^{\nu}\delta_i = \nu\delta$$

(we temporarily ignore the restrictions $z_1 \geq 0$, $z_{\nu+1} \geq 0$, $|\delta_i - \delta_{i-1}| < M z_i$, $i=2,\ldots,\nu$). The inequality $\delta \geq \nu/c$ is included into the set of restrictions because it is necessary and sufficient for consistency of the conditions $\sum_{i=1}^{\nu}\frac{1}{\delta_i} = c$,

$\delta_1 > 0, \ldots, \delta_\nu > 0$, and $\sum_{i=1}^{\nu} \delta_i = \nu\delta$ that the following inequality hold:

$$\min_{\substack{\delta_1,\ldots,\delta\ >0 \\ \sum_{i=1} 1/\delta_i =}} \sum_{i=1}^{\nu} \delta_i = \frac{\nu^2}{c} \leq \nu\delta.$$

To find (5.5), we first consider the problem of finding

$$\max_{\delta,\ldots,\delta\ >0} \sum_{i=1}^{\nu} \delta_i^2 \quad \text{for} \quad \sum_{i=1}^{\nu} \frac{1}{\delta_i} = c, \quad \sum_{i=1}^{\nu} \delta_i = \nu\delta. \tag{5.6}$$

For this problem, we write the Lagrange function and its first two differentials:

$$= \sum_{i=1}^{\nu} \delta_i^2 - \lambda \sum_{i=1}^{\nu} \frac{1}{\delta_i} - \mu \sum_{i=1}^{\nu} \delta_i,$$

$$d = \sum_{i=1}^{\nu} \left(2\delta_i + \frac{\lambda}{\delta_i^2} - \mu\right) d\delta_i, \quad d^2 = \sum_{i=1}^{\nu} \left(2 - \frac{2\lambda}{\delta_i^3}\right) (d\delta_i)^2.$$

To deliver the maximum in (5.6), the point $\delta_0^\nu = (\delta_1^0, \ldots, \delta_\nu^0)$ should solve the following system:

$$2\delta_i^3 - \mu\delta_i^2 = -\lambda, \quad i = 1, \ldots, \nu, \quad \sum_{i=1}^{\nu} \frac{1}{\delta_i} = c, \quad \sum_{i=1}^{\nu} \delta_i = \nu\delta.$$

The equation $2x^3 - \mu x^2 = -\lambda$ has at most two positive roots α and β, $0 < \alpha \leq \beta$. Let $\delta_1^0 = \cdots = \delta_k^0 = \alpha$, $\delta_{k+1}^0 = \cdots = \delta_\nu^0 = \beta$. To find out if δ_0^ν is the point delivering the maximum, consider the quadratic form

$$d^2 = \left(2 - \frac{2\lambda}{\alpha^3}\right) \sum_{i=1}^{k} (d\delta_i)^2 + \left(2 - \frac{2\lambda}{\beta^3}\right) \sum_{i=k+1}^{\nu} (d\delta_i)^2$$

for

$$\sum_{i=1}^{\nu} d\delta_i = 0, \quad \sum_{i=1}^{k} \frac{d\delta_i}{\alpha^2} + \sum_{i=k+1}^{\nu} \frac{d\delta_i}{\beta^2} = 0.$$

The latter equations can hold only in case one of the following two conditions holds:

$$\begin{aligned} &\sum_{i=1}^{\nu} d\delta_i = 0, \quad \alpha = \beta; \\ &\sum_{i=1}^{k} d\delta_i = 0, \quad \sum_{i=k+1}^{\nu} d\delta_i = 0, \quad \alpha < \beta. \end{aligned} \tag{5.7}$$

Taking into account that $0 < \alpha \leq \beta$ and that α and β are roots of the equation $2x^3 - \mu x^2 = -\lambda$, we get

$$\lambda = \frac{2\alpha^2\beta^2}{\alpha + \beta}, \quad 2 - \frac{2\lambda}{\alpha^3} = 2\frac{\alpha(\alpha + \beta) - 2\beta^2}{\alpha(\alpha + \beta)} \leq 0,$$

$$2-\frac{2\lambda}{\beta^3}=2\frac{\beta(\alpha+\beta)-2\alpha^2}{\beta(\alpha+\beta)}\geq 0.$$

Hence, for the maximum to be attained at δ_0^ν, it is necessary that one of the following two conditions hold: either $\alpha=\beta$, or $\alpha<\beta$ and $k=\nu-1$ (otherwise the quadratic form d^2 would be positive definite or indefinite if any of the conditions (5.7) held, and so δ_0^ν could not deliver the maximum). In the both cases, the necessary conditions for the maximum have the form

$$\frac{\nu-1}{\alpha}+\frac{1}{\beta}=c, \quad (\nu-1)\alpha+\beta=\nu\delta, \quad 0<\alpha\leq\beta.$$

Setting $\alpha=\delta-\gamma$, $\beta=\delta+(\nu-1)\gamma$, $\gamma\geq 0$, we see that

$$\gamma(\delta)\stackrel{def}{=}\frac{1}{2c(\nu-1)}[(\nu-2)(c\delta-\nu)+\sqrt{(\nu-2)^2(c\delta-\nu)^2+4c\delta(\nu-1)(c\delta-\nu)}]$$

is a positive root of the equation

$$\frac{\nu-1}{\delta-\gamma}+\frac{1}{\delta+(\nu-1)\gamma}=0.$$

Thus, the problem (5.5) has been reduced to computing

$$\min_{\delta\geq\nu/}\ (\ \delta), \quad \text{where} \quad (\delta)=\frac{M}{4\delta}+\frac{M\delta-\nu(\nu-1)\gamma^2(\delta)}{M}.$$

It is easy to verify that $''(\delta)<0$ for $\delta\geq\nu/c$, and so the minimum is attained at $\delta=\nu/c=\alpha=\beta$ and

$$\min_{x\ ,\delta}\omega(x^\nu,\delta^\nu)=\frac{M}{4\nu}+\frac{\nu}{c}.$$

In the second case, the minimum of the function $\omega(x^\nu,\delta^\nu)$ with respect to x^ν and δ^ν satisfying the conditions (5.3) is attained at the boundary of the set (5.3), that is, we have either $z_1=0$, or $Mz_i=|\delta_i-\delta_{i-1}|$ for some $i\in\{2,\ldots,\nu\}$, or $z_{\nu+1}=0$. In Section 5.4 we prove that in this case we always have $Mz_i=|\delta_i-\delta_{i-1}|$ for some $i\in\{2,\ldots,\nu\}$. This means that the corresponding evaluation of f is not necessary since in the worst case it gives no new information. In this situation, leaving the total number of informational computations intact, we consider the number of points at which f is evaluated not to be greater than ν. Formally, we can say that at some of these ν points evaluations with infinite errors are permitted. Here we refine the above definition of optimality, allowing infinite values of $\delta_1,\ldots,\delta_\nu$.

Combining the both cases, we obtain

$$\min_{x\ ,\delta}\omega(x^\nu,\delta^\nu)=\min_{j\leq\nu}\left(\frac{M}{4j}+\frac{j}{c}\right).$$

It is easily verified that the minimum on the right-hand side is attained at $j=j(\nu)$, where

$$j(\nu)=\max\{j\mid j\leq\nu,\ j(j-1)\leq Mc/4=NM/P\}.$$

Note that the latter inequality is equivalent to the inequality

$$M/(4j)+j/c \leq M/[4(j-1)]+(j-1)/c.$$

Substituting δ_0^ν for δ^ν in the formulas for x_0^ν obtained earlier, and also substituting $j(\nu)$ for ν, we get the following formulas determining the optimal algorithm $(x_0^{j(\nu)}, \delta_0^{j(\nu)})$:

$$x_1^0 = \frac{1}{2j(\nu)}, \quad x_2^0 = \frac{3}{2j(\nu)}, \ldots, x_{j(\nu)}^0 = \frac{2j(\nu)-1}{2j(\nu)},$$

$$\delta_1^0 = \cdots = \delta_{j(\nu)}^0 = \frac{j(\nu)}{c} = \frac{j(\nu)P}{4N}.$$

Disregarding the restrictions

$$z_1 \geq 0, \quad z_{\nu+1} \geq 0, \quad |\delta_i - \delta_{i-1}| < M z_i, \quad i=2,\ldots,\nu,$$

in the original problem does not affect the result, since the minimizing point obtained in this way satisfies all the restrictions of the original problem.

The algorithm $(x_0^{j(\nu)}, \delta_0^{j(\nu)})$ we have just constructed is the main result of this section.

If we still fix the number N of informational computations but do not fix ν at all, then the above derivations yield that the optimal number ν_0 of function evaluations is

$$\nu_0 = \max\{\nu \mid \nu(\nu-1) \leq NM/P\}. \tag{5.8}$$

In this case, the optimal algorithm is given by the formulas

$$x_1^0 = \frac{1}{2\nu_0}, \quad x_2^0 = \frac{3}{2\nu_0}, \ldots, x_{\nu_0}^0 = \frac{2\nu_0-1}{2\nu_0},$$

$$\delta_1^0 = \cdots = \delta_{\nu_0}^0 = \frac{\nu_0}{c} = \frac{\nu_0 P}{4N}.$$

Note once again that, firstly, when we speak about optimal algorithms, we mean optimality in the class of algorithms for iterated integration, and, secondly, the definition of optimal algorithm for iterated integration does not take into account that the set of permissible values of $\delta_1, \ldots, \delta_\nu$ is discrete (in other words, that $n_1, \ldots, n_\nu$ are integer).

5.3. Analogue of sequentially optimal algorithm for iterated integration

In this section, we show how the results on constructing sequentially optimal algorithms (Section 3) can be extended to the problem of computing iterated integrals. We assume here that the accuracy ε to be guaranteed as a result of the computations is fixed.

Firstly, we find the number N of informational computations guaranteeing the prescribed accuracy ε if the optimal nonadaptive algorithm for iterated integration (with non-fixed ν) is used. Since ν is integer, we proceed from (5.8). However, we now neglect the restriction that ν and N should be integer. Then the optimal ν_0 can be obtained from

$$\frac{M}{4\nu_0}+\frac{\nu_0}{c}=\min_{j>0}\left(\frac{M}{4j}+\frac{j}{c}\right),$$

thus $\nu_0=(NM/P)^{1/2}$, and N can be obtained from

$$\min_{x^{\nu_0},\delta^{\nu_0}}\omega(x^{\nu_0},\delta^{\nu_0})=\frac{M}{4\nu_0}+\frac{\nu_0}{c}=\frac{\sqrt{MP}}{2\sqrt{N}}=\varepsilon,$$

thus $N=MP/(2\varepsilon)^2$, $\nu_0=M/(2\varepsilon)$. Hence, the optimal accuracy to be prescribed for computing values of the function $f(x)=\int_0^1 g(x,u)du$ is $\delta_0=\nu_0/c=\varepsilon/2$. We fix this accuracy for all evaluations of f.

Now we proceed to construct an analogue of the sequntially optimal algorithm from Section 3. The first informational computation should be performed at a point

$$x_1\in\left\{\frac{1}{2\nu_0},\frac{3}{2\nu_0},\ldots,\frac{2\nu_0-1}{2\nu_0}\right\}.$$

Suppose we have performed i computations at points $x_1,\ldots,x_i$; let $x_{i1},\ldots,x_{ii}$ be the ordered permutation of $x_1,\ldots,x_i$, and let $y_{ij}=f(x_{ij})$, $j=1,\ldots,i$. Denote by p the minimal number of the remaining evaluations of f allowing us to achieve the accuracy ε. Assume that n_1 out of p computations are to be carried out within $[0,x_{i1})$, n_2 computations – within (x_{i1},x_{i2}), $\ldots$, n_i – within $(x_{i,i-1},x_{ii})$, and n_{i+1} – within $(x_{ii},1]$, where $\sum_{j=1}^{i+1} n_i=p$. The numbers $p,n_1,\ldots,n_{i+1}$ will be determined later.

Consider any interval $(x_{i,j-1},x_{ij})$, $j=2,\ldots,i$. We wish to find the optimal points

$$t_1,\ldots,t_{n_j},\quad t_0\overset{def}{=}x_{i,j-1}<t_1<t_2<\cdots<t_{n_j}<x_{ij}\overset{def}{=}t_{n_j+1},$$

at which to compute values of f, that is, the points delivering the minimum in the formula

$$\min_{t_1,\ldots,t_{n_j}}\ \max_{1,\ldots,\ n_j}\sum_{k=1}^{n_j+1}\left[\frac{M(t_k-t_{k-1})^2}{4}-\frac{(v_k-v_{k-1})^2}{4M}+\delta_0(t_k-t_{k-1})\right]\overset{def}{=}\omega_j,\tag{5.9}$$

where $v_k=f(t_k)$, $k=0,1,\ldots,n_j+1$, and $\delta_0=\varepsilon/2$ is the prescribed accuracy of evaluations of f. The function inside min max in (5.9) is a special case of the expression for $\varepsilon_\nu(w^\nu)$. Restrictions on the variables in (5.9) are analogous to (5.4):

$$|v_k-v_{k-1}|\le M(t_k-t_{k-1})+2\delta_0,\quad k=1,\ldots,n_j+1.$$

It is easy to verify that

$$t_k=x_{i,j-1}+k\frac{x_{ij}-x_{i,j-1}}{n_j+1},$$

$$\omega_j = \frac{M(x_{ij}-x_{i,j-1})^2}{4(n_j+1)} - \frac{(y_{ij}-y_{i,j-1})^2}{4M(n_j+1)} + \delta_0(x_{ij}-x_{i,j-1}).$$

Treating the intervals $[0, x_{i1})$ and $(x_{ii}, 1]$ in a similar way, we come to the conclusion that, to find optimal $n_1, \dots, n_{i+1}$, we have to solve the same problem of integer programming as in the case of exact evaluations of f (see (3.7), (3.9)). Namely, we have to find $n_1, \dots, n_{i+1}$ delivering the minimum in the following formula:

$$\min_{\substack{n_1,\dots,n_{i+1}\in\{0,1,2,\dots\} \\ \sum_{j=1}^{i+1} n_j = p}} \left[\frac{M}{4}\cdot\frac{x_{i1}^2}{n_1+1/2} + \frac{1}{4M}\sum_{j=2}^{i} \frac{(x_{ij}-x_{i,j-1})^2 M^2 - (y_{ij}-y_{i,j-1})^2}{n_j+1} + \frac{M}{4}\cdot\frac{(1-x_{ii})^2}{n_{i+1}+1/2} + \delta_0 \right] \stackrel{def}{=} \omega(p).$$

After the solution has been found, we put x_{i+1} equal to any of the points $t_1, \dots, t_{n_j}$ delivering the minimum in one of the problems (5.9), $j=2,\dots,i$, or to any of the points delivering the minimum in analogous problems for $[0, x_{i+1})$, $(x_{ii}, 1]$. Here, p is the minimal integer such that $\omega(p) \le \varepsilon$.

Observe that we have fixed the same accuracy δ_0 for all evaluations of f. However, it is also possible, when selecting x_{i+1}, to regard errors of the evaluations as parameters to be optimized (it is easy to show that, within every interval $[0, x_{i1}), (x_{i1}, x_{i2}), \dots, (x_{ii}, 1]$, the same accuracy should be prescribed for all evaluations). It would not take much effort to formulate the problems of finding optimal accuracies. However, solving these problems would dramatically increase combinatory complexity of the algorithm and make its application inexpedient. Actually, the approach we have just described enables us to solve the problem of multidimensional integration with the help of algorithms for integrating functions of one variable.

5.4. Auxiliary statements

Here we prove the auxiliary statement we used in the constructions of Section 5.2. We introduce some more notations (in addition to all the notations of Section 5.2). Fix an arbitrary δ^ν satisfying the restrictions

$$\sum_{i=1}^{\nu} \frac{1}{\delta_i} = c, \quad \delta_i > 0, \ i = 1, \dots, \nu,$$

and set

$$z = (z_1, \dots, z_{\nu+1}), \ \omega(z) = \omega(x^\nu, \delta^\nu),$$

$$A=\{z\in S_{\nu+1}\mid z_i\geq a_i,\ i=1,\ldots,\nu+1\},$$

$$A_i=\{z\in A\mid z_i=a_i\},\quad \partial A=\bigcup_{i=1}^{\nu+1}A_i,$$

where

$$S_{\nu+1}=\left\{z\ \middle|\ \sum_{i=1}^{\nu+1}z_i=1\right\},$$

$$a_1=a_{\nu+1}=0,\quad a_i=\frac{|\delta_i-\delta_{i-1}|}{M},\quad i=2,\ldots,\nu.$$

We formulate the statement to be proved as the following lemma.

Lemma 5.1. *If*

$$\min_{z\in A}\omega(z)=\omega(z^*),\quad z^*\in\partial A,$$

then

$$z^*\in\bigcup_{i=2}^{\nu}A_i.$$

Proof. As was shown in Section 5.2, $\min_{z\in\ \ +1}\omega(z)=\omega(z^0)$, where

$$z_1^0=\frac{1}{2\nu}+\frac{\delta-\delta_1}{M},\quad z_i^0=\frac{1}{\nu}+\frac{2\delta-\delta_i-\delta_{i-1}}{M},\quad i=2,\ldots,\nu,$$

$$z_{\nu+1}^0=\frac{1}{2\nu}+\frac{\delta-\delta_\nu}{M},\quad \delta=\sum_{i=1}^{\nu}\frac{\delta_i}{\nu}.$$

Apparently,

$$\begin{array}{lll} z_1^0>a_1 & M+2\nu\delta>2\nu\delta_1, & \\ z_i^0>a_i & M+2\nu\delta>2\nu\max\{\delta_{i-1},\delta_i\}, & i=2,\ldots,\nu, \\ z_{\nu+1}^0>a_{\nu+1} & M+2\nu\delta>2\nu\delta_\nu. & \end{array}\tag{5.10}$$

We set $_0=\{i\mid z_i^0\leq a_i\}$ and show that

$$z^*=\bigcup_{i\in I_0}A_i.\tag{5.11}$$

Indeed, otherwise we have

$$z_i^0>a_i\quad\text{for all}\quad i\in\{i\mid z_i^*=a_i\}.\tag{5.12}$$

Moreover, it is easy to see that, since the function ω is strictly convex in z, the convex set A should not have interior (with respect to $S_{\nu+1}$) points belonging to the set

$$B=\{z\in S_{\nu+1}\mid\omega(z)\leq\omega(z^*)\}.$$

We arrive at a contradiction, since $z^*+\lambda(z^0-z^*)\in B$ for $0\le\lambda\le 1$ and, at the same time, (5.12) yields that $z^*+\lambda(z^0-z^*)$ belongs to the relative interior of A for $\lambda>0$ small enough. This proves (5.11).

Treating the problem of finding

$$\omega(\tilde{z})=\min_{z\in\ \ +1,\ z_i=0}\omega(z),\quad \tilde{z}=(0,\tilde{z}_2,\ldots,\tilde{z}_{\nu+1})\in S_{\nu+1},$$

in the same way as (5.10), we get

$$\begin{aligned}&\tilde{z}_i>a_i \quad M+2\nu\delta>(2\nu-1)\max\{\delta_{i-1},\delta_i\}+\delta_1,\quad i=2,\ldots,\nu,\\ &z^0_{\nu+1}>a_{\nu+1} \quad M+2\nu\delta>(2\nu-1)\delta_\nu+\delta_1.\end{aligned} \tag{5.13}$$

If $\tilde{z}^*\in A_1$, then, similarly to (5.11), we can establish the inclusion

$$z^*=\bigcup_{i\in\tilde{I}}A_i, \tag{5.14}$$

where $\widetilde{\ }=\{i>1\mid \tilde{z}_i\le a_i\}$.

Suppose that $z^*\in A_1\setminus\left(\bigcup_{i=2}^{\nu+1}A_i\right)$. Then $z^0_1\le a_1$ due to (5.11), and $M+2\nu\delta\le 2\nu\delta_1$ by (5.10). On the other hand, in this case $z^*=\tilde{z}$ and, in accordance with (5.13), for $i=2$ we have $M+2\nu\delta>2\nu\delta_1$. This contradiction shows that $z^*\notin A_1\setminus\left(\bigcup_{i=2}^{\nu+1}A_i\right)$. In the same way we prove that $z^*\notin A_{\nu+1}\setminus\left(\bigcup_{i=1}^{\nu}A_i\right)$.

Treating the problem of finding

$$\omega(z)=\min_{z\in\ \ +1,\ z_1=z\ \ +1=0}\omega(z),\quad z=(0,z_2,\ldots,z_\nu,0)\in S_{\nu+1},$$

in the same way as (5.10) and (5.13), we obtain

$$z_i>a_i \quad M+2\nu\delta>2(\nu-1)\max\{\delta_{i-1},\delta_i\}+\delta_1+\delta_\nu,\quad i=2,\ldots,\nu. \tag{5.15}$$

Suppose that $z^*\in(A_1\cap A_{\nu+1})\setminus\left(\bigcup_{i=2}^{\nu}A_i\right)$. Then $\tilde{z}_{\nu+1}\le a_{\nu+1}$ due to (5.14), and $M-2\nu\delta\le(2\nu-1)\delta_\nu+\delta_1$ by (5.13). On the other hand, in this case $z^*=z$ and, in accordance with (5.15), for $i=\nu$ we have $M+2\nu\delta>(2\nu-1)\delta_\nu+\delta_1$. This contradiction shows that $z^*\notin(A_1\cap A_{\nu+1})\setminus\left(\bigcup_{i=2}^{\nu}A_i\right)$.

The only possibility left is that $z^*\in\bigcup_{i=2}^{\nu}A_i$. □

6. COMPUTATION OF MULTIPLE INTEGRALS USING PEANO TYPE DEVELOPMENTS

Consider the problem of integrating a function f over n-dimensional unit cube

$$K=\{u=u(^1,\ldots,u^n)\mid -1/2\le u^i\le 1/2,\ i=1,\ldots,n\}.$$

Let $\phi:[0,1]\rightarrow K$ be a one-valued continuous mapping of $[0,1]$ onto K. Then, under certain restrictions on ϕ, we can replace (exactly or approximately) computation of $\int_K f(u)du$ with computation of $\int_0^1 f(\phi(x))dx$. This idea, for ϕ being the Peano mapping (see, e.g., Luzin [48]), was suggested by Wiener [33]. A similar method for reducing dimension when solving extremal problems was applied by Strongin [78].

In this section, we construct a mapping ϕ_m that approximates the Peano mapping ϕ with accuracy not worse than 2^{-m} for each coordinate, and also show that, with m large enough, $\left|\int_K f(u)du - \int_0^1 f(\phi_m(x))dx\right|$ can be made arbitrarily small. Our method of constructing the mapping ϕ_m differs from that used by Strongin [78] only in some details. However, the structure of ϕ_m is essential for the proof of Theorem 6.1 which is the main result of the section, so we have to repeat some of the constructions from Strongin [78] here.

6.1. Den itions of the mappings ϕ and ϕ_m

At the first stage, the cube is partitioned by coordinate planes into 2^n first partition cubes' (with edge 12). These cubes are enumerated with numbers z_1 from 0 to 2^n-1, and the first partition cube with a number z_1 is denoted by $K(z_1)$. Each of the first partition cubes, in turn, is partitioned into 2^n second partition cubes' (with edge 14) by planes parallel to the coordinate ones and passing through the middle points of the edges of the first partition cubes. The second partition cubes comprising the cube $K(z_1)$ are enumerated with numbers z_2 from 0 to 2^n-1, and the second partition cube with a number z_2 belonging to $K(z_1)$ is denoted by $K(z_1, z_2)$.

Proceeding in this way, we construct mth partition cubes (with edge 2^{-m}) for any m. Denote mth partition cubes by $K(z_1,\dots,z_m)$ is such a way that

$$K(z_1)\supset K(z_1,z_2)\supset\cdots\supset K(z_1,\dots,z_2),\quad 0\le z_j\le 2^n-1,\ j=1,\dots,m.$$

Now, we partition the interval [0,1] into 2^n equal intervals, then each of these intervals into 2^n equal intervals, etc., and enumerate the elements of each partition from left to right with numbers z_j (where j is the number of the partition) from 0 to 2^n-1. Similarly to the first stage, we denote the mth partition intervals by $k(z_1,\dots,z_m)$. Every interval contains its left endpoint; it also contains its right endpoint iff $z_1=\cdots=z_m=2^n-1$. Clearly, the length of $k(z_1,\dots,z_m)$ is equal to 2^{-mn}.

Define a mapping ϕ by the following condition:

$$u=\phi(x)\,[x\quad \in k(z_,\dots,z_m)\Rightarrow u\in K(z_1,\dots,z_m),\ m\ge 1],$$

that is, the point $\phi(x)$ belongs to all the cubes $K(z_1,\dots,z_m)$ such that $x \in k(z_1,\dots,z_m)$. Existence and uniqueness of the point $\phi(x)$ follows from the well-known principle of embedded balls, see Kolmogorov and Fomin, [76]. Strongin [78] has suggested a method of enumerating cubes of each partition with numbers $z_1, z_2, \dots$ which ensures continuity of the mapping ϕ. Assume that in the above construction we used this very method. The mapping ϕ we have obtained is the Peano mapping that determines the Peano curve.

We now proceed to construct the mapping ϕ_m. Clearly, the 2^{mn} centers of the mth partition cubes make up a uniform orthogonal lattice in K with lattice constant 2^{-m} in each coordinate. We enumerate the nodes of this lattice and, therefore, the mth partition cubes, using only one index i. To do this, we first enumerate from left to right with numbers $i=0,1,\dots,2^{mn}-1$ all the intervals making up the mth partition of the segment [0,1]. Next, we denote the cube $K(z_1,\dots,z_m)$ by K_i and its center by v_i if we have assigned the number i to the interval $k(z_1,\dots,z_m)$. Then, as has been shown by Strongin [78], cubes K_i and K_{i+1} are always adjacent, i.e., have a common side.

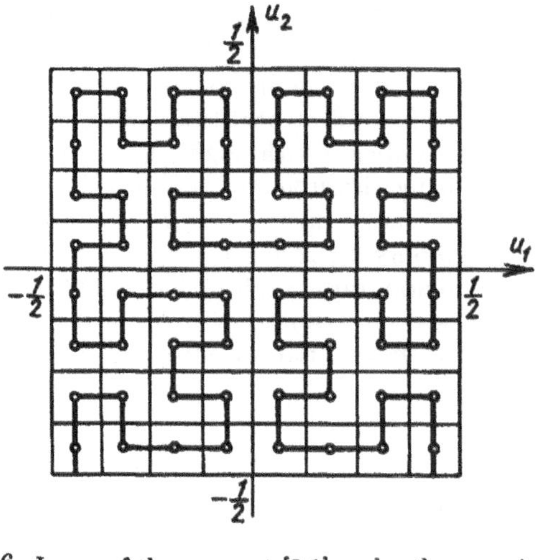

Fig. 6. Image of the segment [0,1] under the mapping ϕ_m

Denote by w_i the point of intersection of the segment linking v_{i-1} and v_i with the common side of the cubes K_{i-1} and K_i, $i=1,\dots,2^{mn}-1$; denote by w_0 the projection of v_0 on the side of K_0 parallel to the common side of K_0 and K_1; and denote by $w_{2^{mn}}$ the projection of v_{2mn-1} on the side of $K_{2^{mn}-1}$ parallel to the common side of $K_2^{mn}-1$ and $K_{2^{mn}-2}$. Consider the following mapping of $[0,1]$ into K:

$$\phi_m(x)=\begin{cases} w_0+(v_0-w_0)x2^{mn+1} & \text{if } 0\le x\le \frac{1}{2^{mn+1}};\\ v_{i-1}+(v_i=v_{i-1})(2^{mn}x-i+1/2) & \\ \qquad \text{if } \frac{i-1/2}{2^{mn}}\le x\le\frac{i+1/2}{2^{mn}},\ i=1,\dots,2^{mn}-1; & \\ v_{2^{mn}-1}+(w_{2^{mn}}-v_{2^{mn}-1})[2^{mn+1}(x-1)+1] & \\ \qquad \text{if } 1-\frac{1}{2^{mn+1}}\le x\le 1. & \end{cases}$$

This mapping transforms a segment $[(i-1/2)2^{-mn},(i+1/2)2^{-mn}]$ into the line segment linking the points v_{i-1} and v_i. Thus, $\{\phi_m(x) \mid x\in[0,1]\}$ is a piecewise linear curve successively connecting centers of adjacent mth partition cubes in the order of their numbers. It is easy to see that $\phi_m(x)$ differs from $\phi(x)$ by at most 2^{-m} in each coordinate for any $x\in[0,1]$. Fig. 6 corresponds to $n=2$, $m=3$.

6.2. Main result

Theorem 6.1 (Glinkin and Sukharev [74]). *Let*

$$|f(x)-f(x')|\le M\|x-x'\|,\quad x,x'\in K, \tag{6.1}$$

where $\|\cdot\|$ is the Euclidean norm. Then

$$\left|\int_K f(u)du-\int_0^1 f(\phi_m(x))dx\right|\le M\sqrt{n}\cdot 2^{-m}. \tag{6.2}$$

Proof. Denote by p_i the inverse image of the point v_i and by q_i the inverse image of the point w_i under the mapping ϕ_m; $i=0,1,\dots,2^{mn-1}$; $j=0,1,\dots,2^{mn}$. Fix an arbitrary $i\in\{0,1,\dots,2^{mn-1}\}$. The part of the polygonal line ϕ_m lying in the cube K_i is described by the formula

$$\phi_m(x)=\begin{cases} w_i+(v_i-w_i)(x-q_i)2^{mn+1} & \text{if } x\in[q_i,p_i],\\ v_i+(w_{i+1}-v_i)(x-p_i)2^{mn+1} & \text{if } x\in[p_i,q_{i+1}]. \end{cases}$$

Compare the integrals $\int_{K_i} f(u)du$ and $\int_{q_i}^{q_{i+1}} f(\phi(x))dx$. Since f is continuous, we have

$$\int_{K_i} f(u)du=f(\xi_i)\mu(K_i)=f(\xi_i)2^{-mn},$$

where $\mu(K_i)$ is the volume of the cube K_i and $\xi_i \in K_i$. On the other hand,

$$\int_{q_i}^{q_{i+1}} f(\phi_m(x))dx = \int_{q_i}^{p_i} f(\phi_m(x))dx + \int_{p_i}^{q_{i+1}} f(\phi_m(x))dx$$

$$= f(\eta_i)(p_i - q_i) + f(\zeta_i)(q_{i+1} - p_i) = 2^{-mn}[f(\eta_i) + f(\zeta_i)],$$

where $\eta_i \in K_i$ and $\zeta_i \in K_i$. From here, taking into account (6.1) and the fact that the diameter of K_i is equal to $2^{-m}\sqrt{N}$, we get

$$\left| \int_{K_i} f(u)du - \int_{q_i}^{q_{i+1}} f(\phi_m(x))dx \right| = 2^{-mn-1}|2f(\xi_i) - f(\eta_i) - f(\zeta_i)|$$

$$\leq 2^{-mn-1}(|f(\xi_i) - f(\eta_i)| + |f(\xi_i) - f(\zeta_i)|) \leq 2^{-mn-m} M\sqrt{n}.$$

To complete the proof, it suffices to sum up the obtained inequalities for $i = 0, 1, \ldots, 2^{mn-1}$. □

Remark. If the norm in (6.1) is an arbitrary (not necessarily Euclidean) norm in n-dimensional coordinate space, then Theorem 6.1 still holds provided that we replace the right-hand side of (6.2) by Md_m, where d_m is the diameter of the mth partition cubes (the metric we use for computing the diameter is determined by the norm that figures in (6.1)). □

6.3. Optimal quadrature formula for the class of functions satisfying the Hölder condition

Consider the problem of integrating $g(x) = f(\phi_m(x))$. Strongin [78] has shown that the function g on $[0, 1]$ satisfies the Hölder condition with the exponent $1/n$:

$$|g(x) - g(x')| \leq M_0 |x - x'|^{1/n}, \quad x, x' \in [0, 1], \tag{6.3}$$

where $M_0 = 4M\sqrt{n}$. Now we apply the results of Section 1 to obtain the optimal on the class (6.3) quadrature formula.

Theorem 6.2 (Glinkin and Sukharev [74]). *The optimal on the class (6.3) quadrature formula with N nodes has the nodes $x_i^0 = h(2^{-1/n} + i - 1)$, $i = 1, \ldots, N$, where $h = (N - 1 + 2^{1-1/n})^{-1}$, and the weights $p_1^0 = p_N^0 = (2^{-1/n} + 1/2)h$, $p_2^0 = \cdots = p_{N-1}^0 = h$. The result guaranteed by this quadrature is equal to*

$$\frac{M_0 h^{1+1/n}(N - 1 + 2^{1-1/n^2})}{2^{1/n}(1 + 1/n)}.$$

Proof. By the corollary of Theorem 1.2, the best guaranteed result is equal to

$$M_0 \min_{x_1,\dots,x_N \in [0,1]} \int_0^1 \min_{i=1,\dots,N} |x-x_i|^{1/n} dx, \tag{6.4}$$

and the optimal nodes are the points delivering the outer minimum in (6.4). Set $z_1 = x_z$; $z_i = x_i - x_{i-1}$, $i=2,\dots,N$; $z_{N+1} = 1 - x_N$. Then we have

$$\int_0^1 \min_{i=1,\dots,N} |x-x_i|^{1/n} dx = \int_{x_1-z_1}^{x_1+z_2/2} |x-x_1|^{1/n} dx$$

$$+ \sum_{i=2}^{N-1} \int_{x_i-z_i/2}^{x_i+z_{i+1}/2} |x-x_i|^{1/n} dx + \int_{x_N-z_N/2}^{x_N+z_{N+1}} |x-x_N|^{1/n} dx \tag{6.5}$$

$$= \frac{2^{1/n} z_1^{1+1/n} + \sum_{i=2}^{N} z_i^{1+1/n} + 2^{1/n} z_{N+1}^{1+1/n}}{2^{1/n}(1+1/n)}.$$

Applying Lagrange's method of undetermined multipliers to minimization of (6.5) under the condition $\sum_{i=1}^{N+1} z_i = 1$, $z_i \geq 0$, $i=1,\dots,N+1$, we find the optimal nodes and the best guaranteed result. The optimal weights, due to Theorem 1.2, are given by the formulas

$$p_1^0 = z_1^0 + \frac{z_2^0}{2}, \quad p_i^0 = \frac{z_i^0 + z_{i+1}^0}{2}, \; i=2,\dots,N, \quad p_N^0 = \frac{z_N^0}{2} + z_{N+1}^0,$$

where $z_1^0, \dots, z_{N+1}^0$ deliver the conditional minimum in (6.5). □

6. . Remark on program implementation of algorithms for computing multiple integrals

We could suggest a number of algorithms for integrating functions of many variables, all of them based on replacing computation of $\int_K f(u)du$ by computation of $\int_0^1 f(\phi(x))dx$ and different only in what method is used for integrating functions of one variable. As to simplicity of program implementation, the most easily implemented algorithms are those in which the integral $\int_0^1 f(\phi_m(x))dx$ is computed with the help of the optimal quadrature formula derived in Section 6.3.

Theorem 6.1 gives the estimate $M\sqrt{n}2^{-n}$ for the irrevocable error of any of these algorithms. Numerical tests carried out on BESM-6 computer have shown that the time required for computation of $\phi_m(x)$ for various values of m and n within the limits $1 \leq n \leq 10$, $4 \leq m \leq 20$ amounts to a few hundredths of a second. Moreover, if we choose m too large in order to reduce the irrevocable error, it can lead the number 2^{mn} out of the range of numbers admitting computer representation. When choosing m for a specific problem, we should take this into account as well as the characteristics of the problem.

CHAPTER 3

RECOVERY OF FUNCTIONS FROM THEIR VALUES

In this chapter, we deal with the problem of approximating functions on the basis of their values at a finite number of points. We derive optimal deterministic nonadaptive algorithms and show that adaption does not help improve the result guaranteed in the class of all nonadaptive algorithms. We also obtain a sequentially optimal algorithm for functions of one variable.

1. OPTIMAL NONADAPTIVE ALGORITHMS

Problems of optimal recovery of a function on the basis of some information on this function at a number of points were studied in many papers. In most of them, various classes of differentiable (Berezovskii and Ivanov [77], Bojanov [75, 86], Bojanov and Chernogorov [77], Gaffney [77, 78], Gaffney and Powell [76], Ivanov [77], Korneichuk [86], Micchelli and Rivlin [77], Micchelli, Rivlin and Winograd [76], Traub and Woźniakowski [80]) or analytic (Golomb [77], Osipenko [72, 76], Traub and Woźniakowski [80]) functions were considered. Both cases of exact and approximate (Marchuk [76], Marchuk and Osipenko [75], Micchelli [78]) information on the function were analyzed. Computational aspects of the promblem were considered, for instance, by De Boor [77].

In most of the papers mentioned, the points at which values of the function (or some other information) are computed are assumed to be fixed and only the terminal operation is to be chosen (in other words, it is the problem of interpolating some given data). In our setting, we focus on optimal choice of the points for informational computations.

1.1. Statement of the problem

The problem of recovering a function from its values naturally arises in the folowing situation.

Suppose we have to compute (at least approximately) the values of a function $f \in F$ at many different points of the set K, and we do not know in advance what points they will be. Our computational resources are limited, so the number N of evaluations (informational computations) is fixed. We choose $x^N=(x_1,\dots,x_N)$, compute $y^N=(y_1,\dots,y_N)$, where $y_j=f(x_j)$, $j=1,\dots,N$, and, on the basis of the information $z^N=(x^N,y^N)$, select a function β: $K\to\mathbb{R}$. After that, evaluations of f are replaced by evaluations of β. The function β is said to recover the function f from the information z^N. A natural way of estimating the approximation error is $\sup_{x\in K}|f(x)-\beta(x)|$.

The above situation fits into the computation model described in Chapter 1 if we set

$$S(f)=f,\quad B=\{\beta\mid\,:\,K\to\mathbb{R}\},$$

$$\varepsilon(\alpha,f)=\gamma(S(f),\beta)=\sup_{x\in K}|f(x)-\beta(x)|. \tag{1.1}$$

Here $\alpha=(\tilde{x}_1,\dots,\tilde{x}_N,\tilde{\beta})$, the function $\beta=\tilde{\beta}(z^N)\in B$ is computed by formulas (3.2), Chapter 1, with $x_j(f)=f(x_j)$, $X_j=K$, $j=1,\dots,N$, and all terminal operations are permitted.

Due to Theorem 4.3 of Chapter 1, constructing an optimal (by error) nonadaptive algorithm amounts to computing

$$v_N\overset{def}{=}\inf_{x^N\in X^N}\ \sup_{y^N\in\{y^N\mid F(z^N)\neq\emptyset\}}\ \inf_{\beta\in B}\ \sup_{j\in F(z^N)}\ \sup_{x\in K}|f(x)-\beta(x)|, \tag{1.2}$$

obtaining the vector x_0^N that delivers the outer infimum, and finding the central terminal operation $\tilde{\beta}_*$ (for z^N fixed, the element $\tilde{\beta}_*(z^N)$ delivers the inner infimum in (1.2)).

1.2. Reduction of the problem of constructing an optimal nonadaptive algorithm to the problem of optimal covering for K

In this section, we assume that $F=F_\rho$, where F_ρ is a functional class determined by some quasi-metric ρ. The below lower and upper envelopes ϕ_{1N} and ϕ_{2N} for functions from $F(z^N)$ are defined by formulas (1.2), Chapter 2, with $i=N$ (see also Lemmas 1.1 and 1.2, Chapter 2), and the notions of radius $R(x^N)$ of the covering determined by centers $x_1,\dots,x_N$ and radius R_N of optimal covering are defined by (2.2) and (2.3), Chapter 2. We also put

$$\phi_N(x)=\frac{1}{2}(\phi_{1N}(x)+\phi_{2N}(x)),\quad x\in K. \tag{1.3}$$

Fig.3 (see p. 79) shows the graph of ϕ_N in the case of one variable for $\rho(u,v)=M|u-v|$.

Theorem 1.1 (Sukharev [78]). *The central terminal operation for approximation of a function belonging to F_ρ from its values is given by the formula*

$$\tilde{\beta}_*(z^N) = \phi_N.$$

The result guaranteed by a nonadaptive algorithm $(x^N, \tilde{\beta}_*)$ is equal to the radius $R(x^N)$ of the covering determined by the centers $x_1, \ldots, x_N$. The best guaranteed result v_N is equal to the radius R_N of optimal covering. An algorithm $(x_0^N, \tilde{\beta}_*)$ is optimal (by error) in the class of all nonadaptive algorithms iff the vector x_0^N determines an optimal covering of K.

Proof. To find the central terminal operation, we make use of the formula

$$\begin{aligned}\inf_{\beta\in B} \sup_{f\in F(z^N)} \sup_{x\in K} |f(x)-\beta(x)| &= \inf_{\beta\in B} \sup_{x\in K} \sup_{f\in F(z^N)} |f(x)-\beta(x)| \\ &= \sup_{x\in K} \inf_{r\in R} \sup_{f\in F(z^N)} |f(x)-r| \\ &= \sup_{x\in K} \inf_{r\in R} \max\{\phi_{2N}(x)-r,\ r-\phi_{1N}(x)\} = \sup_{x\in K} \frac{1}{2}[\phi_{2N}(x)-\phi_{1N}(x)].\end{aligned} \tag{1.4}$$

The first and the fourth equations are trivial, the second one follows from Lemma 4.1 of Chapter 1, and the third one – from Lemmas 1.1 and 1.2 of Chapter 2. The infimum in r is attained at $r=\phi_n(x)$ for any $x\in K$. Hence, due to the second statement of Lemma 4.1, Chapter 1, and the definition (see (4.4), Chapter 1), the central terminal operation is $\tilde{\beta}_*(z^N)=\phi_N$.

We now show that

$$\sup_{y^N\in\{y^N|\ F(z^N)\neq\emptyset\}} \frac{1}{2}[\phi_{2N}(x)-\phi_{1N}(x)] = \min_{i=1,\ldots,N} \rho(x_i, x). \tag{1.5}$$

We have

$$\begin{aligned}&\sup_{y^N\in\{y^N|\ F(z^N)\neq\emptyset\}} \frac{1}{2}[\phi_{2N}(x)-\phi_{1N}(x)] \\ &= \frac{1}{2} \sup_{y^N\in\{y^N|\ F(z^N)\neq\emptyset\}} \left[\min_{i=1,\ldots,N}(y_j+\rho(x,x_j)) - \max_{i=1,\ldots,N}(y_j-\rho(x,x_j))\right] \\ &\leq \frac{1}{2} \sup_{y^N\in\{y^N|\ F(z^N)\neq\emptyset\}} \min_{i=1,\ldots,N}[(y_j+\rho(x,x_j))-(y_j-\rho(x,x_j))] \\ &= \min_{i=1,\ldots,N} \rho(x,x_j).\end{aligned}$$

On the other hand, taking into account that $(0,\ldots,0)\in\{y^N|F(z^N)\neq\emptyset\}$, we get

$$\begin{aligned}&\sup_{y^N\in\{y^N|\ F(z^N)\neq\emptyset\}} \frac{1}{2}[\phi_{2N}(x)-\phi_{1N}(x)] \\ &\geq \frac{1}{2}\left[\min_{i=1,\ldots,N}(0+\rho(x,x_j)) - \max_{i=1,\ldots,N}(0-\rho(x,x_j))\right] = \min_{i=1,\ldots,N} \rho(x,x_j).\end{aligned}$$

This proves (1.5).

We now rewrite the formula for the result guaranteed by the algorithm $(x_0^N, \widetilde{\beta}_*)$. Due to (1.4) and (1.5) we have

$$\sup_{y^N \in \{y^N | F(z^N) \neq \emptyset\}} \inf_{\beta \in B} \sup_{f \in F(z^N)} \sup_{x \in K} |f(x) - \beta(x)|$$

$$= \sup_{y^N \in \{y^N | F(z^N) \neq \emptyset\}} \sup_{x \in K} \frac{1}{2}[\phi_{2N}(x) - \phi_{1N}(x)]$$

$$= \sup_{x \in K} \sup_{y^N \in \{y^N | F(z^N) \neq \emptyset\}} \frac{1}{2}[\phi_{2N}(x) - \phi_{1N}(x)]$$

$$= \sup_{x \in K} \min_{i=1,\dots,N} \rho(x_i, x) = R(x^N).$$

The rest of the theorem is now obvious. □

1.3. On the problem of optimal covering

We consider some issues connected to optimal covering of the set K with a given number of ρ-balls of the same radius. We start with the following remark.

If K is a compact set and ρ is a continuous function, then there exists an optimal covering of K with a given number N of ρ-balls of the same radius.

This assertion follows from continuity of the functions $\min_{i=1,\dots,N} \rho(x_i, x)$ and $R(x^N) = \max_{x \in K} \min_{i=1,\dots,N} \rho(x_i, x)$ in x and x^N respectively, which implies that the minimum of $R(x^N)$ is attained at some x_0^N. We give the solution to the problem of constructing an optimal covering for some specific ρ and K.

Let

$$u = (u^1, \dots, u^n), \quad M > 0, \quad k_1 \geq 0, \dots, k_n \geq 0,$$

$$K = \{u \mid a^i \leq u^i \leq b^i, \ i = 1, \dots, n\},$$

$$\rho(u, v) = M \max_{i=1,\dots,n} k_i |u^i - v^i|. \tag{1.6}$$

Apparently, in this case the ρ-ball $\{u \mid \rho(z, u) \geq r\}$ of radius r with center z is the coordinate parallelepiped

$$\left\{ u \;\middle|\; z^i - \frac{r}{Mk_i} \leq u^i \leq z^i + \frac{r}{Mk_i}, \ i = 1, \dots, n \right\}.$$

Denote $c^i = b^i - a^i$ and $d^i = 2/(Mk_i)$, $i = 1, \dots, n$.

The problem is to cover a coordinate parallelepiped with edges $c^1, \dots, c^n$ by N coordinate parallelepipeds with edges $rd^1, \dots, rd^n$ in such a way that

r is minimal. Clearly, the minimal value of r is equal to the radius R_N of optimal covering. Thus,

$$R_N = \min \left\{ r \;\middle|\; \prod_{i=1}^{n} \left\lceil \frac{c^i}{rd^i} \right\rceil \le N \right\}. \tag{1.7}$$

We now describe an algorithm for solving the problem (1.6) based on (1.7). Let $\bar{r} = \left(\frac{1}{N} \prod_{i=1}^{n} \frac{c^i}{d^i} \right)^n$. Apparently, $R_N \ge \bar{r}$. Moreover, (1.7) implies that at least one of the numbers $c^i/(R_N d^i)$, $i=1,\dots,n$, must be integer. Set $m_i = \lceil c^i/(\bar{r} d^i) \rceil$ and enumerate in the increasing order the roots of the following $m_1 + \dots + m_n$ equations:

$$\frac{c^1}{rd^1} = m_1, \quad \frac{c^1}{rd^1} = m_1 - 1, \quad \dots \quad \frac{c^1}{rd^1} = 1,$$

$$\dots\dots\dots\dots\dots\dots\dots\dots\dots\dots$$

$$\frac{c^n}{rd^n} = m_n, \quad \frac{c^n}{rd^n} = m_n - 1, \quad \dots \quad \frac{c^n}{rd^n} = 1.$$

We get $m_1 + \dots + m_n$ numbers $r_1 \le r_2 \le \dots \le r_{m_1 + \dots + m_n}$. Now we start to check the condition $\prod_{i=1}^{n} \lceil c^i/(r_j d^i) \rceil \le N$ for the numbers r_j as j increases. If this condition gets satisfied for the first time at $j = j_0$, then $R_N = r_{j_0}$. So, we have solved the problem in the case (1.6).

If K is a coordinate cube and in (1.6) we have $k_1 = \dots = k_n = 1$, then it is easy to obtain an optimal covering in the explicit form.

Theorem 1.2. *Let*

$$K = \{u \mid 0 \le u^i \le 1, \; i=1,\dots,n\}, \quad \rho(u,v) = M \max_{i=1,\dots,n} |u^i - v^i|.$$

Then $R_N = M/(2m)$, *where* $m = \lfloor \sqrt[n]{N} \rfloor$, m^n *out of* N *centers* $x_1^0, \dots, x_N^0$ *of an optimal covering coincide with the points* $(j_1/(2m), \dots j_n/(2m))$, $j_1, \dots, j_n \in \{1, 2, \dots, 2m-1\}$, *and the rest* $N - m^n$ *centers are arbitrary.*

Proof. First, assume that $N = m^n$ for an integer m. Then the hyperplanes $u^i = j/m$, $i = 1, \dots, n$, $j = 1, \dots, m-1$, partition K into m^n coordinate cubes with edge $1/m$. Denote their centers by $x_1^0, \dots x_N^0$. Clearly, the vector $x_0^N = (x_1^0, \dots x_N^0)$ determines an optimal covering with the radius $R_N = M/(2m) = M(2\sqrt[n]{N})$.

Now let $m^n < N < (m+1)^n$, where m is integer. We show that in this case the minimal length of edges of the cubes covering K is also equal to $1/m$. This amounts to proving the following: if K is covered by coordinate cubes with edge $l < 1/m$, then at least $(m+1)^n$ cubes are required.

We prove this statement by induction on dimension n of the space. In one dimension it is trivial. Suppose it holds for $\mathbb{R}^{n-1}$ and prove it for $\mathbb{R}^n$. We draw hyperplanes

$$u^1 = l + \varepsilon, \quad u^1 = 2(l+\varepsilon), \dots, u^1 = m(l+\varepsilon),$$

where $\varepsilon > 0$ is such that $l+\varepsilon < 1/m$. Denote by $T_1, T_2, \ldots T_m$ the cross-sections of K by these hyperplanes. Denote by T_0 the side of K lying in the hyperplane $u^1 = 0$.

Let K be covered by coordinate cubes with edge $l < 1/m$. Then the side T_0 is covered by $(n-1)$-dimensional cubes (cross-sections or sides of the corresponding n-dimensional cubes) with edge $l < 1/m$. According to the induction hypothesis, the number of these $(n-1)$-dimensional cubes is not less than $(m+1)^{n-1}$, and, obviously, none of the corresponding n-dimensional cubes intersects with the cross-section T_1. Repeating the argument about the covering of T_0 for the cross-sections $T_1, \ldots, T_m$, we prove that K is covered by at least $(m+1)(m+1)^{n-1} = (m+1)^n$ coordinate cubes with edge l. □

Corollary. *If $n=1$, $K=[0,1]$, $\rho(u,v) = M|u-v|$, then for recovery of a function belonging to the class F_ρ from its values we have $v_N = M/(2N)$, and the algorithm $(1/(2N), 3/(2N), \ldots, (2N-1)/(2N))$ is optimal in the class of all nonadaptive algorithms.* □

Assume that $\rho(u,v) = \|u-v\|$, K is a bounded set with a boundary of zero measure, and we have a least dense covering of the space $\mathbb{R}^n$ by translations of the ball $\Omega(1) = \{x \mid \|x\| \leq 1\}$. Then, with the help of the construction described in Theorem 2.2 of Chapter 2, we can easily form an asymptotically optimal sequence of algorithms.

We can also obtain an optimal covering through numerical solution of the problem

$$R(x^N) = \sup_{a \in K} \min_{i=1,\ldots,N} \rho(x_i, a) \to \min_{x_1,\ldots,x_N \in K}.$$

For instance, in this way an optimal covering of the square K was derived for the Euclidean metric ρ and $N \leq 15$ by Brusov and Piyavskii [71].

1.4. Coincidence of the best guaranteed results in the classes of adaptive and nonadaptive algorithms

In this section we show that, for recovery of functions belonging to F_ρ from their values, the best guaranteed result v_N in the class of nonadaptive algorithms coincides with the best guaranteed result ε_N in the class of all adaptive algorithms. Several coincidence results of this type were obtained in Section 5 of Chapter 1, where S was assumed to be a functional. We could extend those results to the case $S(f) = f$ and thus establish coincidence of v_N and ε_N. However, here we prove this fact independently by reasoning of some other type.

Suppose that the central terminal operation $\tilde{\beta}_*(z^N)=\phi_N$ is used (see (1.2) and Theorem 1.1), and consider the class X^N of nonadaptive algorithms (see (3.8), Chapter 1) and the class $\tilde{X}^N$ of adaptive algorithms (see (3.10), Chapter 1). In our case $X^N=K^N$. For estimating efficiency of algorithms, we use the criterion (4.14), Chapter 1, i.e.,

$$\varepsilon(x^N,f)=\sup_{x\in K}|f(x)-\phi_N(x)|, \quad \varepsilon(\tilde{x}^N,f)=\varepsilon(x^N,f), \tag{1.8}$$

where in the first equation $\phi_N=(\phi_{1N}+\phi_{2N})/2$, $y_1=f(x_i)$, $i=1,\dots,N$,

$$\phi_{1N}(x)=\max_{i=1,\dots,N}\{y_i-\rho(x,x_i)\},$$

$$\phi_{2N}(x)=\min_{i=1,\dots,N}\{y_i+\rho(x,x_i)\},$$

and in the second equation the vector x^N is determined by the algorithm $\tilde{x}^N$ and the function f in accordance with (3.2), Chapter 1.

Theorem 1.3. *For recovery of functions belonging to F_ρ from their values, the best guaranteed results in the classes X^N and $\tilde{X}^N$ coincide,* i.e.,

$$\inf_{\tilde{x}^N\in\tilde{X}^N}\sup_{f\in F}\varepsilon(\tilde{x}^N,f)=\inf_{x^N\in X^N}\sup_{f\in F}\varepsilon(x^N,f), \tag{1.9}$$

and every algorithm optimal in X^N is also optimal in $\tilde{X}^N$.

Proof. Denote the left-hand side of (1.9) by ε_N and the right-hand side by v_N. Since $X^N\subset\tilde{X}^N$, we have $\varepsilon_N\leq v_N$. To prove the theorem, it suffices to show that $\varepsilon\geq v_N$.

Denote an algorithm δ-optimal in $\tilde{X}^N$ by $\tilde{x}^N_\delta$, $\delta>0$. Then

$$\sup_{f\in F_\rho}\varepsilon(\tilde{x}^N_\delta,f)<\varepsilon_N+\delta.$$

Let $\tilde{x}^N_\delta(\mathbf{0})=x^N_\delta=(x_{1\delta},\dots x_{N\delta})$, where $\mathbf{0}$ is a function identically equal to zero on K and $\tilde{x}^N_\delta(\mathbf{0})$ denotes the vector obtained by (3.2), Chapter 1, for $\tilde{x}^N=\tilde{x}^N_\delta$ and $f=\mathbf{0}$. Due to Theorem 1.1,

$$v_N=R_N=\inf_{x^N\in K^N}\sup_{x\in K}\min_{i=1,\dots,N}\rho(x_i,x),$$

therefore,

$$\sup_{x\in K}\min_{i=1,\dots,N}\rho(x_{i\delta},x)\geq v_N.$$

Hence there is a point $a_\delta\in K$ such that

$$\min_{i=1,\dots,N}\rho(x_{i\delta},a_\delta)\geq v_N-\delta. \tag{1.10}$$

Consider the function

$$g_\delta\overset{def}{=}\begin{cases} v_N-\delta-\rho(x,a_\delta) & \text{if } \rho(x,a_\delta)\leq v_N-\delta,\\ 0 & \text{if } \rho(x,a_\delta)\geq v_N-\delta.\end{cases}$$

Apparently, $g_\delta(x)=\max\{v_N-\delta-\rho(x,a_\delta),0\}$. From Lemmas 2.1 and 2.7 of Chapter 1 we conclude that $g_\delta\in F_\rho$. It follows from (1.10) that $g_\delta(x_{1\delta})=0$,

$i=1,\dots,N$, hence $\tilde{x}_\delta^N(g_\delta)=\tilde{x}_\delta^N(0)=x_\delta^N$, where $\tilde{x}_\delta^N(g_\delta)$ denotes the vector obtained by (3.2), Chapter 1, for $\tilde{x}^N=\tilde{x}_\delta^N$ and $f=g_\delta$. Finally, due to (1.8),

$$\begin{aligned}\varepsilon_N &> \sup_{f\in F_p} \varepsilon(\tilde{x}_\delta^N,f)-\delta \geq \varepsilon(\tilde{x}_\delta^N,g_\delta)-\delta=\varepsilon(x_\delta^N,g_\delta)-\delta\\ &=\sup_{x\in K}|g_\delta(x)-\phi_N(x)|-\delta=\sup_{x\in K} g_\delta(x)-\delta=v_N-2\delta,\end{aligned}$$

where $\phi_N(x)\equiv 0$ since $y_i=g_\delta(x_{i\delta})=0$, $i=1,\dots,N$. Taking the limit as $\delta\to 0$, we get $\varepsilon_N\geq v_N$. □

2. SEQUENTIALLY OPTIMAL AND ONE-STEP OPTIMAL RECOVERY ALGORITHMS

Let

$$F=\{f \mid |f(u)-f(v)|\leq M|u-v|,\ u,v\in[a,b]\}. \tag{2.1}$$

As before, we assume that informational computations are evaluations of the unknown function at some points of $K=[a,b]$. Theorem 1.1 gives the central terminal operation in the case where $\rho(u,v)=M|u-v|$ (the case under consideration), and the corollary of Theorem 1.2 gives an optimal (by error) nonadaptive recovery algorithm. The main purpose of this section is to construct a sequentially optimal (by error) algorithm for the class (2.1) of one-variable functions satisfying the Lipschitz condition. Having solved this problem, we will easily construct a one-step optimal algorithm and also sequentially optimal algorithm (counting informational computations).

2.1. Optimal error algorithms for subclasses of the functional class (2.1)

When constructing sequentially optimal (by error) algorithms, we have to derive optimal algorithms for subclasses of functions from the class (2.1) with fixed values at one or both endpoints of $[a,b]$. The solution to this problem is given by Lemmas 2.1 and 2.2 (and also Theorem 3.1) below, where the number of informational computations is assumed to be equal to n, $x_0=a$, $x_{n+1}=b$, and F denotes the functional class (2.1). In this case $F(x_{n+1},y_{n+1})$, $F(x_0,y_0)$, and $F(x_0,x_{n+1},y_0,y_{n+1})$ are subclasses of F containing functions with fixed values at the right, left, and both endpoints of $[a,b]$ respectively.

In the formulations of Lemmas 2.1 and 2.2 it is not stated explicitly what terminal operation is used. Actually, it is always the central terminal operation, easily obtained with the help of Theorem 1.1 if we take into account *a priori* information on the function values at the endpoints of $[a, b]$. Namely,

$$\tilde{\beta}_*(x^n, x_{n+1}, y^n, y_{n+1})$$
$$= \frac{1}{2}\left(\max_{j=1,\dots,n+1}\{y_j - M|x-x_j|\} + \min_{j=1,\dots,n+1}\{y_j + M|x-x_j|\}\right),$$

$$\tilde{\beta}_*(x_0, x^n, y_0, y^n)$$
$$= \frac{1}{2}\left(\max_{j=0,1,\dots,n}\{y_j - M|x-x_j|\} + \min_{j=0,1,\dots,n}\{y_j + M|x-x_j|\}\right),$$

$$\tilde{\beta}_*(x_0, x^n, x_{n+1}, y_0, y^n, y_{n+1})$$
$$= \frac{1}{2}\left(\max_{j=0,1,\dots,n+1}\{y_j - M|x-x_j|\} + \min_{j=0,1,\dots,n+1}\{y_j + M|x-x_j|\}\right)$$

are the central terminal operations for the classes $F(x_{n+1}, y_{n+1})$, $F(x_0, y_0)$, and $F(x_0, x_{n+1}, y_0, y_{n+1})$ respectively. In what follows, we assume that we deal with these terminal operations and with the algorithmic classes X^N (see (3.8), Chapter 1) and $\tilde{X}^N$ (see (3.10), Chapter 1).

Lemma 2.1. *The algorithm*

$$\left(a + \frac{b-a}{2n+1}, a + 3\frac{b-a}{2n+1}, \dots, a + (2n-1)\frac{b-a}{2n+1}\right) \in X^n$$

is optimal (by error) on $F(x_{n+1}, y_{n+1})$ *both in the class* X^N *of all nonadaptive algorithms and in the class* $\tilde{X}^N$ *of all adaptive algorithms.*

The algorithm

$$\left(a + 2\frac{b-a}{2n+1}, a + 4\frac{b-a}{2n+1}, \dots, a + 2n\frac{b-a}{2n+1}\right) \in X^n$$

is optimal (by error) on $F(x_0, y_0)$ *both in the class* X^N *of all nonadaptive algorithms and in the class* $\tilde{X}^N$ *of all adaptive algorithms.*

The best guaranteed result in the both cases is equal to $M(b-a)/(2n+1)$. □

The lemma is easily proved by the same reasoning we used when proving Theorems 1.1 and 1.3. Here we have to take into account that the worst "nature's" behaviour from "the computer's" point of view is selecting $f(x_j) = y_{n+1}$ for the class $F(x_{n+1}, y_{n+1})$ and selecting $f(x_j) = y_0$ for the class $F(x_0, y_0)$, where x_j, $j = 1, \dots, n$, are the points of function evaluations. We also observe that the class $F(x_{n+1}, y_{n+1})$ is symmetric with respect to $f(x) \equiv y_{n+1}$, and $F(x_0, y_0)$ is symmetric with respect to $f(x) \equiv y_0$.

Thus, the optimal algorithm prescribes that the first evaluation be performed at any of the points

$$\left\{ a + r\frac{b-a}{2n+1} \;\middle|\; \begin{array}{l} r = 1, 3, \dots, 2n-1 \ \text{ for the class } \ F(b, y_b), \\ r = 2, 4, \dots, 2n \ \text{ for the class } \ F(a, y_a) \end{array} \right\}. \tag{2.2}$$

The situation with the class $F(x_0, x_{n+1}, y_0, y_{n+1})$, which is not symmetric if $y_0 \neq y_{n+1}$, is much more complicated.

We use our old notation $\varepsilon_N(z^i)$ for the best guaranteed result (see (6.7) and (6.8), Chapter 1) and denote the best guaranteed result on $F(x_0, x_{n+1}, y_0, y_{n+1})$ in the class of all adaptive algorithms with n informational computations by $\varepsilon_{n+1}(x_0, x_{n+1}, y_0, y_{n+1})$. It is sufficient to consider the case $x_0 = a = 0$, $x_{n+1} = b = 1$. Clearly, the best guaranteed result in the situation $0, 1, y_0, y_{n+1}$ depends only on $|y_{n+1} - y_0|$, or on $l = -|y_{n+1} - y_0|/M + 1$ (the second quantity corresponds to the first one in a one-to-one manner), see Fig.7. The Lipschitz constant M and the number of informational computations n are assumed to be fixed. Denote the best guaranteed result by $(M/2)W_n(l)$; in other words, set

$$W_n(l) = \frac{2}{M}\varepsilon_{n+2}(0, 1, y_0, y_{n+1}).$$

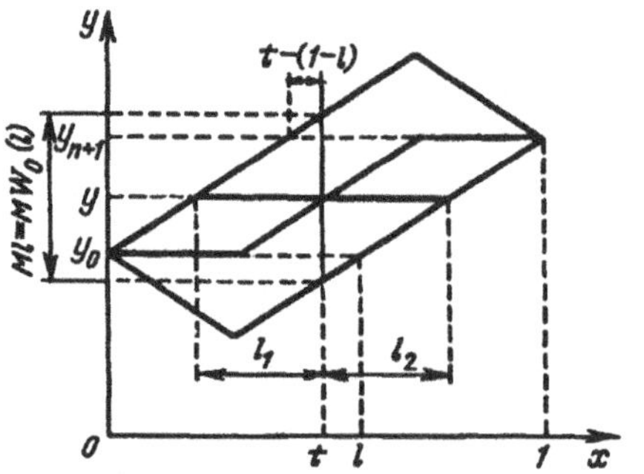

Fig. 7. On finding the best guaranteed result of recovering a function with fixed values at the endpoints of [0, 1]

The situation where we know the values of the function at the endpoints of $[a, b]$ can be reduced to the case under consideration using the substitution $x' = (x - a)/(b - a)$, $y' = y$. If $f(x)$ satisfies the Lipschitz condition on $[a, b]$ with a constant M, then $f(a + x'(b - a))$ obviously satisfies the Lipschitz condition in x' on [0,1] with the constant $M(b - a)$. Therefore,

$$\varepsilon_{n+2}(a, b, y_0, y_{n+1}) = \frac{M}{2}(b - a)W_n(l'), \tag{2.3}$$

where

$$l' = \frac{-|y_{n+1} - y_0|/M + b - a}{b - a}.$$

Lemma 2.2. *The best guaranteed result on $F(0, 1, y_0, y_{n+1})$ in the* *class of all adaptive algorithms satisfies the recursion equation*

$$W_n(l)=\min_{0<t<1}\ \max_{\substack{l_1,l_2\\ 0\le l_1\le t,\\ 0\le l_2\le 1-t\\ l_1+l_2=l}}\ \min_{j=0,1,\ldots,n-1}\ \max\left\{tW_j\left(\frac{l_1}{t}\right),(1-t)W_{n-1-j}\left(\frac{l_2}{1-t}\right)\right\} \tag{2.4}$$

with

$$W_0(t)=l. \tag{2.5}$$

The first computation of an optimal on $F(0,1,y_0,y_{n+1})$ adaptive algorithm is to be performed at the point delivering the outer minimum on the right-hand side of (2.4).

Proof. Suppose at the first step "the computer" selects a point $t\in[0,1]$. We have

$$f(t)=y\in A(t)\stackrel{def}{=}[\phi_{12}(t),\phi_{22}(t)],$$

where $\phi_{12}(t)=\max\{y_0-M|x|,\ \ y_{n+1}-M|x-1|\}$ and $\phi_{22}(t)=\min\{y_0+M|x|,y_{n+1}+M|x-1|\}$ are the lower and upper envelopes respectively constructed on the basis of the information $0,1,y_0,y_{n+1}$. Now we repeat the argument of Section 3.2, Chapter 2. Observe that, for the class (2.1), the values f takes at any points within one of the intervals $(0,t)$, $(t,1)$ do not change the range of possible values of f at any point within the other interval. (This range is bounded by the lower and upper envelopes ϕ_{13} and ϕ_{23} constructed on the basis of the information $0,1,t,y_0,y_{n+1},y$.) Therefore, the optimal strategy for "the computer" after the first step is to choose the optimal number $j\in\{0,1,\ldots,n-1\}$ of computations to be performed within $(0,t)$ (the rest $n-1-j$ computations will be performed within $(t,1)$) and then, at the rest $n-1$ steps, to select the points of informational computations in the optimal way.

According to (2.3), the best guaranteed result after the first step has the form

$$\varepsilon_{n+2}(0,1,t,y_0,y_{n+1},y)=\frac{M}{2}\min_{j=0,1,\ldots,n-1}\max\left\{tW_j\left(\frac{-|y-y_0|/M+t}{t}\right),\right.$$
$$\left.(1-t)W_{n-1-j}\left(\frac{-|y-y_{n+1}|/M+1-t}{1-t}\right)\right\}. \tag{2.6}$$

Since an optimal strategy in a multistage decision process consists of an optimal decision at the first stage and optimal behaviour at all the next stages, we can write the following equation for the best result guaranteed on $F(0,1,y_0,y_{n+1})$:

$$W_n(l)=\min_{0<t<1}\max_{y\in A(t)}\min_{j=0,1,\ldots,n-1}\max\left\{tW_j\left(\frac{-|y-y_0|/M+t}{t}\right),\right.$$
$$\left.(1-t)W_{n-1-j}\left(\frac{-|y-y_{n+1}|/M+1-t}{1-t}\right)\right\}. \tag{2.7}$$

Moreover, as we can see from Fig.7,

$$W_0(l)=l. \tag{2.8}$$

The recursion equation (2.7) expresses the well-known optimality principle of dynamic programming, see Bellman [57], Bellman and Dreyfus [62].

To prove the lemma, it now suffices to show that equations (2.4) and (2.7) are equivalent. In order to do this, we show that the function $W_n(l)$ is nondecreasing with respect to l for any n.

For $n=0$, this follows from (2.8). Assume that $W_j(l)$ is nondecreasing in l for $j=1,\dots,n-1$. Fix some $t\in(0,1)$. To be definite, assume that $y_0\leq y_{n+1}$ (otherwise in what follows $[y_0,y_{n+1}]$ should be replaced by $[y_{n+1},y_0]$). Using (2.6) and the assumption that the functions W_j, $j=1,\dots,n-1$, are nondecreasing, we obtain

$$\begin{aligned}&\varepsilon_{n+2}(0,1,t,y_0,y_{n+1},y)\leq\varepsilon_{n+2}(0,1,t,y_0,y_{n+1},y_0)\quad\text{for}\quad y\leq y_0,\\&\varepsilon_{n+2}(0,1,t,y_0,y_{n+1},y)\leq\varepsilon_{n+2}(0,1,t,y_0,y_{n+1},y_{n+1})\quad\text{for}\quad y\geq y_0.\end{aligned} \tag{2.9}$$

Therefore,

$$W_n(l)=\frac{2}{M}\min_{0<t<1}\max_{y\in N(t)}\varepsilon_{n+2}(0,1,t,y_0,y_{n+1},y),$$

where $N(t)=A(t)\cap[y_0,y_{n+1}]$. Set $l_1=-|y-y_0|/M+t$, $l_2=-|y-y_{n+1}|/M+1-t$. Then for $y\in N(t)$ we have $l_1+l_2=l$. As y runs over the set $N(t)$, l_1 runs over the interval $[\max\{0,t-(1-l)\},\min\{t,l\}]$ (see Fig.7) and l_2 runs over the interval $[\max\{0,l-t\},\min\{1-t,l\}]$. It is easy to see that the restrictions on l_1 and l_2 as y takes values in $N(t)$ look as follows: $0\leq l_1\leq t$, $0\leq l_2\leq 1-t$, $l_1+l_2=l$. So, we have established equivalence of the equations (2.4) and (2.7). However, in our argument we have made use of the inequalities (2.9) derived under the assumption that W_j is nondecreasing.

To complete the proof of the lemma, we now have to complete the induction proof of $W_n(l)$ being nondecreasing in l. For any fixed t, the function inside the outer minimum on the right-hand side of (2.4) (and, therefore, the function $W_n(l)$) is nondecreasing in l. This follows from the fact that the functions W_j, $j=0,\dots,n-1$, are nondecreasing by the induction hypothesis. □

The main difficulty we meet with when constructing a sequentially optimal algorithm is to solve explicitly the system of equations (2.4), (2.5). A sequentially optimal algorithm requiring numerical solution of the system (2.4), (2.5) at every step would have inadequately high combinatory complexity. An explicit solution of (2.4), (2.5) will be given in Section 3. Due to Theorem 3.1,

$$W_n(l)=\begin{cases} l/k, & a_k(n)\leq l\leq b_k(n),\\ b_k(n)/k, & b_k(n)\leq l\leq a_{k+1}(n),\end{cases}\qquad k=1,\dots,\lfloor n/2\rfloor+1,$$

where

$$a_1(n)=0,\quad a_k(n)=\left(\frac{1}{2}\right)^{m_{k-1}(n)-1}\frac{k}{k+s_{k-1}(n)},$$
$$k=2,\dots,\lfloor n/2\rfloor+1,$$

$$a_{\lfloor n/2\rfloor+2}(n)=1, \quad b_k(n)=\left(\frac{1}{2}\right)^{m_k(n)-1}\frac{k}{k+1+s_{k-1}(n)},$$
$$k=1,\ldots,\lfloor n/2\rfloor+1;$$

here $m_k(n)$ is the quotient and $s_k(n)$ is the remainder on dividing n by k. The outer minimum in (2.4) is attained at t equal to one of the following numbers:

$$T_{n,r}(l)=\begin{cases}\dfrac{r}{k+s_{k-1}(n)} & \text{if } b_{k-1}(n)\le l\le b_k(n),\\ \text{any number from } [0,1] & \text{if } 0\le l\le b_1(n),\end{cases} \tag{2.10}$$

where $b_{\lfloor n/2\rfloor+2}(n)=1$, $r=1,\ldots,k+s_{k-1}(n)-1$.

2.2. Sequentially optimal (by error) algorithm

For simplicity of the notation, let $K=[0,1]$. From the definition of sequentially optimal algorithm, Theorem 1.3 and corollary of Theorem 1.2 we get

$$x_1\in\left\{\frac{1}{2N},\frac{3}{2N},\ldots,\frac{2N-1}{2N}\right\}, \tag{2.11}$$

that is, any of the points (2.11) can be selected for the first informational computation.

Suppose i computations have been performed at points $x_1,\ldots,x_i$, and $x_{i1},\ldots,x_{ii}$ is the ordered permutation of $x_1,\ldots,x_i$, i.e., $x_{i1}<\cdots<x_{ii}$. Analogously to the problem of numerical integration, constructing a sequentially optimal algorithm amounts to distributing the rest $N-i$ informational computations in the optimal way among the intervals

$$[0,x),\ (x_{i1},x_{i2}),\ \ldots,\ (x_{i,i-1},x_{ii}),\ (x_{ii},1], \tag{2.12}$$

constructing optimal adaptive algorithms for the functional classes that correspond to these intervals, and putting x_{i+1} equal to the point of the first informational computation prescribed by any of these algorithms.

According to Lemmas 2.1 and 2.2, this requires that we find a set of numbers $n_1^i,\ldots,n_{i+1}^i$ (the jth number in this set is the number of informational computations to be performed within the jth interval in (2.12)) delivering the minimum in the problem of obtaining

$$\varepsilon_N(z^i)=\frac{M}{2}\min_{\substack{n_1,\ldots,n_{i+1}\in\{0,1,2,\ldots\}\\ \sum_{j=1}^{i+1}n_j=N-i}}\max\left\{\frac{2x_{i1}}{2n_1+1},(x_{i2}-x_{i1})W_{n_2}(l_{i2}),\ldots,\right.$$
$$\left.(x_{ii}-x_{i,i-1})W_{n_i}(l_{ii}),\frac{2(1-x_{ii})}{2n_{i+1}+1}\right\}, \tag{2.13}$$

where

$$l_{ij} = \frac{-|f(x_{ij}) - f(x_{i,j-1})|/M + x_{ij} - x_{i,j-1}}{x_{ij} - x_{i,j-1}}.$$

After the solution $n_1^i, \ldots, n_{i+1}^i$ has been found, we can put x_{i+1} equal to any point from the union of the sets given by (2.2) and (2.10), i.e.,

$$\begin{aligned} x_{i+1} \in & \left\{ x_{i1} \frac{r_1}{2n_1^i + 1} \;\middle|\; r_1 = 1, 3, \ldots, 2n_1^i - 1 \right\} \\ & \bigcup \left(\bigcup_{j=2}^{i} \{ x_{i,j-1} + (x_{ij} - x_{i,j-1}) T_{n_j^i, r_j}(l_{ij}) \mid r_j = 1, \ldots, k + s_{k-1}(n_j^i) - 1 \} \right) \\ & \bigcup \left\{ x_{ii} + (1 - x_{ii}) \frac{r_{i+1}}{2n_{i+1}^i + 1} \;\middle|\; r_{i+1} = 2, 4, \ldots, 2n_{i+1}^i \right\}, \end{aligned} \tag{2.14}$$

where $i = 1, \ldots, N-1$; the number k, according to (2.10), is determined by the condition $b_{k-1}(n_j^i) \le l_{ij} \le b_k(n_j^i)$; finally, we assume that if some of the numbers $n_1^i, \ldots, n_{i+1}^i$ are equal to zero then the corresponding sets are empty.

Lemmas 2.1 and 2.2 together with the above construction give a complete proof of the following theorem.

Theorem 2.1 (Sukharev [78]). *The algorithm* (2.11), (2.14) *is sequentially optimal (by error) on the class* (2.1) *with* $[a, b] = [0, 1]$. □

2.3. Solution of auxiliary integer optimization problems

We now proceed to solving the sequence of problems (2.13). We introduce, for fixed x_{ij} and l_{ij}, the following notations[1]:

$$\alpha_1(n_1) = \frac{2x_{i1}}{2n_1 + 1}, \quad \alpha_j(n_j) = (x_{ij} - x_{i,j-1}) W_{n_j}(l_{ij}), \quad j = 2, \ldots, i,$$

$$\alpha_{i+1}(n_{i+1}) = \frac{2(1 - x_{ii})}{2n_{i+1} + 1}. \tag{2.15}$$

Rewrite the problem (2.13) in the form

$$\max_{i=1,\ldots,i+1} \alpha_i(n_i) \to \min_{\substack{n_1, \ldots, n_{i+1} \in \{0,1,2,\ldots\} \\ \sum_{j=1}^{i+1} n_j = N - i}}. \tag{2.16}$$

Let $n_1^{i-1}, \ldots, n_i^{i-1}$ be the solution to the problem of finding $\varepsilon_N(z^{i-1})$, which arises when computing x_i, and let, for definiteness, $x_i \in (x_{i-1,j-1},$

---

[1] Though the functions $\alpha_1, \ldots, \alpha_{i+1}$ depend on the number of step, for simplicity of notation we do not introduce additional subscripts.

$x_{i-1,j}$). Then $x_{i-1,1}=x_{i1},\ldots, x_{i-1,j-1}=x_{i,j-1}, x_i=x_{ij}, x_{i-1,j}=x_{i,j+1},\ldots,$ $x_{i-1,i-1}=x_{ii}$. As an initial approximation to the solution $n_1^i,\ldots,n_{i+1}^i$ we choose the set of integers $n_1^{i0},\ldots,n_{j+1}^{i0}$ defined in the following way:

$$n_1^{i0}=n_1^{i-1},\ \ldots,\ n_{j-1}^{i0}=n_{j-1}^{i-1},\ n_{j+2}^{i0}=n_{j+1}^{i-1},\ \ldots,\ n_{i+1}^{i0}=n_i^{i-1}.$$

The numbers n_j^{i0}, n_{j+1}^{i0} are defined as the minimal integers such that

$$\alpha_j(n_j^{i0})\le\frac{2}{M}\varepsilon_N(z^{i-1}),\quad \alpha_{j+1}(n_{j+1}^{i0})\le\frac{2}{M}\varepsilon_N(z^{i-1}).$$

We assign to the variables $n_1,\ldots,n_{i+1}$ initial values $n_1^{i0},\ldots,n_{i+1}^{i0}$ so that $n_1:=n_1^{i0},\ldots,n_{i+1}:=n_{i+1}^{i0}$. Set $\Delta n^i=n_j^{i-1}-1-n_j^{i0}-n_{j+1}^{i0}$ and perform Δn^i times the operator $n_s:=n_s+1$, where $s\in\{1,\ldots,i+1\}$ is some (say, the least) subscript such that $\alpha_s(n_s)=\max\limits_{t=1,\ldots,i+1}\alpha_t(n_t)$.

It is easy to see that the obtained values of $n_1,\ldots,n_{i+1}$ coincide with the solution $n_1^i,\ldots,n_{i+1}^i$ we are seeking and

$$\varepsilon_N(z^i)=\frac{M}{2}\max_{t=1,\ldots,i+1}\alpha_i(n_t^i). \tag{2.17}$$

Note that if $i=N$, then (2.17) immediately yields the following formula for the error estimate (i.e., *a posteriori* result guaranteed by the algorithm after all informational computations have been performed):

$$\varepsilon_N(z^N)=\frac{M}{2}\max_{t=1,\ldots,N+1}\alpha_t(0). \tag{2.18}$$

2.4. One-step optimal algorithm

To select x_1 in the optimal way, we have to solve the problem of minimizing the result (2.18) guaranteed after the first step:

$$\varepsilon_1(z^1)=\max\{Mx_1, M(1-x_1)\}\to\min,\quad x_1\in[0,1].$$

The solution is

$$x_1=1/2. \tag{2.19}$$

Solving the problem (2.16) for $N-i=1$, we get from (2.14)

$$x_{i+1}=\begin{cases} x_{i1}/3 & \text{if } \alpha_1(0)=m,\\ (x_{i,k-1}+x_{ik})/2 & \text{if } \alpha_k(0)=m,\ k\in\{2,\ldots,i\},\\ x_{ii}+2(1-x_{ii})/3 & \text{if } \alpha_{i+1}(0)=m,\end{cases} \tag{2.20}$$

where $m=\max\limits_{t=1,\ldots,i+1}\alpha_t(0)$ and the numbers $\alpha_t(0)$ are defined by (2.15). Thus we have proved the following theorem.

Theorem 2.2. *The algorithm* (2.19), (2.20) *is one-step optimal on the class* (2.1) *with* $[a,b]=[0,1]$. □

2.5. Sequentially optimal (counting informational computations) algorithm

The setting with a fixed accuracy ε in 'simpler' than the setting with a fixed number N of informational computations since constructing a sequentially optimal (counting informational computations) algorithm does not require solving integer programming problems of the type (2.16).

By virtue of Theorem 1.3, the best guaranteed result in the class of all adaptive algorithms ε_N coincides with the best guaranteed result in the class of all nonadaptive algorithms v_N, and, due to the corollary of Theorem 1.2, $v_N = M/(2N)$. Therefore,

$$N_\varepsilon = \min\{N \mid \varepsilon_N \le \varepsilon\} = \min\left\{N \,\middle|\, \frac{M}{2N} \le \varepsilon\right\} = \left\lceil \frac{M}{2\varepsilon} \right\rceil .$$

In accordance with the definitions of optimal (counting informational computations) algorithm (see Section 6.4, Chapter 1), choice of x_1 in a sequentially optimal algorithm is given by (2.11) for $N = N_\varepsilon$.

After i steps the stopping criterion (4.20), Chapter 1, is applied. In the problem being considered, due to (2.18), satisfying the stopping criterion means satisfying the inequality

$$\varepsilon_i(z^i) = \frac{M}{2} \max_{t=1,\dots,i+1} \alpha_t(0) \le \varepsilon.$$

If the stopping criterion is not satisfied, then the integer numbers $n_1^i, \dots, n_{i+1}^i$ are determined by the condition $M\alpha_t(n_t^i)/2 \le \varepsilon$ in such a way that either $n_t^i = 0$ or $M\alpha_t(n_t^i - 1)/2 > \varepsilon$, $t = 1, \dots, i+1$; after that x_{i+1} is chosen by formula (2.14).

2.6. Construction scheme and implementation of sequentially optimal algorithms

Note that we constructed sequentially optimal algorithms according to the same scheme we used for integration problems. The key points of the scheme were described in Section 3.7, Chapter 2. That section also gives the characteristics of computation models to which this scheme applies. Our considerations regarding the integration problem remain valid for the problem of recovering functions from their values. However, for the recovery problem, obtaining optimal algorithms for functional classes determined by some fixed information at the endpoints of the segment may turn out to be much more difficult than for the integration problem.

When implementing sequentially optimal algorithms, we should make proper use of freedom in choice of the next informational computation point given by formulas (2.11) and (2.14). This point can be selected randomly with equal probability from the set of all possible points, or we can select the point from this set closest to the point prescribed by the one-step optimal algorithm, etc.

We present the results of testing the version of sequentially optimal (by error) algorithm that selects from the set of all possible points (2.14) the point closest to the point prescribed by the one-step optimal algorithm. This version was applied to a random test function from the family $f(x)= c_0+c_1x+c_2x^2/2+\cdots+c_nx^n/n!$, where $x\in[a,b]=[-2,2]$ and $c_0,c_1,\ldots,c_n$ are random numbers uniformly distributed over $[-3,3]$; $n=4$. The algorithm estimated the Lipschitz constant M by the formula

$$M=\max_{j=0,1,\ldots,160}|f'(-2+j/40)|.$$

Table 4

N	0.10	0.30	0.50	0.80
v_N	0.6117	0.2039	0.1223	0.0765
$\varepsilon_N(z^N)$	0.5942	0.1594	0.0967	0.0543
$\varepsilon_N(z^N)/v_N$	0.9715	0.7820	0.7904	0.7106

The first line of Table 4 gives the number N of informational computations. The second line gives the accuracy $v_N=M(b-a)/(2N)$ guaranteed by the optimal in X^N and $\tilde{X}^N$ algorithm

$$\left(a+\frac{b-a}{2N},a+3\frac{b-a}{2N},\ldots,a+(2N-1)\frac{b-a}{2N}\right)$$

(this accuracy coincides with *a posteriori* accuracy guaranteed on completion of all informational computations). The third line contains *a posteriori* accuracy guaranteed on completion of all informational computations by the sequentially optimal (by error) algorithm; the fourth line gives the ratio of the numbers from the third and the second lines.

Table 4 shows that for the problem of recovering functions from their values, as well as for the integration problem, the sequentially optimal algorithm has considerable advantage over the optimal algorithm.

3. SOLUTION OF A MULTISTEP ANTAGONISTIC GAME RELATED TO THE PROBLEM OF OPTIMAL RECOVERY

In Section 2, we applied the optimality principle for multistage decision processes to construct a sequentially optimal algorithm for recovery of a function from its values. This enabled us to reduce the problem of constructing an algorithm to the problem of solving the following system of recursion equations (Lemma 2.2):

$$W_n(l) = \min_{0<t<1} \max_{\substack{l_1,l_2 \\ 0\le l_1\le t,\ 0\le l_2\le 1-t}} \min_{j=0,1,\ldots,n-1} \max\left\{ tW_j\left(\frac{l_1}{t}\right),\right.$$

$$\left.(1-t)W_{n-1-j}\left(\frac{l_2}{1-t}\right)\right\}, \tag{3.1}$$

$$W_0(l) = l. \tag{3.2}$$

In this section we find the solution of the system (3.1), (3.2) in an explicit form. The solution is obtained though reducing the problem to a multistep antagonistic game with complete information and finding the saddle point of the game.

Solution of the system (3.1), (3.2) and of the corresponding game is of interest in itself since theory of systems of recursion equations has been developed considerably only for the linear case (see, e.g., Milne-Thomson [33] and Ostrovskii [63]), and we have but a few examples of nonlinear systems being solved explicitly (see Bellman [57], Bellman and Dreyfus [62], Chernous'ko [70a]), as well as of multistep games with complete information and a continuum of alternatives for every player being solved explicitly (see Vorob'yov and Vrublevskaya [67]).

3.1. Reduction of the problem to an antagonistic game, formulation and discussion of the main theorem

The form of equations (3.1) and (3.2) suggests that $W_n(l)$ is the value of the following multistep antagonistic game (which is called *recovery game*).

At the first step, the minimizing player chooses a point $x_1 \in (0,1)$ (point t in the notation of (3.1)), thus partitioning the segment [0,1] into segments $[0, x_1]$, $[x_1, 1]$, and the maximizing player, knowing x_1, assigns nonnegative numbers l_1 and l_2 to these two segments in such a way that the number assigned to a segment does not exceed its length and $l_1 + l_2 = l$, where $l \in [0,1]$ is some *a priori* fixed number known to the both players.

At the second step, the minimizing player chooses a point $x_2 \in (0, x_1) \cup (x_1, 1)$, thus partitioning one of the segments of the first step into two, and the maximizing player, knowing x_1 and x_2, assigns nonnegative numbers to these two segments in such a way that each number does not exceed the length of the corresponding segment and their sum is equal to the number that was assigned to the union of these segments at the first step, and so on.

After all the n steps have been performed, the maximizing player gets from the minimizing player maximum of the numbers assigned to the $n+1$ segments into which the segment [0,1] has been partitioned.

The solution of this game is given by the following theorem, which is the main result of Sections 2 and 3.

Theorem 3.1 (Sukharev [77]). *The solution $W_n(l)$ of the above game depends on n and l in the following way:*

$$W_n(l)=\begin{cases} l/k, & a_k(n)\le l\le b_k(n),\\ b_k(n)/k, & b_k(n)\le l\le a_{k+1}(n),\ k=1,\dots,\lfloor n/2\rfloor+1,\end{cases} \tag{3.3}$$

where

$$a_1(n)=0,\quad a_k(n)=\left(\frac{1}{2}\right)^{m_{k-1}(n)-1}\frac{k}{k+s_{k-1}(n)},\quad k=2,\dots,\left\lfloor\frac{n}{2}\right\rfloor+1,$$

$$a_{\lfloor n/2\rfloor}=1;\quad b_k(n)=\left(\frac{1}{2}\right)^{m_k(n)-1}\frac{k}{k+1+s_k(n)},\quad k=1,\dots,\left\lfloor\frac{n}{2}\right\rfloor+1;$$

$m_k(n)$ is the quotient and $s_k(n)$ is the remainder on dividing n by k.

The optimal choice for the minimizing player at the first step (delivering the outer minimum in (3.1)) can be any of the points

$$T_{n,r}(l)=\begin{cases}\dfrac{r}{k+s_{k-1}(n)} & \text{if } b_{k-1}(n)\le l\le b_k(n),\ k=2,\dots,\left\lfloor\frac{n}{2}\right\rfloor+2,\\ \text{any number from } [0,1] & \text{if } 0\le l\le b_1(n),\end{cases} \tag{3.4}$$

where $b_{\lfloor n/2\rfloor+2}(n)=1$, $r=1,\dots,k+s_{k-1}(n)-1$.

The optimal choice for the maximizing player at the first step (delivering the maximum with respect to l_1 and l_2 in (3.1)) is the pair of numbers $l_1(x_1)$, $l_2(x_1)$[1] *such that $l_2(x_1)=l-l_1(x_1)$ and*

$$l_2(x_1)=l-l_1(x_1),$$

$$l_1(x_1)=\begin{cases} 0 & \text{if } l\le a_2(n),\ x_1\le 1/2,\\ 0 & \text{if } (3.6),\ x_1\le\dfrac{1}{k+1+s_k(n)},\\ (t-1)\dfrac{l}{k} & \text{if } (3.6),\ (3.7),\ s_k(n)\le k-t+1,\\ \dfrac{t-1}{2}\cdot\dfrac{l}{k} & \text{if } (3.6),\ (3.7),\ t-1 \text{ is even},\ s_k(n)\le\dfrac{t-1}{2},\\ \dfrac{t}{2}\cdot\dfrac{l}{k} & \text{if } (3.6),\ (3.7),\ t \text{ is even},\ s_k(n)\le\dfrac{t}{2}-1,\\ l-l_1(1-x_1) & \text{if } x_1\ge 1/2,\end{cases} \tag{3.5}$$

where (3.6) and (3.7) are the following conditions:

$$a_k(n)\le l\le a_{k+1}(n),\quad k=2,\dots,\left\lfloor\frac{n}{2}\right\rfloor+1, \tag{3.6}$$

---

[1] For unification of notation, we always replace the symbol t in (3.1) by x_1 when dealing with the recovery game.

$$\frac{t-1}{k+1+s_k(n)} \le x_1 \le \frac{t}{k+1+s_k(n)}, \quad 1 \le t-1 \le \frac{k+1+s_k(n)}{2}. \tag{3.7}$$

Before proving the theorem, we make the following observation. Knowing optimal choices of the first step (3.4) and (3.5) for the both players, it is easy to construct by induction optimal strategies for the players in the recovery game. Moreover, these strategies will be sequentially optimal, that is, not only will they guarantee the result $W_n(l)$ to the both players, but also punish the adversary in the optimal way for mistakes s/he commits if the strategy s/he uses is not optimal.

To obtain such strategies, it is sufficient to be able to find after i steps an optimal distribution of the rest $n-i$ points to be selected by the minimizing player among the intervals (x_{i0}, x_{i1}), (x_{i1}, x_{i2}), $\ldots$, $(x_{i,i-1}, x_{ii})$, $(x_{ii}, x_{i,i+1})$, where $x_{i0}=0$, $x_{i,i+1}=1$, and $x_{i1}, \ldots, x_{ii}$ is the ordered permutation of the numbers $x_1, \ldots, x_i$ already selected by the minimizing player. To distribute these points among the intervals, we have to find a set of nonnegative integers $n_1, \ldots, n_{i+1}$ delivering the minimum in the following formula:

$$\min_{\substack{n_1,\ldots,n_{i+1} \\ n_1+\cdots+n_{i+1}=n-i}} \max_{j=1,\ldots,i+1} (x_{ij} - x_{i,j-1}) W_{n_j}\left(\frac{l_{ij}}{x_{ij} - x_{i,j-1}}\right),$$

where l_{ij} is the number assigned by the maximizing player to the segment $[x_{i,j-1}, x_{ij}]$. An algorithm for solving a similar problem was given in Section 2.3. If $n_1^i, \ldots, n_{i+1}^i$ is the solution for the problem, then, for the minimizing player's strategy to remain seguentially optimal, at the $(i+1)$st step s/he has to put x_{i+1} equal to any point $x_{i,j-1} + (x_{ij} - x_{i,j-1}) T_{n_j^i r_j}\left(\frac{l_{ij}}{x_{ij} - x_{i,j-1}}\right)$, where the r_j's are given by (3.4) and $j=1,\ldots,i+1$ are such that $n_j^i > 0$. For the maximizing player, it is sufficient to assign the numbers

$$(x_{ij} - x_{i,j-1}) l_1 \left(\frac{x_{i+1} - x_{i,j-1}}{x_{ij} - x_{i,j-1}}\right),$$

$$l_{ij} - (x_{ij} - x_{i,j-1}) l_1 \left(\frac{x_{i+1} - x_{i,j-1}}{x_{ij} - x_{i,j-1}}\right)$$

to the segments $[x_{i,j-1}, x_{i+1}]$ and $[x_{i+1}, x_{ij}]$ respectively. Here we have to put $n = n_j^i$ and $l = l_{ij}/(x_{ij} - x_{i,j-1})$ when using (3.5)-(3.7).

The proof of Theorem 3.1 is given in Sections 3.2 and 3.3 below. In Section 3.2 we show that, if at the first step the minimizing player uses the strategy described in Theorem 3.1, this ensures that his/her loss will not be greater than $W_n(l)$ (provided that his/her subsequent actions are optimal). In Section 3.3 we show that, if at the first step the maximizing player uses the strategy described in Theorem 3.1, this ensures that his/her gain will not be less than $W_n(l)$ (provided that his/her subsequent actions are optimal). This proves the theorem, since these two facts together mean that the choices prescribed by Theorem 3.1 for the both players at the first step are optimal and $W_n(l)$ given by (3.3) is the value of the game.

3.2. Proof of Theorem 3.1: guaranteed result for the minimizing player

In this section we show that, if at the first step of the recovery game the minimizing player chooses any of the points (3.4) as x_1, this ensures that his/her loss will not be greater than (3.3) (provided that his/her subsequent actions are optimal).

For $n=1$ the statement is trivial since

$$W_1(l)=\begin{cases} l, & l\leq 1/2, \\ 1/2, & j\geq 1/2, \end{cases} \quad T_{1,1}(l)=\begin{cases} \text{any number from } [0,1], & l\leq 1/2, \\ 1/2, & l\geq 1/2. \end{cases}$$

Suppose it holds for the 2-,...,$(n-1)$-step games. We now prove it for the n-step game.

For $l\leq b_1(n)$ there is nothing to prove. Let

$$a_k(n)\leq l\leq b_k(n), \quad k=2,\ldots,\lfloor n/2\rfloor+1, \tag{3.8}$$

or

$$\left(\frac{1}{2}\right)^{m_{k-1}(n)-1}\frac{1}{k+s_{k-1}(n)}\leq\frac{l}{k}\leq\left(\frac{1}{2}\right)^{m_k(n)-1}\frac{1}{k+1+s_k(n)},$$

$$k=2,\ldots,\lfloor n/2\rfloor+1. \tag{3.9}$$

It suffices to demonstrate that if at the first step the minimizing player chooses $x_1=\dfrac{r}{k+s_{k-1}(n)}$ (where r is any of the numbers $1,\ldots,k+s_{k-1}(n)-1$), this ensures that his/her loss will not be greater than $W_n(l)=l/k$, no matter what l_1 the maximizing player chooses (the latter equation holds due to (3.3) and (3.8)).

If $l_1\leq l/k$, then, to ensure that his/her loss will not be greater than l/k, the minimizing player has to allocate all the points $x_2,\ldots,x_n$ within the segment $[x_1,1]$, and so, taking the induction hypothesis into account, we have to prove that in this case

$$(1-x_1)W_{n-1}\left(\frac{l_2}{1-x_1}\right)\leq\frac{l}{k}.$$

We prove this inequality in the worst for the minimizing player case $l_2=l$, $r=1$; in this situation it takes the form

$$\frac{k+s_{k-1}(n)-1}{k+s_{k-1}(n)}W_{n-1}\left(l\frac{k+s_{k-1}(n)}{k+s_{k-1}(n)-1}\right)\leq\frac{l}{k}.$$

For the latter inequality to hold, it is necessary and sufficient that

$$l\frac{k+s_{k-1}(n)}{k+s_{k-1}(n)-1}\geq a_k(n-1).$$

This is obviously true due to (3.9) if

$$\left(\frac{1}{2}\right)^{m_{k-1}(n)-1} \frac{k}{k+s_{k-1}(n)} \cdot \frac{k_{k-1}(n)}{k+s_{k-1}(n)-1}$$
$$\geq \left(\frac{1}{2}\right)^{m_{k-1}(n-1)-1} \frac{k}{k+s_{k-1}(n-1)},$$

which is equivalent to

$$k+s_{k-1}(n-1) \geq 2^{m_{k-1}(n)-m_{k-1}(n-1)}(k+s_{k-1}(n)-1). \qquad (3.10)$$

In fact, the left- and right-hand sides of (3.10) are always equal, since, if $m_{k-1}(n)=m_{k-1}(n-1)+1$, then $s_{k-1}(n-1)=k-2$, $s_{k-1}(n)=0$, and if $m_{k-1}(n)=m_{k-1}(n-1)$, then $s_{k-1}(n)=s_{k-1}(n-1)+1$.

Now consider the case

$$(t-1)\frac{l}{k} \leq l_1 \leq t\frac{l}{k}, \quad t=2,\ldots,k-1 \qquad (3.11)$$

(the case $t=k$ can be reduced to the case $t=1$ since it does not matter which endpoint of the segment we take). To ensure that his/her loss will not be greater than l/k for any l_1 satisfying (3.11), it is sufficient that the minimizing player choose n_1 out of the rest $n-1$ points within the interval $(0,x_1)$, where n_1, due to the induction hypothesis, is determined by the conditions

$$x_1 W_{n_1}\left(t\frac{l}{k}\cdot\frac{1}{x_1}\right) \leq \frac{l}{k}, \quad x_1 W_{n_1-1}\left(t\frac{l}{k}\cdot\frac{1}{x_1}\right) > \frac{l}{k},$$

which are equivalent to the inequalities

$$t\frac{l}{k}\cdot\frac{1}{x_1} \geq a_t(n_1), \qquad (3.12)$$

$$t\frac{l}{k}\cdot\frac{1}{x_1} < a_t(n-1) \qquad (3.13)$$

respectively. Since $l-l_1 \leq (k+1-t)l/k$, it suffices to show that we have

$$(1-x_1)W_{n-1-n_1}\left(\frac{k+1-t}{1-x_1}\cdot\frac{l}{k}\right) \leq \frac{l}{k},$$

or, equivalently,

$$\frac{k+1-t}{1-x_1}\cdot\frac{l}{k} \geq a_{k+1-t}(n-1-n_1).$$

Due to (3.9), establishing this inequality amounts to checking that

$$\frac{k+1-t}{1-x_1}\left(\frac{1}{2}\right)^{m_{k-1}(n)-1} \frac{1}{k+s_{k-1}(n)} \geq a_{k+1-t}(n-1-n_1).$$

Substituting $x_1 = \dfrac{r}{k+s_{k-1}(n)}$ and using the expression for $a_{k+1-t}(n-1-n_1)$, we get

$$k+s_{k-1}(n)-r \leq 2^{m_{k-1}(n-1-n_1)-m_{k-1}(n)}[k+1-t+s_{k-1}(n-1-n_1)]. \qquad (3.14)$$

This inequality we have to prove.

For the proof we use the following corollary of (3.13) and (3.9):

$$t\left(\frac{1}{2}\right)^{m_{k-1}(n)-1}\frac{1}{k+s_{k-1}(n)}\cdot\frac{k+s_{k-1}(n)}{r}$$
$$<\left(\frac{1}{2}\right)^{m_{t-1}(n_1-1)-1}\frac{t}{t+s_{t-1}(n_1-1)},$$

which can be written as

$$2^{m_{t-1}(n_1-1)}[t+s_{t-1}(n_1-1)]<r2^{m_{k-1}(n)}. \tag{3.15}$$

Assume that

$$\frac{1}{2}<\frac{t-1}{r}<r. \tag{3.16}$$

Then $0\leq r-t\leq t-3$, and so $n_1^0=1+(t-1)m_{k-1}(n)+r-t$ satisfies the equations

$$2^{m_{t-1}(n_1-1)}=2^{m_{k-1}(n)},\quad t+s_{t-1}(n_1-1)=r,$$

and, therefore, the equation

$$2^{m_{t-1}(n_1-1)}[t+s_{t-1}(n_1-1)]=r2^{m_{k-1}(n)}.$$

Since (3.15) follows from (3.13) and the left-hand side of (3.15) is increasing with respect to n_1, the number n_1 determined by (3.12) and (3.13) satisfies the inequality

$$n_1\leq(t-1)m_{k-1}(n)+r-t. \tag{3.17}$$

In view of (3.17),

$$\begin{aligned}n-n_1-1&\geq n-(t-1)m_{k-1}(n)-r+t-1\\&=n-\frac{t-1}{k-1}n+\frac{t-1}{k-1}s_{k-1}(n)-r+t-1\\&=\frac{(k-t)n}{k-1}+\frac{t-1}{k-1}s_{k-1}(n)+t-r-1\\&=(k-t)\left[m_{k-1}(n)+\frac{s_{k-1}(n)}{k-1}\right]+\frac{t-1}{k-1}s_{k-1}(n)+t-r-1\\&=(k-t)m_{k-1}(n)+s_{k-1}(n)+t-r-1,\end{aligned}$$

hence

$$\frac{n-n_1-1}{k-t}\geq m_{k-1}(n)+\frac{s_{k-1}(n)+t-r-1}{k-t} \tag{3.18}$$

and

$$m_{k-t}(n-n_1-1)-m_{k-1}(n)\geq\frac{s_{k-1}(n)-s_{k-t}(n-n_1-1)+t-r-1}{k-t}. \tag{3.19}$$

Taking into account that, by the definitions of x_1 and $s_{k-t}(n-n_1-1)$, we have $r\leq k+s_{k-1}(n)-1$ and $s_{k-t}(n-n_1-1)\leq k-t-1$, we derive from (3.19)

$$\begin{aligned}&m_{k-t}(n-n_1-1)-m_{k-1}(n)\\&\quad\geq\frac{s_{k-1}(n)-(k-t-1)+t-r-1}{k-t}\geq-2+\frac{1}{k-t},\end{aligned}$$

and, since the left-hand side is integer, $m_{k-t}(n-n_1-1)-m_{k-1}(n)\geq -1$.

Let

$$m_{k-t}(n-n_1-1)-m_{k-1}(n)=-1; \tag{3.20}$$

then (3.19) yields

$$s_{k-1}(n)-s_{k-t}(n-n_1-1)+t-r-1\leq t-k, \tag{3.21}$$

and the inequality (3.14) to be proved has the form

$$2[k+s_{k-1}(n)-r]\leq k+1-t+s_{k-1}(n-1-n_1). \tag{3.22}$$

Observe that, in the case (3.20), it follows from (3.18) that $s_{k-1}(n))+t-r-1\leq 0$, i.e., $s_{k-1}(n)+t-r\leq 0$, and make use of (3.21). Then (3.22) becomes obvious since

$$\begin{aligned}&2[k+s_{k-1}(n)-r]-[k+1-t+s_{k-1}(n-1-n_1))]\\&\quad\equiv[s_{k-1}(n)-s_{k-t}(n-n_1-1)+t-r-1]+k+s_{k-1}(n)-r\\&\quad\leq t-k+k+s_{k-1}(n)-r=s_{k-1}(n)+t-r\leq 0.\end{aligned}$$

Let

$$m_{k-t}(n-n_1-1)-m_{k-1}(n)=0. \tag{3.23}$$

Then (3.19) yields $s_{k-1}(n)-s_{k-t}(n-n_1-1)+t-r-1\leq 0$, and in the case (3.23) this is exactly the inequality (3.14) being proved.

Finally, let

$$m_{k-t}(n-n_1-1)-m_{k-1}(n)=q\geq 1. \tag{3.24}$$

Then (3.19) implies

$$s_{k-1}(n)-r\leq 1-t+s_{k-t}(n-n_1-1)+q(k-t).$$

Moreover, we have $q(k-t)\leq(2^q-1)[k+1-t+s_{k-t}(n-n_1-1)]$ since $q\leq 2^q-1$ for $g\geq 1$; therefore,

$$\begin{aligned}k+s_{k-1}(n)-r&\leq k+1-t+s_{k-t}(n-n_1-1)+q(k-t)\\&\leq 2^q[k+1-t+s_{k-t}(n-n_1-1)].\end{aligned}$$

Thus, we have proved (3.14) under the assumption (3.24), and so we are through with the case (3.16).

We now proceed to the case

$$(t-1)/r\leq 1/2 \quad\text{or}\quad (t-1)/r\geq 1. \tag{3.25}$$

We start with two auxiliary inequalities.

Firstly, under the assumption (3.25) we have

$$(t-1)\log_2\frac{t-1}{r}\geq t-r-1. \tag{3.26}$$

To prove this, put $(t-1)/r=\tau$ and consider the function $z(\tau)=\tau\log_2\tau-\tau+1$. We have to show that $z(\tau)\geq 0$ if $0<\tau\leq 1/2$ or $\tau\geq 1$, but this is obvious since the function $z(\tau)$ is convex for $\tau>0$ and $z(1/2)=z(1)=0$.

Secondly, we have

$$(t-1)\log_2(t+\alpha)-\alpha\geq(t-1)\log_2(t-1)+1 \quad\text{for}\quad 0\leq\alpha\leq t-2. \tag{3.27}$$

To prove this, consider the concave function $z(\alpha)=(t-1)\log_2(t+\alpha)-\alpha$ and verify that $z(0)\geq z(t-2)$. Then we have $\min_{0\leq\alpha\leq t-2} z(\alpha)=z(t-2)=(t-1)\log_2(t-1)+1$ and (3.27) is proven. So it is sufficient to check that $z(0)\geq z(t-2)$, i.e., $(t-1)\log_2[t/(t-1)]\geq 1$. Set $\beta=1/(t-1)$; then $0\leq\beta\leq 1$ (since $t\geq 2$ due to (3.11)), and the inequality to be verified takes the form $\log_2(1+\beta)-\beta\geq 0$. But this is obvious for $0\leq\beta\leq 1$, since the function $\log_2(1+\beta)-\beta$ is concave and equal to zero at $\beta=0$ and $\beta=1$.

Now we make some equivalent transformations of inequality (3.15). We have

$$m_{t-1}(n_1-1)+\log_2[t+s_{t-1}(n_1-1)]<m_{k-1}(n)+\log_2 r,$$

hence

$$\frac{n_1-1}{t-1}+\frac{s_{t-1}(n_1-1)}{t-1}<\frac{n}{k-1}-\frac{s_{k-1}(n)}{k-1}+\log_2\frac{r}{t+s_{t-1}(n_1-1)}.$$

After multiplying the both parts of the inequality by $t-1$, subtracting from n and then dividing by $k-t$, we get

$$\frac{n-n_1-1}{k-t}>\frac{n}{k-1}+\frac{(t-1)\log_2\frac{t+s_{t-1}(n_1-1)}{r}-s_{t-1}(n_1-1)+\frac{t-1}{k-1}s_{k-1}(n)-2}{k-t}$$
$$=m_{k-1}(n)+\frac{(t-1)\log_2\frac{t+s_{t-1}(n_1-1)}{r}-s_{t-1}(n_1-1)+s_{k-1}(n)-2}{k-t}.$$

Using (3.27) for $\alpha=s_{t-1}(n_1-1)$ and then (3.26) gives

$$m_{k-t}(n-n_1-1)-m_{k-1}(n)$$
$$>\frac{(t-1)\log_2\frac{s_{t-1}(n_1-1)}{r}-s_{t-1}(n_1-1)-s_{k-1}(n)-2}{k-t}+\frac{-s_{k-t}(n-n_1-1)}{k-t}$$
$$\geq\frac{(t-1)\log_2\frac{t-1}{r}-1s_{k-1}(n)-s_{k-t}(n-n_1-1)}{k-t}$$
$$\geq\frac{t-r-2+s_{k-1}(n)-s_{k-t}(n-n_1-1)}{k-t}. \tag{3.28}$$

Continuing (3.28), we obtain

$$m_{k-t}(n-n_1-1)-m_{k-1}(n)>\frac{t-r-2+s_{k-1}(n)-(k-t-1)}{k-t}$$
$$=-1+\frac{t-r-1+s_{k-1}(n)}{k-t}. \tag{3.29}$$

From here, recalling that $r\leq k+s_{k-1}(n)-1$, we get the inequality

$$m_{k-t}(n-n_1-1)-m_{k-1}(n)>-1+\frac{t-k}{k-t}=-2. \tag{3.30}$$

Let

$$m_{k-t}(n-n_1-1)-m_{k-1}(n)=-1. \tag{3.31}$$

Then, in view of (3.28),

$$-1>[t-r-2+s_{k-1}(n)-s_{k-t}(n-n_1-1)]/(k-t),$$

or

$$k-r-q+s_{k-1}(n)-s_{k-1}(n-n_1-1)\leq 0, \tag{3.32}$$

and, by (3.29),

$$-1>-1+(t-r-1+s_{k-1}(n))/(k-t),$$

or

$$t-r+s_{k-1}(n)\leq 0. \tag{3.33}$$

Summing inequalities (3.32) and (3.33), we get

$$2[k+s_{k-1}(n)-r]\leq k+1-t+s_{k-t}(n-n_1-1),$$

and in the case (3.31) this is exactly the inequality (3.31) being proved.

Let

$$m_{k-t}(n-n_1-1)-m_{k-1}(n)=0. \tag{3.34}$$

Then, by virtue of (3.28), we have

$$0>t-r-2+s_{k-1}(n)-s_{k-t}(n-n_1-1),$$

hence

$$k+s_{k-1}(n)-r\leq k+1-t+s_{k-1}(n-n_1-1),$$

and in the case (3.34) this is the inequality (3.14) being proved.

Finally, let

$$m_{k-t}(n-n_1-1)-m_{k-1}(n)=q\geq 1. \tag{3.35}$$

Then (3.28) yields

$$q>\frac{t-r--2+s_{k-1}(n)-s_{k-t}(n-n_1-1)}{k-t},$$

therefore,

$$s_{k-1}(n)-r\leq 1-t+s_{k-t}(n-n_1-1)+q(k-t). \tag{3.36}$$

Since $q\leq 2^q-1$ for $q\geq 1$, we have

$$q(k-t)\leq(2^q-1)[k+1-t+s_{k-t}(n-n_1-1)]. \tag{3.37}$$

Using (3.36) and (3.37), we get

$$\begin{aligned}k+s_{k-1}(n)-r&\leq k+1-t+s_{k-t}(n-n_1-1)+q(k-t)\\&\leq 2^q[k+1-t+s_{k-t}(n-n_1-1)],\end{aligned}$$

and in the case (3.35) this coincides with the inequality (3.14) being proved.

According to (3.30), we have exhausted all the possibilities and covered the situation (3.25) completely, thus covering the case (3.8).

Now, if we have

$$b_k(n)\leq l\leq a_{k+1}(n),\quad k=1,\ldots,\lfloor n/2\rfloor, \tag{3.38}$$

then, choosing at the first step $T_{n,r}(l)=r/(k+1+s_k(n))$, $k=1,\ldots,k+s_k(n)$, the minimizing player can ensure that his/her loss will not be greater than $W_n(l)=a_{k+1}(n)/(k+1)$, as we have already proved for $l=a_{k+1}(n)$. If (3.38) holds, then, all the more, the loss does not exceed $W_n(l)=a_{k+1}(n)/(k+1)=b_n(n)/k$.

Finally, if

$$l \geq b_{\lfloor n/2 \rfloor + 1}, \tag{3.39}$$

then the minimizing player can ensure that his/her loss will not be greater than $W_n(l) = \frac{b_{\lfloor n/2 \rfloor + 1}}{\lfloor n/2 \rfloor + 1} = \frac{1}{n+1}$ by choosing at the rth step the point $T_{n,r}(l) = r/(n+1)$, $r = 1, \ldots \lfloor n/2 \rfloor + 1 + s_{\lfloor n/2 \rfloor + 1}(n) = n$, regardless of the maximizing player's actions. Now we have considered all the possibilities $l \leq b_1(n)$, (3.8), (3.38) and (3.39) and thus completed the proof of the statement opening Section 3.2.

3.3. Proof of Theorem 3.1: guaranteed result for the maximizing player

In this section we show that, if at the first step the maximizing player assigns the numbers $l_1(x_1)$ and $l - l_1(x_1)$ (where $l_1(x_1)$ is given by (3.5)) to the segments $[0, x_1]$ and $[x_1, 1]$ respectively, this ensures that his/her gain will not be less than (3.3), provided that his/her subsequent actions are optimal.

First, we make sure that (3.5) fully determines a strategy of the maximizing player at the first step (though this strategy may not be unique).

Assume that

$$k - t + 2 \leq s_k(n) \leq \frac{t-1}{2} - 1, \quad t-1 \text{ is even}, \tag{3.40}$$

and so $l_1(x_1)$ is not determined. From the second two-sided inequality in (3.7) and the left inequality in (3.40) we get $2t - s_k(n) - 3 \leq k \leq t + s_k(n) - 2$, hence $t - 1 \leq 2s_k(n)$. But this contradicts the right inequality in (3.40), therefore, (3.40) is not possible.

Assume that

$$k - t + 2 \leq s_k(n) \leq \frac{t}{2} - 2, \quad t \text{ is even}, \tag{3.41}$$

and so $l_1(x_1)$ is not determined. Then, again, we have $t - 1 \leq 2s_k(n)$, which contradicts the right inequality in (3.41); therefore, (3.41) is not possible either.

Thus, for any $x_1 \leq 1/2$ the function $l_1(x_1)$ is determined by the first five lines in (3.5), and for $x_1 \geq 1/2$ it is determined by the sixth line in (3.5).

The conditions $0 \leq l_1(x_1) \leq x_1$ and $0 \leq l - l_1(x_1) \leq 1 - x_1$ to be satisfied by the maximizing player's strategy are also easy to check.

If $l \leq b_1(n) = (1/2)^n$, then the maximizing player can ensure that s/he will get $W_n(l) = l$ by assigning l to the larger segment and 0 to the smaller segment at every step. Since the maximizing player guarantees himself (herself) the gain $b_1(n)$ if $l = b_1(n)$, s/he will, all the more, get $b_1(n)$ if $l \geq b_1(n)$. This justifies the first line in (3.5) (note that we need not justufy the sixth line separately once the first five lines have been justified). Thus, we have proved our statement for $n=1$.

Suppose the statement also holds for the $2-,\dots,(n-1)$-step games. We now prove it for the n-step game.

It is sufficient to give the proof for the case (3.8), since if the statement is proven for this case and $b_k(n)\le l\le a_{k+1}(n)$, $k=2,\dots,\lfloor n/2\rfloor+1$, then the maximizing player will, all the more, get $b_k(n)/k$ if $l\ge b_k(n)$.

So, we assume that in the situation (3.8) we have $x_1\le 1/(k+1+s_k(n))$. In accordance with the induction hypothesis, it suffices to show that $(1-x_1)W_{n-1}(l/(1-x_1))\ge W_n(l)=l/k$. Furthermore, it is sufficient to check the inequality in the worst for the maximizing player case $x_1=1/(k+1+s_k(n))$:

$$\frac{k+s_k(n)}{k+1+s_k(n)}W_{n-1}\left(l\frac{k+1+s_k(n)}{k+s_k(n)}\right)\ge\frac{l}{k}.$$

Due to the definition of W_{n-1}, this inequality is equivalent to the inequality $l(k+1+s_k(n))/(k+s_k(n))\le b_k(n-1)$, and to prove the latter inequality in the case (3.8) it is sufficient to show that $b_k(n)(k+1+s_k(n))/(k+s_k(n))\le b_k(n-1)$, or, in other words, $(k+s_k(n))2^{m_k(n)}\ge(k+1+s_k(n-1))2^{m_k(n-1)}$. In fact, the left- and right-hand sides of the latter formula are equal; if $m_k(n)=m_k(n-1)+1$, then $s_k(n-1)=k-1$, $s_k(n)=0$; and if $m_k(n)=m_k(n-1)$, then $s_k(n)=s_k(n-1)+1$. Thus, we have justified the second line in (3.5).

Next, we have to consider the case (3.7) under the conditions (3.8). We show that, if

$$x_1W_{n_1}\left(\frac{l_1(x_1)}{x_1}\right)<\frac{l}{k},\tag{3.42}$$

then

$$(1-x_1)W_{n-n_1-1}\left(\frac{l-l_1(x_1)}{1-x_1}\right)\ge\frac{l}{k}.\tag{3.43}$$

Due to the induction hypothesis, this means the following: if the minimizing player allocates within the segment $[0,x_1]$ such a number n_1 out of the rest $n-1$ points that the maximizing player cannot guarantee that s/he gets $W_n(l)=l/k$ on $[x_1,1]$ (this is (3.42)), then the maximizing player can guarantee that s/he gets at least $W_n(l)=l/k$ on $[x_1,1]$ (this is (3.43)), and, therefore, in the whole game. So, this is the fact we wish to prove.

Making the substitution $l_1(x_1)=(t-1)l/k$ in (3.42), we obtain

$$x_1W_{n_1}\left(\frac{t-1}{x_1}\cdot\frac{l}{k}\right)<\frac{l}{k}.$$

In view of the definition of the function W_{n_1}, this inequality is equivalent to

$$\frac{t-1}{x_1}\cdot\frac{l}{k}>b_{t-1}(n_1).$$

Using the right-hand side of (3.9) and the left-hand side of (3.7), we derive

$$(t-1)\left(\frac{1}{2}\right)^{m_1(n)-1}\frac{1}{k+1+s_k(n)}\cdot\frac{k+1+s_k(n)}{t-1}$$
$$>\left(\frac{1}{2}\right)^{m_{t-1}(n_1)-1}\frac{t-1}{t+s_{t-1}(n_1)}.$$

Equivalent transformations give successively

$$m_{t-1}(n_1)>m_k(n)+\log_2\frac{t-1}{t+s_{t-1}(n_1)},$$

$$n_1-s_{t-1}(n_1)>(t-1)m_k(n)+(t-1)\log_2\frac{t-1}{t+s_{t-1}(n_1)}.$$

Observing that $(t-1)/(t+s_{t-1}(n_1))\geq 1/2$, we continue:

$$n_1-s_{t-1}(n_1)>(t-1)m_k(n)+(t-1)\log_2\frac{1}{2}=(t-1)(m_k(n)-1).$$

Recalling now the meaning of the quantities participating in the inequality, we conclude

$$n_1\geq(t-1)m_k(n). \tag{3.44}$$

Thus, we have derived (3.44) from (3.42). Making the substitution $l_1(x_1)=(t-1)l/k$ in the inequality (3.43) being proved and replacing it with an equivalent one, we get successively

$$(1-x_1)W_{n-n_1-1}\left(\frac{k-tr+1}{1-x_1}\cdot\frac{l}{k}\right)\geq\frac{l}{k},$$

$$\frac{k-t+1}{1-x_1}\cdot\frac{l}{k}\leq b_{k-t+1}(n-n_1-1).$$

We make the latter inequality stronger using the right inequalities in (3.7) and (3.9), the inequality $n-n_1-1\leq n-(t-1)m_k(n)-1$ which follows from (3.44), and the fact that the function $b_k(n)$ is decreasing with respect to n:

$$(k-t+1)\left(\frac{1}{2}\right)^{m_k(n)-1}\frac{1}{k+1+s_k(n)}\left(1-\frac{t}{k+1+s_k(n)}\right)^{-1}$$
$$\leq\left(\frac{1}{2}\right)^{m_{k-t+1}(n-(t-1)m_k(n)-1)}\frac{k-t+1}{k-t+2+s_{k-t+1}(n-(t-1)m_k(n)-1)},$$

or

$$\left(\frac{1}{2}\right)^{m_k(n)}\frac{1}{k-t+1+s_k(n)}$$
$$\leq\left(\frac{1}{2}\right)^{m_{k-t+1}(n-(t-1)m_k(n)-1)}\frac{1}{k-t+2+s_{k-t+1}(n-(t-1)m_k(n)-1)}. \tag{3.45}$$

We show that, in fact, if $s_k(n)<k-t+1$, then the left- and right-hand sides of (3.45) are equal, which justifies (3.43) and the third line in (3.5). Observe that

$$\frac{n-(t-1)m_k(n)-1}{k-t+1}=\frac{n-(t-1)(n/k-s_k(n)/k)-1}{k-t+1}$$
$$=\frac{n}{k}+\frac{(t-1)\frac{s_k(n)}{k}-1}{k-t+1}=m_k(n)+\frac{s_k(n)}{k}+\frac{(t-1)s_k(n)}{k(k-t+1)}-\frac{1}{k-t+1}$$
$$=m_k(n)+\frac{s_k(n)-1}{k-t+1}.$$

If $s_k(n)=0$, then $m_{k-t+1}(n-(t-1)m_k(n)-1)=m_k(n)-1$ and $s_{k-t+1}(n-(t-1)m_k(n)-1)=k-t$; on the other hand, if $1\le s_k(n)\le k-t+1$, then $m_{k-t+1}(n-(t-1)m_k(n)-1)=m_k(n)$ and $s_{k-t+1}(n-(t-1)m_k(n)-1)=s_k(n)-1$. In the both cases, the equality of the two sides of (3.45) is obvious.

The cases corresponding to the fourth and the fifth lines in (3.5) are dealt with in a similar way.

We now make the substitution $l_1(x_1)=\frac{t-1}{2}\cdot\frac{l}{k}$. The obtained inequality yields

$$n_1>r[m_k(n)+1]+s_r(n_1)+r\log_2\frac{r}{r+1+s_r(n_1)}\ge rm_k(n)+s_r(n_1),$$

where $r=(t-1)/2$ and the number $(t-1)$ is even, which implies $n_1\ge r[m_k(n)+1]$, hence $n-n_1-1\le n-\frac{t-1}{2}m_k(n)-\frac{t+1}{2}$. Replacing (3.43) in the case $l_1(x_1)=\frac{t-1}{2}\cdot\frac{l}{k}$ being considered with an equivalent inequality and making it stronger with the help of the latter of the obtained inequalities and the right inequalities in (3.7) and (3.9), we get

$$\left(\frac{1}{2}\right)^{m_k(n)}\frac{1}{k-t+1+s_k(n)}\le\left(\frac{1}{2}\right)^{m_{k-\frac{t-1}{2}}\left(n-\frac{t-1}{2}m_k(n)-\frac{t+1}{2}\right)}$$
$$\times\frac{1}{k-\frac{t-3}{2}+s_{k-\frac{t-1}{2}}\left(n-\frac{t-1}{2}m_k(n)-\frac{t+1}{2}\right)}. \qquad (3.46)$$

Now observe that

$$\frac{n-\frac{t-1}{2}m_k(n)-\frac{t+1}{2}}{k-\frac{t-1}{2}}=m_k(n)+\frac{s_k(n)-\frac{t+1}{2}}{k-\frac{t-1}{2}}.$$

For $t-1$ even and $s_k(n)\ge(t+1)/2$, this yields

$$m_{k-\frac{t-1}{2}}\left(n-\frac{t-1}{2}m_k(n)-\frac{t+1}{2}\right)=m_k(n),$$
$$s_{k-\frac{t-1}{2}}\left(n-\frac{t-1}{2}m_k(n)-\frac{t+1}{2}\right)=s_k(n)-\frac{t+1}{2}.$$

On the other hand, if $s_k(n)=(t-1)/2$, then

$$m_{k-\frac{t-1}{2}}\left(n-\frac{t-1}{2}m_k(n)-\frac{t+1}{2}\right)=m_k(n)-1,$$

$$s_{k-\frac{t-1}{2}}\left(n-\frac{t-1}{2}m_k(n)-\frac{t+1}{2}\right)=k-\frac{t+1}{2}.$$

In the both cases, the left- and right-hand sides of (3.46) are equal, which justifies the fourth line in (3.5).

Finally, making the substitution $l_1(x_1)=\frac{t}{2}\cdot\frac{l}{k}$ in (3.42), we derive

$$n_1>r(m_k(n)+1)+s_r(n_1)+r\log_2\frac{r-1/2}{r+1+s_r(n_1)},$$

where now we have $r=t/2$ (t is even). If $s_r(n_1)\leq r-2$, then $\frac{r-1/2}{r+1+s_r(n_1)}\geq\frac{1}{2}$, hence $n_1-s_r(n_1)\geq rm_k(n)$; so, we have

$$n\geq r(m_k(n)+1) \text{ if } s_r(n_1)\leq r-2. \tag{3.47}$$

On the other hand, if $s_r(n_1)=r-1$, then $r\log_2\frac{r-1/2}{r+1+s_r(n_1)}=r\log_2\frac{r-1/2}{2r}\geq -r-1$. Indeed, dividing both sides of the inequality by r and putting $\alpha=1/(2r)$ and $z(\alpha)=\log_2(1-\alpha)$, we can rewrite it in the form $z(\alpha)\geq 0$ for $0\leq\alpha\leq 1/2$. But the latter inequality is valid since $z(0)=z(1/2)=0$ and the function z is concave. Thus, $n_1>r(m_k(n)+1)-2$, and we have

$$n_1\geq r(m_k(n)+1)-1 \text{ if } s_r(n_1)=r-1. \tag{3.48}$$

Combining (3.47) and (3.48), we get $n_1\geq r(m_k(n)+1)-1$, hence $n-n_1-1\leq n-r(m_k(n)+1)$. Transforming now (3.43) and making it stronger, we obtain

$$\left(\frac{1}{2}\right)^{m_k(n)}\frac{1}{k+1-2r+s_k(n)}\leq\frac{1}{2}^{m_{k-r}(n-rm_k(n)-r)}\frac{1}{k-r+1+s_{k-r}(n-rm_k(n)-r)}. \tag{3.49}$$

Observe that

$$\frac{n-rm_k(n)-r}{k-r}=m_k(n)+\frac{s_k(n)-r}{k-r},$$

which yields that, from t even and $s_k(n)=r-1$, we have

$$m_{k-r}(n-rm_k(n)-r)=m_k(n)-1,\quad s_{k-r}(n-rm_k(n)-r)=k-r-1,$$

and for $s_k(n)\geq r$

$$m_{k-r}(n-rm_k(n)-r)=m_k(n),\quad s_{k-r}(n-rm_k(n)-r)=s_k(n)-r.$$

It is easy to see that in the both cases the left- and right-hand sides of (3.49) are equal, which justifies the fifth line in (3.5).

This completes the proof of the statement opening Section 3.3 and, in view of the last paragraph of Section 3.1, also the proof of Theorem 3.1. □

CHAPTER 4

SEARCH FOR THE GLOBAL EXTREMUM

In this chapter, we deal with problems of optimal search for the maximum of a function of one or more variables. For functional classes determined by quasi-metrics, we derive optimal deterministic and stochastic nonadaptive algorithms, as well as one-step optimal deterministic and stochastic algorithms and a sequentially optimal deterministic algorithm.

We start with a short historical survey.

The first papers on optimal (in the framework of the minimax approach) algorithms for seeking extremum were devoted to unimodal functions, i.e., functions with only one maximum. Now the paper by Kiefer [53] is well-known where an optimal adaptive algorithm for finding the extremum of a unimodal function was derived. The same result was obtained by Johnson [55]. These papers were followed by Afanas'ev [74], Afanas'ev and Novikov [77], Avriel and Wilde [66, 68], Beamer and Wilde [69, 70, 71], Chernous'ko [70a], Gal [71], Karp and Miranker [68], Kiefer [57], Kononov and Biryukova [75], Oliver and Wilde [64], Rubal'skii [82], Shapiro and Wilde [74a, b], Tarasova [84], Traub and Woźniakowski [80], Witzgall [72], etc. Some of the results were extended to multimensional case by Gal [72], Kaupe [64], Krolak [66,68], Newman [65], Sugie [64], Wilde [65], Wilde and Beightler [67], Wilde and Sanchez-Anton [71a, b].

The problem of constructing optimal methods for maximization of concave (minimization of convex) functions of one variable are discussed in Afanas'ev [74], Chernous'ko [70b], Sonnevend [77], Witzgall [72]. An optimal adaptive algorithm has been constructed only in the setting where the criterion is the accuracy of computation of the maximizing point, see Witzgall [72]. In the setting where the criterion is the accuracy of computation of the maximal value of the function, only a one-step optimal algorithm has been obtained, see Chernous'ko [70b].

Many papers are devoted to optimization of methods for finding the maximum of a concave (minimum of a convex) function of several variables. It has been shown that the minimal number of informational computations required to guarantee a given accuracy ε has the same order of magnitude as $\ln(1/\varepsilon)$, both in the case where at every step only a value of the function being maximized is computed, and in the case where the gradient (subgradient) is also computed. The lower bound of the order is attained for the centroid method (see Levin [65]) and the ellipsoid method (see Nemirovsky and Yudin [79]).

Optimal algorithms for classes of differentiable functions were studied by Ganshin [76], Chuyan [84, 86], Girlin [78], Ivanov [72c], Ivanov et al. [79], Pevnyi [82], Zaliznyak and Ligin [78], etc. Ivanov [72c] deals with the class of functions m times differentiable in $\mathbb{R}^n$ and satisfying the condition

$$\left| \frac{\partial^m}{\partial t^m} f(x+te)\bigg|_{t=t_1} - \frac{\partial^m}{\partial t^m} f(x+te)\bigg|_{t=t_2} \right| \le M|t_1-t_2|,$$

where $\|e\|=1$ and the constant M is indepent of x and e. He shows that the best accuracy attainable with a given number N of informational computations is $O(N^{-(m+1)/n})$. This result is valid in the case where at every step the values of the function and all its partial derivatives upto the order m are computed, as well as in the case where only function values are computed. Earlier, similar estimates were obtained for classes of m times differentiable functions for other problems of numerical analysis by Bakhvalov [59], Kolmogorov and Tikhomirov [59], and Vitushkin [59]. Kolmogorov and Tikhomirov [59] have constructed optimal by order algorithms for the class of analytic functions and some other functional classes.

1. ON THE CHOICE OF STARTING POINTS FOR LOCAL OPTIMIZATION METHODS

In practice of computations, it is common to use some local optimization method many times in order to find the global maximum of a function f. The starting points are selected either at random according to a given probability distribution, or in a deterministic way so that they are distributed uniformly in some sense over the domain of maximization $K \subset \mathbb{R}^n$. In this section, we discuss optimal choice of starting points.

1.1. Optimal deterministic and stochastic methods for selection of starting points

Suppose we have to select N starting points $x_1, \ldots, x_N$. It is natural to choose them in such a way that the radius $R(x^N) = \sup_{x\in K} \min_{i=1,\ldots,N} \rho(x, x_i)$ of the covering determined by the centers $x^N = (x_1, \ldots, x_N)$ (here ρ is some metric; see (2.2), Chapter 2) is minimal. This choice guarantees that the distance from an arbitrary point $x \in K$ to the closest starting point does not exceed the radius $R_N = \inf_{x_1,\ldots,x_N\in K} \sup_{x\in K} \min_{i=1,\ldots,N} \rho(x, x_i)$ of optimal covering. This distance cannot be reduced for all points from K simultaneously.

The set of starting points delivering the infimum in the expression for R_N is called an *optimal deterministic method for selection of starting points* or *optimal deterministic configuration of starting points.* Thus, optimal deterministic configurations coincide with sets of centers of optimal coverings for K (about optimal coverings see Section 2.1 of Chapter 2 and Section 1.3 of Chapter 3).

Stochastic methods for choosing starting points are determined by probability measures on $\sigma^N \in \Sigma^N$, where Σ^N is the set of all probability measures on the σ-algebra of Borel subsets of K^N. Application of a stochastic method σ^N consists in random choice of a configuration $(x_1, \dots, x_N)$ of starting points in accordance with the probability measure σ^N.

An optimal stochastic method for selection of starting points is the method determined by a probability measure that minimizes the function

$$\bar{R}(\sigma^N) \stackrel{def}{=} \sup_{x \in K} \int_{K^N} \min_{i=1,\dots,N} \rho(x, x_i) \sigma^N [dx^N]. \tag{1.1}$$

Thus, if we apply an optimal stochastic method, then the average distance from an arbitrary point $x \in K$ to the closest starting point does not exceed

$$\bar{R}_N \stackrel{def}{=} \inf_{\sigma^N \in \Sigma^N} \bar{R}(\sigma^N). \tag{1.2}$$

This average distance cannot be reduced for all points from K simultaneously.

Comparison of stochastic and deterministic methods for selection of starting points was carried out, e.g., by Aird and Rice [77] and Anderssen and Bloomfield [75], but, in all the cases they considered, at least one of the methods being compared was not optimal.

1.2. Comparison of optimal stochastic and deterministic methods for selection of starting points

We carry out the comparison for the setting where

$$K = \{u = (u^1, \dots, u^n) \mid 0 \le u^i \le 1,\ i = 1, \dots, n\},$$
$$\rho(u, v) = \max_{i=1,\dots,n} |u^i - v^i|, \quad N = p^n, \quad p \text{ is integer}. \tag{1.3}$$

Theorem 1.1 (Sukharev [71]). *Under assumptions* (1.3) *the following inequalities hold:*

$$\frac{n}{n+1} R_N \le \bar{R}_N \le R_N. \tag{1.4}$$

Moreover, the lower bound is asymptotically sharp, i.e.,

$$\lim_{p \to \infty} \frac{\frac{n}{n+1} R_{p^n}}{\bar{R}_{p^n}} = 1. \tag{1.5}$$

Proof. The right inequality in (1.4) is obvious since Σ^N contains any atomic probability measure concentrated at a single point from K^N.

In order to prove the left inequality, we use Lemma 2.4 of Chapter 2. Taking into account that in the case being considered $R_N = r_N = 1/(2p)$ and $\mu(K) = 1$, we have

$$\inf_{x^N \in K^N} \int_K \min_{i=1,\dots,N} \rho(x, x_i)dx = \frac{n}{n+1} R_N,$$

hence

$$\int_k \min_{i=1,\dots,N} \rho(x, x_i)dx \geq \frac{n}{n+1} R_N, \quad x^N \in K^N. \tag{1.6}$$

Integrating the two parts of (1.6) with respect to an arbitrary probability measure $\sigma^N \in \Sigma^N$, we get

$$\int_K \left(\int_{K^N} \min_{i=1,\dots,N} \rho(x, x_i) \sigma^N\{ds^N\} \right) ds \geq \frac{n}{n+1} R_N$$

(we can change the order of integration since the integrand is continuous). This inequality yields $\bar{R}(\sigma^N) \geq \frac{n}{n+1} R_N$. From here we immediately get the left inequality in (1.4) since $\sigma^N \in \Sigma^N$ is arbitrary.

Now we proceed to prove (1.5). For this, we introduce into consideration the function

$$\phi(n) = \frac{n^2 + \sqrt{3n^2 + 4n + 1} - 1}{2(n^2 + n)}, \quad n = 1, 2, \dots.$$

It is easily verified that $1/2 < \phi(n) < 1/2 + 1/[2(n+1)]$.

Observe that in what follows we can replace $\phi(n)$ by $1/2$. This will transform the stochastic method σ_*^N we consider below into a worse (i.e., giving a greater value of the function $\bar{R}$) method. However, the proof of (1.5) remains valid for this method, too. Our choice of a more complicated method is caused by our wish to construct as efficient an algorithm as possible (an optimal stochastic method for selection of starting points for $n > 1$ has not been obtained yet).

Set $\tau = 1/(p - 2 + 2\phi(n))$ and consider the cube

$$K(\tau) = \left\{ u = (u^1, \dots, u^n) \;\middle|\; -\frac{p\tau - 1}{2} \leq u^i \leq 1 + \frac{p\tau - 1}{2}, \; i = 1, \dots, n \right\}$$

with edge $p\tau$. Partition each of the edges of $K(\tau)$ into p parts and draw hyperplanes parallel to the coordinate hyperplanes. Thus the cube $K(\tau)$ will be partitioned into $N = p^N$ cubes with edge τ. Denote the centers of these cubes by $x_1^*, \dots, x_N^*$, and put $x_*^N = (x_1^*, \dots, x_N^*)$. Let θ be a random vector uniformly distributed over the cube $K_0 = \{\theta = (\theta^1, \dots, \theta^n) \mid |\theta^i| \leq \tau/2, \; i = 1, \dots, n\}$; let $x_i^*(\theta)$ be the projection of the vector $x_i^* + \theta$ on the cube K,

$i=1,\ldots,N$; and let $x_*^N(\theta)=(x_1^*(\theta),\ldots,x_N^*(\theta))$. Denote the probability measure corresponding to the random vector $x_*^N(\theta)$ by σ_*^N.

Compute

$$\bar{R}(\sigma_*^N)=\max_{a\in K}\int\limits_{K^N}\min_{i=1,\ldots,N}\rho(a,x_i)\sigma_*^N\{ds^N\}.$$

Fig.8 illustrates the case $n=2$, $p=5$; here K is the square bounded by a heavy line, and $K(\tau)$ is the larger square.

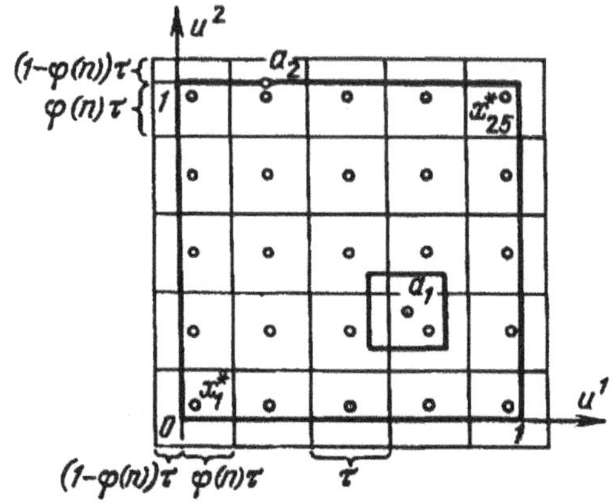

Fig. 8. To the definition of the stochastic method δ_*^N

Assume that $a=x_j^*$ and x_j^* is not the center of any of the boundary cubes. We have

$$\int\limits_{K^N}\min_{i=1,\ldots,N}\rho(x_j^*,x_i)\sigma_*^N\{dx^N\}=\frac{1}{\tau^n}\int\limits_{K_0}\rho(x_j^*,x_j^*(\theta))d\theta$$

$$=\frac{1}{\tau^n}\int\limits_{K_0}\rho(0,\theta)d\theta=\frac{n}{n+1}\cdot\frac{\tau}{2}. \tag{1.7}$$

Let now $a=a_1$, and let the coordinate cube with edge τ and center a_1 lie within K. It is easy to see (Fig.8) that in this case, too,

$$\int\limits_{K^N}\min_{i=1,\ldots,N}\rho(a_1,x_i)\sigma_*^N\{dx^N\}=\frac{n}{n+1}\cdot\frac{\tau}{2}. \tag{1.8}$$

Let, finally, the point $a=a_2$ be such that the coordinate cube with edge τ and center a_2 does not belong to K entirely. It is sufficient to consider the case where a_2 lies on the boundary of K and all the coordinates of a_2 except one are greater than 0 and less than 1. Assume without loss of generality that $a_2 = x_i^* + a'$, where $a' = (0, \ldots, 0, \tau/2 - (1-\phi(n))\tau)$. This situation is shown in Fig.8. We have

$$\int_{K^N} \min_{i=1,\ldots,N} \rho(a_2, x_i)\sigma_*^N\{dx^N\}$$

$$= \int_{\substack{|\theta^i|\le\tau/2,\ i=1,\ldots,n-1 \\ -\tau/2\le\theta^n\le\tau/2+(1-\phi(n))\tau}} \rho(\theta, a')\frac{1}{\tau^n}d\theta + (1-\phi(n))\int_{K_0'} \rho'(0,\theta')\frac{1}{\tau^{n-1}}d\theta'$$

$$= \frac{n}{n+1}\cdot\frac{\tau}{2}, \tag{1.9}$$

where

$$K_0' = \{\theta' = (\theta^1, \ldots, \theta^{n-1}) \,|\, |\theta^i| \le \tau/2, \quad i = 1, \ldots, n-1\},$$

$$\rho'(u', v') = \max_{i=1,\ldots,n-1} |u^i - v^i|, \quad u' = (u^1, \ldots, u^{n-1}), \quad v' = (v^1, \ldots, v^{n-1}).$$

Now on the basis of (1.7)-(1.9) we can conclude

$$\bar{R}(\sigma_*^N) = \frac{n}{n+1}\cdot\frac{1}{2(p-2+2\phi(n))}. \tag{1.10}$$

Taking into account that $R_{p^n} = 1/(2p)$, we get from (1.4) and (1.10)

$$1 \ge \frac{\frac{n}{n+1}R_{p^n}}{\bar{R}_{p^n}} \ge \frac{\frac{n}{n+1}R_{p^n}}{\bar{R}(\sigma_*^{p^n})} = \frac{p-2+2\phi(n)}{p}. \tag{1.11}$$

This immediately yields (1.5), which completes the proof of the theorem. □

Remark 1. It is easy to see that

$$\frac{p-2+2\phi(n)}{p} = \frac{\frac{n}{n+1}R_{p^n}}{\bar{R}(\sigma_*^{p^n})} \le \frac{\bar{R}_{p^n}}{\bar{R}(\sigma_*^{p^n})} \le 1.$$

This implies that, as p tends to infinity, the sequence of stochastic methods $\sigma_*^{p^n}$ gives a sequence of guaranteed results equivalent to the optimal one. □

Remark 2. Theorem 1.1 shows that, under assumptions (1.3), the relative advantage of using stochastic methods for selection of starting points instead of deterministic ones is not large and goes to zero as the dimension of the space grows. □

2. OPTIMAL NONADAPTIVE SEARCH FOR A FUNCTIONAL CLASS DETERMINED BY A QUASI-METRIC

In this section, we deal with the problem of optimal deterministic and stochastic nonadaptive search for the global extremum (i.e., global optimization) of a function from the class F_ρ, where ρ is an arbitrary quasi-metric. Here and in the subsequent sections of this chapter we assume that

$S(f)=\sup_{x\in K} f(x)$, $B=\mathbb{R}$, and also, unless otherwise stated, informational computations are evaluations of the function f and the only terminal operation permitted is $\tilde{\beta}(x^N, y^N)=\max_{i=1,\dots,N} y_i$ (see (3.6), Chapter 1). Thus criterion (4.14) from Chapter 1 takes the form

$$\varepsilon(x^N, f)=\sup_{x\in K} f(x)-\max_{i=1,\dots,N} f(x_i), \tag{2.1}$$

where $\varepsilon(\tilde{x}^N, f)$ is given by the second formula in (4.14), Chapter 1. In the remainder of this chapter we deal with the algorithmic classes X^N (see (3.8), Chapter 1) and $\tilde{X}^N$ (see (3.10), Chapter 1) without speaking about the terminal operation any more.

2.1. Reduction of the problem of constructing an optimal deterministic nonadaptive algorithm to the problem of optimal covering for the domain K

Denote the best guaranteed in the algorithmic class X^N result by v_N:

$$v_N=\inf_{x^N\in x^N}\sup_{f\in F_\rho}\varepsilon(x^N, f).$$

The theorem below is formulated in terms of the radii $R(x^N)$ and R_N defined by formulas (2.2) and (2.3) of Chapter 2 respectively.

Theorem 2.1 (Sukharev [71]). *The result guaranteed by an algorithm x^N on the functional class F_ρ is equal to the radius $R(x^N)$ of the covering determined by the centers $x_1, \dots, x_N$. The best guaranteed result v_N is equal to the radius of optimal covering R_N. An algorithm x_0^N is optimal by error on F_ρ in the class X^N iff the vector x_0^N determines an optimal covering of K.*

Proof. For the function h_a introduced in Lemma 2.1 of Chapter 1, we have

$$\varepsilon(x^N, h_a)=\sup_{x\in K} h_a(x)-\max_{i=1,\dots,N} h_a(x_i)=\min_{i=1,\dots,N}\rho(x_i, a).$$

Let $f\in F_\rho$, $\delta>0$, $f(x_\delta)\geq\sup_{x\in K} f(x)-\delta$. Then

$$\begin{aligned}\varepsilon(x^N, h_{x_\delta})&=\min_{i=1,\dots,N}\rho(x_i, x_\delta)=\rho(x_{i_\delta}, x_\delta)\geq f(x_\delta)-f(x_{i_\delta})\\&\geq f(x_\delta)-\max_{i=1,\dots,N} f(x_i)\geq\sup_{x\in K} f(x)-\max_{i=1,\dots,N} f(x_i)-\delta\\&=\varepsilon(x^N, f)-\delta.\end{aligned}\tag{2.3}$$

Therefore, $\sup_{a\in K}\varepsilon(x^N, h_a)\geq\varepsilon(x^N, f)-\delta$. Taking into account that $f\in F_\rho$ is arbitrary and letting δ tend to zero, we derive

$$\sup_{a\in K}\varepsilon(x^N, h_a)\geq\sup_{f\in F_\rho}\varepsilon(x^N, f).$$

The opposite inequality follows from Lemma 2.1 of Chapter 1. Combining this with (2.2), we get

$$\sup_{f\in F_\rho} \varepsilon(x^N, f) = \sup_{a\in K} \varepsilon(x^N, h_a) = \sup_{a\in K} \min_{i=1,\ldots,N} \rho(x_i, a) = R(x^N).$$

The rest of the theorem is now obvious. □

Remark 1. The proof of Theorem 2.1 shows that if $\bar{F} \subset F_\rho$ and $h_a \in \bar{F}$ for all $a \in K$, then an algorithm x_0^N optimal on the class F_ρ is also optimal on the subclass $\bar{F}$ of F_ρ. The best guaranteed results for F_ρ and $\bar{F}$ coincide. □

Remark 2. Optimal on F_ρ nonadaptive deterministic algorithms for recovery of functions and for global optimization coincide. In the both cases, the problem is reduced to constructing an optimal covering of K. Thus, if $n=1$, $K=[0,1]$, and $\rho(u,v)=M|u-v|$, then, as well as for the problem of recovering a function $f \in F_\rho$ from its values, the algorithm $(1/(2N), 3/(2N), \ldots, (2N-1)/(2N))$ is optimal in the class of all nonadaptive algorithms, and $v_N = M/(2N)$ (see the corollary of Theorem 1.2, Chapter 3). Below we show that optimal nonadaptive stochastic algorithms and sequentially optimal algorithms for the recovery problem and the global optimization problem are significantly different. □

Remark 3. Recall that, for the functional class F_ρ, an optimal nonadaptive algorithm is also optimal in the class of all adaptive algorithms (see the corollary of Theorem 5.3, Chapter 1). This fact can also be proved by another method (see Sukharev [71]), similar to the method we used for the proof of Theorem 1.3 in Chapter 3. □

2.2. Reduction of the problem of constructing an optimal stochastic nonadaptive algorithm to an antagonistic game

In view of definition (7.8), Chapter 1, obtaining an optimal by error stochastic algorithm calls for minimizing the function $\sup_{f\in F_\rho} \int_{K^N} \varepsilon(x^N, f)\sigma^N\{dx^N\}$ with respect to $\sigma^N \in \Sigma^N$. Here, as before, Σ^N is the set of all probability measures on the σ-algebra of Borel subsets of K^N (K is assumed to be either an open set or the closure of an open set).

Lemma 2.1. *Let K be a compact set, and let ρ be a quasi-metric* *continuous on* $K\times K$. *Then, for any* $\sigma^N\in\Sigma^N$,

$$\sup_{f\in F_\rho}\int\limits_{K^N}\varepsilon(X^N,f)\sigma^N\{dx^N\}=\sup_{a\in K}\int\limits_{K^N}\varepsilon(X^N,h_a)\sigma^N\{dx^N\}$$

$$=\sup_{a\in K}\int\limits_{K^N}\min_{i=1,\dots,N}\rho(X_i,a)\sigma^N\{dx^N\}.$$

Proof. Let $f\in F_\rho$. Since ρ is continuous, f is also continuous. Hence, there is a point $x_0\in K$ such that $f(x_0)=\max_{x\in K}f(x)$. Due to (2.3), we have $\varepsilon(x^N,f)\le\varepsilon(x^N,h_{x_0})$ for an arbitrary $x^N\in K^N$. Therefore,

$$\int\limits_{K^N}\varepsilon(X^N,f)\sigma^N\{dx^N\}\le\int\limits_{K^N}\varepsilon(X^N,h_{x_0})\sigma^N\{dx^N\},$$

which yields

$$\sup_{f\in F_\rho}\int\limits_{K^N}\varepsilon(X^N,f)\sigma^N\{dx^N\}\le\sup_{a\in K}\int\limits_{K^N}\varepsilon(X^N,h_a)\sigma^N\{dx^N\}.$$

The opposite inequality is obvious since $h_a\in F_\rho$ for $a\in K$. This proves the first equation of the lemma. The second equation follows from (2.2). □

Corollary. *An optimal (by error) stochastic nonadaptive algorithm of global optimization coincides with an optimal mixed strategy of the minimizing player in the antagonistic game with the payoff function* $\min_{i=1,\dots,N}\rho(x_i,a)$. *Pure strategies of the minimizing player in this game are vectors* $x^N\in K^N$, *pure strategies of the maximizing player are points* $a\in K$. *The best guaranteed result for stochastic nonadaptive algorithms is equal to the value of the game.*

2.3. Construction of an optimal stochastic nonadaptive algorithm in one dimension

In one dimension, we can give a complete solution to the problem of obtaining an optimal (by error) stochastic nonadaptive algorithm. The algorithm appears to be rather simple.

Theorem 2.2 (Sukharev [71]). *Let* $n=1$, $K=[0,1]$, $\rho(u,v)=M|u-v|$. *Then the stochastic nonadaptive algorithm* σ_0^N *assigning probabilities* $1/2$ *to the vectors*

$$x_1^N=\left(0,\frac{2}{2N-1},\dots,\frac{2N-2}{2N-1}\right),x_2^N=\left(\frac{1}{2N-1},\frac{3}{2N-1},\dots,1\right),$$

is optimal by error on the class F_ρ. The best guaranteed result is equal to $M/(4N-2)$.

Proof. In accordance with the corollary of Lemma 2.1, it is sufficient to prove that σ_0^N is an optimal mixed strategy of the minimizing player in the game with the payoff function

$$\eta(x^N, a) \overset{def}{=} \min_{i=1,\dots,N} |x_i - a|,$$

and that $s_N = 1/(4N-2)$ is the value of the game. First, we show that the strategy σ_0^N guarantees the result s_N to the minimizing player.

Let $a \in K = [0,1]$. Then $a = a' + j/(2N-1)$, $j \in \{0, 1, \dots, 2N-1\}$, $0 \leq a' \leq 1/(2N-1)$, and

$$\bar{\eta}(\sigma_0^N, a) = \frac{1}{2}\eta(x_1^N, a) + \frac{1}{2}\eta(x_2^N, a) = \frac{1}{2}a' + \frac{1}{2}\left(\frac{1}{2N-1} - a'\right) = s_N,$$

where

$$\bar{\eta}(\sigma^N, a) = \int_{K^N} \eta(x^N, a)\sigma^N\{dx\}, \quad \sigma^N \in \Sigma^N.$$

Thus, the strategy σ_0^N does guarantee the result s_N.

Denote by ω_0 the mixed strategy of the maximizing player assigning equal probabilities $1/(2N)$ to the $2N$ points $0, 1/(2N-1), 2/(2N-1), \dots, (2N-1)/(2N-1) = 1$. We show that the result ω_0 guarantees to the maximizing player is not worse than s_N.

Let $x^N = (x_1, \dots, x_i, \dots, x_N) \in K^N$, and let $j/(2N-1) < x_i < (j+1)/(2N-1)$ for some integer j. Then at least one of the following two inequalities is valid:

$$\begin{aligned} &\bar{\eta}\left(\left(x_1, \dots, \frac{j}{2N-1}, \dots, x_N\right), \omega_0\right) \leq \bar{\eta}(x^N, \omega_0), \\ &\bar{\eta}\left(\left(x_1, \dots, \frac{j+1}{2N-1}, \dots, x_N\right), \omega_0\right) \leq \bar{\eta}(x^N, \omega_0), \end{aligned} \tag{2.4}$$

where

$$\bar{\eta}(x^N, \omega) = \int_K \eta(x^N, a)\omega\{da\}, \quad \omega \in \sum,$$

and $\sum$ is the set of all probability measures on the σ-algebra of Borel subsets of K. Indeed, we have

$$\bar{\eta}(x^N, \omega_0) = \frac{1}{2N}\sum_{\alpha=0}^{2N-1} \eta\left(x^N, \frac{\alpha}{2N-1}\right) = \frac{1}{2N}\sum_{\alpha=0}^{2N-1} \min\left|x_\beta - \frac{\alpha}{2N-1}\right|.$$

Put

$$M(i) = \left\{\frac{m}{2N-1} \;\middle|\; \left|x_i - \frac{m}{2N-1}\right| = \min_{j=1,\dots,N}\left|x_j - \frac{m}{2N-1}\right|, \right.$$
$$\left. 0 \leq m \leq 2N-1\right\};$$

that is, $M(i)$ contains all the points from $\left\{\frac{0}{2N-1}, \frac{1}{2N-1}, \ldots, \frac{2N-1}{2N-1}\right\}$ for which x_i is the closest point from the set $\{x_1, \ldots, x_N\}$. Then, if there are more points from $M(i)$ to the left of x_i than to the right, the first inequality in (2.4) holds, otherwise the second one holds. If there are as many points from $M(i)$ to the left of x_i as to the right, then the two inequalities (2.4) are both valid.

Repeating this argument for all $i=1,\ldots,N$ gives

$$\bar{\eta}\left(\left(\frac{j_1}{2N-1}, \ldots, \frac{j_N}{2N-1}\right), \omega_0\right) \leq \bar{\eta}(x^N, \omega_0),$$

where $j_1, \ldots, j_N \in \{0, 1, \ldots, 2N-1\}$. Finally,

$$\bar{\eta}\left(\left(\frac{j_1}{2N-1}, \ldots, \frac{j_N}{2N-1}\right), \omega_0\right)$$
$$= \frac{1}{2N} \sum_{\alpha=0}^{2N-1} \min_{\beta=1,\ldots,N} \left| \frac{j_\beta}{2N-1} - \frac{\alpha}{2N-1} \right| \geq \frac{1}{2N} \cdot \frac{N}{2N-1} = s_N.$$

Since $x^N \in K^N$ is arbitrary, the latter two inequalities imply that the result guaranteed by the strategy ω_0 is not worse than s_N. □

Remark 1. The result $M/(4N-2)$ guaranteed by the optimal (by error) stochastic nonadaptive algorithm is almost twice as good as the result $M/(2N)$ guaranteed by the optimal (by error) deterministic nonadaptive algorithm. □

Remark 2. For the problem of recovering a function from its values, the stochastic nonadaptive algorithm σ_0^N is not optimal by error. To see this, it suffices to consider the function $f(x) = \min\{x, |x-2/3|\}$ on the segment $K=[0,1]$ for $N=2$. It is easy to verify that for this function $\int_{K^N} \varepsilon(x^N, f)\sigma_0^N\{dx^N\} = M/3$, where $\varepsilon(x^N, f)$ is given by (1.8), Chapter 3. Therefore, the result guaranteed by the algorithm σ_0^N is not better than $M/3$, while the optimal (by error) deterministic nonadaptive algorithm $x_0^N = (1/4, 3/4)$ guarantees the result $M/4$. □

2.4. Comparison of the best guaranteed results in the classes of deterministic and stochastic nonadaptive algorithms in higher dimensions

Theorem 2.3. *Let $K = \{u = (u^1, \ldots, u^n) | 0 \leq u^i \leq 1, i = 1, \ldots, n\}$, let $\rho(u,v) = M \max_{i=1,\ldots,N} |u^i - v^i|$, and let $N = p^n$, where p is integer. Then the best* *guaranteed results in the classes K^N and Σ^N of deterministic and stochastic nonadaptive algorithms respectively are related in the following way:*

$$\frac{n}{n+1}\min_{x^N\in K^N}\max_{f\in F_\rho}\varepsilon(x^N,f)\le\min_{\sigma^N\in\Sigma^N}\max_{f\in F_\rho}\int_{K^N}\varepsilon(x^N,f)\sigma^N\{dx^N\}$$
$$\le\min_{x^N\in K^N}\max_{f\in F_\rho}\varepsilon(x^N,f).$$

Moreover, the lower bound is asymptotically sharp, i.e.,

$$\lim_{p\to\infty}\frac{\frac{n}{n+1}\min_{x^{p^n}\in K^{p^n}}\max_{f\in F_\rho}\varepsilon(x^{p^n},f)}{\min_{\sigma^{p^n}\in\sum^{p^n}}\max_{f\in F_\rho}\int_{K^{p^n}}\varepsilon(x^{p^n},f)\sigma^{p^n}\{dx^{p^n}\}}=1.$$

Proof. Noticing that the function ρ in (1.3) and the function ρ in the theorem differ only by a constant factor M and taking Theorem 2.1 and Lemma 2.1 into account, we see that Theorem 2.3 is just Theorem 1.1 formulated in other terms. □

Remark 1. The sequence of the results guaranteed by the stochastic nonadaptive algorithms (stochastic methods in terms of Theorem 1.1) $\sigma_*^{p^n}$ which were constructed in the course of proving Theorem 1.1 is equivalent to the sequence of the best guaranteed results as $p\to\infty$ (see Remark 1 after Theorem 1.1):

$$\max_{f\in F_\rho}\int_{k^{p^n}}\varepsilon\left(x^{p^n},f\right)\sigma_*^{p^n}\left\{dx^{p^n}\right\}\sim\min_{\sigma^{p^n}\in\sum^{p^n}}\max_{f\in F_\rho}\int_{K^{p^n}}\varepsilon\left(x^{p^n},f\right)\sigma^{p^n}\left\{dx^{p^n}\right\}.$$

□

Remark 2. Theorem 2.3 shows that, in this case, the relative advantage of using stochastic algorithms instead of deterministic ones is not large. It can be expressed by the coefficient $n/(n+1)$ which tends to 1 as the dimension of the space tends to infinity (cf. Remark 2 after Theorem 1.1). □

3. REDUCTION OF THE PROBLEM OF CONSTRUCTING A SEQUENTIALLY OPTIMAL ALGORITHM FOR A FUNCTIONAL CLASS DETERMINED BY A QUASI-METRIC TO A SERIES OF PROBLEMS OF OPTIMAL COVERING

In this section, we clear up the structure of sequentially optimal algorithms. We show that constructing an algorithm sequentially optimal on

F_ρ can be reduced to solving a series of special type problems of optimal covering. When speaking about the functional class F_ρ, we will assume that K is a compact set and the quasi-metric ρ is continuous on $K \times K$ without mentioning this in the formulations of the lemmas explicitly. We also assume that $F = F_\rho$ unless otherwise stated. In particular, we have

$$F(z^i) = \{f \in F_\rho | f(x_j) = y_j, \quad j = 1, \ldots, i\},$$

and in all the constructions below the situation x^i is assumed to be realizable, that is, $F(z^i) \neq 0$.

3.1. Auxiliary statements

Denote

$$\bar{y}_i = \max_{j=1,\ldots,i} y_j,$$

$$K_i(t) = \{x | \phi_{2i}(x) \leq t\} = \bigcup_{j=1}^{i} \{x | y_j + \rho(x, x_j) \leq t\} = \bigcup_{j+1}^{i} S(x_j, t - y_j),$$

$$h_i(x) = \min\{\phi_{2i}(x), \bar{y}_i\},$$

where the upper envelope ϕ_{2i} for the functions from $F(z^i)$ is given by (1.3), Chapter 2, and $S(v, R) = \{u | \rho(u, v) \leq R\}$ is the ρ-ball of radius R with center v (if $R < 0$, then $S(v, R) = \emptyset$)[1]. We will also use the notation $CA = K \setminus A$ for $A \subset K$ and denote the closure of A by $\bar{A}$. It is easy to see that

$$h_i(x) = \begin{cases} \phi_{2i}(x) & \text{for } x \in K_i(\bar{y}_i), \\ \bar{y}_i & \text{for } x \in CK_i(\bar{y}_i). \end{cases}$$

Due to Lemma 2.7 of Chapter 1, we have $h_i \in F(z^i)$. Let $a \in CK_i(\bar{y}_i + r) \neq \emptyset$, $r \geq 0$. Then

$$\rho(a, x_j) \geq \bar{y}_i - y_j + r, \quad j - 1, \ldots, i. \tag{3.1}$$

Note that at the points from the set $C\bar{K}_i(\bar{y}_i + r)$ (and only at them) functions from $F(z^i)$ can take values $y \geq \bar{y}_i + r$. Denote $g_i(x) = \max\{h_i(x), \bar{y}_i + r - \rho(a, x)\}$.

Lemma 3.1. *We have*

$$g_i(x) = \begin{cases} \bar{y}_i + r - \rho(a, x), & \rho(a, x) \leq r, \\ h_i(x), & \rho(a, x) \geq r, \end{cases} \qquad g_i \in F(z^i).$$

---

[1] As well as for the functions ϕ_{1i} and ϕ_{2i} introduced earlier, for simplicity of notation we do not list $z^i = (x_1, \ldots, x_i, y_1, \ldots, y_i)$ among the arguments and the subscripts of $K_i(t)$ and $h_i(t)$. The subscript i reminds the reader of the dependence on z^i. The same goes for their functions $g_i(x)$, $R_i(x_{i+1}, \ldots, x_N)$ and R_{iN} introduced below.

Proof. For $\rho(a,x)\le r$ we have $\bar{y}_i+r-\rho(a,x)\ge\bar{y}_i\ge h_i(x)$, thus $g_i(x)=\bar{y}_i+r-\rho(a,x)$. If $\rho(a,x)\ge r$, then

$$\bar{y}_i-(\bar{y}_i+r-\rho(a,x))\ge 0, \tag{3.2}$$

$$\begin{aligned}\phi_{2i}(x)-(\bar{y}_i+r-\rho(a,x))&=\min_{j=1,\dots,i}\{y_i+\rho(x,x_j)\}-(\bar{y}_i+r-\rho(a,x))\\ &=y_s+\rho(x,x_s)-(\bar{y}_i+r-\rho(a,x))\ge y_s-\bar{y}_i-r+\rho(a,x_s)\ge 0,\end{aligned} \tag{3.3}$$

where the last inequality holds due to (3.1) and the next to last - due to the properties of a quasi-metric (2.3) and (2.4), Chapter 1. Inequalities (3.2) and (3.3) yield $h_i(x)\ge\bar{y}_i+r-\rho(a,x_i)$, hence $g_i(x)=h_i(x)$.

Finally, due to Lemma 2.7 of Chapter 1 we have $g_i\in F$. We also have $g_i(x_j)=h_i(x_j)=y_j$ since $\rho(a,x_j)\ge r$, $j=1,\dots,i$ (see (3.1)). Therefore, $g_i\in F(z^i)$.

Denote

$$R_i(x_{i+1},\dots,x_N)=\min\left\{r\;\middle|\;r\ge 0,\ \bigcup_{j=i+1}^{N}S(x_j,r)\supset\overline{CK_i(\bar{y}_i+r)}\right\}.$$

To make sure that $R_i(x_{i+1},\dots,x_N)$ exists and, incidentally, to obtain its formula, we perform the following transformations:

$$\begin{aligned}R_i(x_{i+1},\dots,x_N)&=\min\left\{r\;\middle|\;r\ge 0,\ \bigcup_{j=i+1}^{N}S(x_j,r)\cup K(\bar{y}_i+r)\supset K\right\}\\ &=\min\left\{r\;\middle|\;r\ge 0,\ \bigcup_{j=i+1}^{N}S(x_j,r)\cup\left[\bigcup_{j=1}^{i}S(x_j,\bar{y}_i-y_j+r)\right]\supset K\right\}\\ &=\min\left\{r\;\middle|\;r\ge 0,\ \sup_{x\in K}\min\left[\min_{j=i+1,\dots,N}\rho(x,x_j),\right.\right.\\ &\qquad\left.\left.\min_{j=1,\dots,i}(\rho(x,x_j)-\bar{y}_i+y_j)\right]\ge r\right\}\\ &=\sup_{x\in K}\min\left\{\min_{j=i+1,\dots,N}\rho(x,x_j),\ \min_{j=1,\dots,i}(\rho(x,x_j)-\bar{y}_i+y_j)\right\}.\end{aligned} \tag{3.4}$$

Note that the last expression is nonnegative since for $x=x_{i_0}$, with z^i being realizable, we have $\rho(x,x_j)-\bar{y}_i+y_j\ge 0$, $j=1,\dots,i$, where $i_0\in\{1,\dots,i\}$ is determined by the condition $y_{i_0}=\bar{y}_i$.

Now we prove the following (almost trivial) statement.

Lemma 3.2. *Let a vector $(x_{i+1},\dots,x_N)$ determine a covering of $A\subset K$ of radius r, i.e.,*

$$\sup_{x\in A}\min_{j=i+1,\dots,N}\rho(x,x_j)=r. \tag{3.5}$$

Then there is a point $a \in \bar{A}$ *such that* $\min_{j=i+1,\dots,N} \rho(x_j, a) = r$.

Proof. By virtue of (3.5), for any positive integer k there is a point $a_k \in A$ such that $\min_{j=i+1,\dots,N} \rho(x_j, a_k) > r - 1/k$. The sequence of a_k has a limit point $a \in \bar{A}$. For this point we have

$$\rho(x_j, a) \geq \rho(x_j, a_k) - \rho(a, a_k) - \rho(a, a_k) > r - 1/k - \rho(a, a_k),$$
$$j = i+1, \dots, N.$$

Since k is arbitrary, the latter inequality yields $\min_{j=i+1,\dots,N} \rho(x_j, a) \geq r$. The opposite inequality is also valid since the radius of the covering of $\bar{A}$ determined by the centers $x_1, \dots, x_N$, as well as the radius of the covering of A, is equal to r. □

The following result is an analogue of Theorem 2.1 for the class $F(z^i)$. For estimating efficiency of nonadaptive algorithms $(x_{i+1}, \dots, x_N)$ we use the criterion $\varepsilon(x^i, x_{i+1}, \dots, x_N, f)$.

Lemma 3.3. *The result guaranteed by a nonadaptive algorithm* $(x_{i+1}, \dots, x_N)$ *on the subclass* $F(z^i)$ *of the class* F_ρ *is equal to* $R_i(x_{i+1}, \dots, x_N)$, *i.e.*

$$\sup_{f \in F(z^i)} \varepsilon(x^i, x_{i+1}, \dots, x_N, f) = R_i(x_{i+1}, \dots, x_N).$$

Proof. Let

$$f \in F(z^i), \quad \max_{x \in K} f(x) = f(x_0), \quad x_0 \in K, \quad r \geq R_i(x_{i+1}, \dots, x_N).$$

If $x_0 \in K_i(\bar{y}_i + r)$, then $x_0 \in S(x_j, \bar{y}_i - y_j + r)$, $j \in \{1, \dots, i\}$, and, therefore, $f(x_0) \leq f(x_j) + \rho(x_0, x_j) \leq y_j + (\bar{y}_i - y_j + r) = \bar{y}_i + r$. Hence,

$$\varepsilon(x^i, x_{i+1}, \dots, x_N, f) = f(x_0) - \max\{\bar{y}_i, f(x_{i+1}), \dots, f(x_N)\}$$
$$\leq f(x_0) - \bar{y}_i \leq r.$$

On the other hand, if $x_0 \in CK_i(\bar{y}_i + r)$, then $x_0 \in S(x_j, r)$, $j \in \{i+1, \dots, N\}$, and, therefore,

$$\varepsilon(x^i, x_{i+1}, \dots, x_N, f) \leq f(x_0) - f(x_j) \leq \rho(x_0, x_j) \leq r.$$

Hence, in any case we have $\varepsilon(x^i, x_{i+1}, \dots, x_N, f) \leq r$. Making use of the fact that $f \in F(z^i)$ and $r \geq R_i(x_{i+1}, \dots, x_N)$ are arbitrary, we get

$$\sup_{f \in F(z^i)} \varepsilon(x^i, x_{i+1}, \dots, x_N, f) \leq R_i(x_{i+1}, \dots, x_N). \tag{3.6}$$

Let now $r < R_i(x_{i+1}, \dots, x_N)$. Then the radius of the covering of $CK_i(\bar{y}_i + r)$ with the centers $x_{i+1}, \dots, x_N$ is greater than r, and, by Lemma 3.2, there is $a \in \overline{CK_i(\bar{y}_i + r)}$ such that

$$\min_{j=i+1,\dots,N} \rho(x_j, a) > r. \tag{3.7}$$

For these a and r, consider the function g_i defined before Lemma 3.1. In view of (3.7) and Lemma 3.1,

$$\varepsilon(x^i, x_{i+1}, \ldots, x_N, g_i) = g_i(a) - \max\{\bar{y}_i, g_i(x_{i+1}), \ldots, g_i(x_N)\} \\ = \bar{y}_i + r - \bar{y}_i = r. \tag{3.8}$$

Since $r < R_i(x_{i+1}, \ldots, x_N)$ is arbitrary, (3.8) yields

$$\sup_{f \in F(z^i)} \varepsilon(x^i, x_{i+1}, \ldots, x_N, f) \geq R_i(x_{i+1}, \ldots, x_N). \tag{3.9}$$

Inequalities (3.6) and (3.9) prove the lemma. □

3.2. Sequentially optimal (by error) algorithm

Denote

$$R_{iN} = \min_{x_{i+1}, \ldots, x_N \in K} R_i(x_{i+1}, \ldots, x_N). \tag{3.10}$$

The minimum on the right-hand side of (3.10) is attained since K is a compact set and the function $R_i(x_{i+1}, \ldots, x_N)$ is continuous with respect to the set of variables $x_{i+1}, \ldots, x_N$ (its continuity easily follows from (3.4)). We can say that R_{iN} is the minimal number r such that the radius of optimal covering of $CK_i(\bar{y}_i + r)$ with $N-i$ centers does not exceed r.

Theorem 3.1 (Sukharev [72]). *Assume that K is a compact set and ρ is a quasi-metric continuous on $K \times K$. Let $\tilde{x}_1^0$ be any center of an optimal covering of K with N centers, and let $\tilde{x}_{i+1}^0(z^i)$ be any center of a covering of $CK_i(\bar{y}_i + R_{iN})$ with $N-i$ centers such that its radius is not greater than R_{iN}, $i = 1, \ldots, N-1$ (in other words, $\tilde{x}_{i+1}^0(z^i)$ is one of the points delivering the minimum in* (3.10)).

Then the algorithm $\tilde{x}_0^N = (\tilde{x}_1^0, \ldots, \tilde{x}_N^0)$ is sequentially optimal (by error) on the class F_ρ, and the result guaranteed by this algorithm after i steps in a situation z^i is equal to R_{iN}.

Proof. Selection of the above point $\tilde{x}_1^0$ for the next step of a sequentially optimal algorithm is justified by the fact that a sequentially optimal algorithm is optimal in $\tilde{X}^N$ (the structure of an algorithm optimal in $\tilde{X}^N$ is given by Theorem 2.1 and Remark 3 after it).

By construction, the result guaranteed by the algorithm $\tilde{x}_0^N$ after i steps in a situation z^i is equal to R_{iN}. In accordance with the definition of sequentially optimal algorithm (see (6.6), Chapter 1), to prove the theorem it suffices to show that

$$\inf_{\tilde{x}_{i+1}, \ldots, \tilde{x}_N} \sup_{f \in F(z^i)} \varepsilon(x^i, \tilde{x}_{i+1}, \ldots, \tilde{x}_N, f) = R_{iN}. \tag{3.11}$$

Let $\tilde{x}_{i+1},\ldots,\tilde{x}_N$ be arbitrary mappings of the form (3.1), Chapter 1, let $r<R_{iN}$, and let the vector $(x^i,x_{i+1},\ldots,x_N)$ be obtained through applying (see (3.2), Chapter 1) the algorithm $(x^i,\tilde{x}_{i+1},\ldots,\tilde{x}_N)$ to the function h_i defined at the beginning of Section 3.1. Then we have $r<R_i(x_{i+1},\ldots,x_N)$. Define a and g_i in the same way as in the proof of Lemma 3.3. Observe that $g_i(x_j)=h_i(x_j)$, $j=1,\ldots,N$, hence application of the algorithm $(x^i,\tilde{x}_{i+1},\ldots,\tilde{x}_N)$ to the function g_i gives the same vector $(x^i,x_{i+1},\ldots,x_N)$ as its application to the function h_i. Therefore, taking (3.8) into account, we get

$$\varepsilon(x^i,\tilde{x}_{i+1},\ldots,\tilde{x}_N,g_i)=\varepsilon(x^i,x_{i+1},\ldots,x_N,g_i)=r.$$

Since $\tilde{x}_{i+1},\ldots,\tilde{x}_N$ are arbitrary, we have

$$\inf_{\tilde{x}_{i+1},\ldots,\tilde{x}_N}\ \sup_{f\in F(z^i)} \varepsilon(x^i,\tilde{x}_{i+1},\ldots,\tilde{x}_N,f)\geq r,$$

and since $r<R_{iN}$ is arbitrary, we also have

$$\inf_{\tilde{x}_{i+1},\ldots,\tilde{x}_N}\ \sup_{f\in F(z^i)} \varepsilon(x^i,\tilde{x}_{i+1},\ldots,\tilde{x}_N,f)\geq R_{iN}. \tag{3.12}$$

On the other hand, Lemma 3.3 immediately yields

$$\inf_{x_{i+1},\ldots,x_N}\ \sup_{f\in F(z^i)} \varepsilon(x^i,x_{i+1},\ldots,x_N,f)\geq R_{iN}. \tag{3.13}$$

Apparently,

$$\inf_{\tilde{x}_{i+1},\ldots,\tilde{x}_N}\ \sup_{f\in F(z^i)} \varepsilon(x^i,\tilde{x}_{i+1},\ldots,\tilde{x}_N,f)$$
$$\leq \inf_{x_{i+1},\ldots,x_N}\ \sup_{f\in F(z^i)} \varepsilon(x^i,x_{i+1},\ldots,x_N,f).$$

From here, making use of (3.13), we derive the inequality opposite to (3.12), which completes the proof of (3.11).□

Remark. It is easy to verify that, for any y_1, the point $\tilde{x}_2^0(z^1)$ is one of the $N-1$ centers (different from the center $\tilde{x}_1^0$) of an optimal covering of K with N centers. Moreover, if $y_{i+1}=\bar{y}_i$, then the point $\tilde{x}_{i+2}^0(z^{i+1})$ is one of the $N-i-1$ centers (different from the center $\tilde{x}_{i+1}^0(z^i)$) of the optimal covering with $N-i$ centers that was constructed when selecting $\tilde{x}_{i+1}^0(z^i)$. In this case, the best guaranteed results after i steps and after $i+1$ steps coincide. We can say that $y_{i+1}=\bar{y}_i$ is the worst for "the computer" behaviour of the function f at the $(i+1)$st step.□

Theorem 3.1 reduces the problem of constructing a sequentially optimal algorithm to a series of N special type problems of optimal covering. Solutions to these problems for some special cases are given in Section 4.

3.4. Sequentially optimal (counting informational computations) algorithm

Assume now that the problem should be solved with a prescribed accuracy ε. We construct a sequentially optimal (counting informational computations) global optimization algorithm and give all the necessary proofs.

We can select any center of any covering of K with the minimal number of ρ-balls of radius ε as the point of the first informational computation. This follows directly from the definition of sequentially optimal algorithm (see (6.10), Chapter 1).

Suppose we have performed i informational computations. In view of (3.11),

$$\inf_{\tilde{x}_{i+1},\ldots,\tilde{x}_{i+s}} \sup_{f\in F(z^i)} \varepsilon(x^i,\tilde{x}_{i+1},\ldots,\tilde{x}_{i+s},f)=R_{i,i+s}.$$

Making use of (6.8) from Chapter 1 and (3.10), we can rewrite $N_\varepsilon(z^i)$ defined by (6.9), Chapter 1, in the following form:

$$\begin{aligned} N_\varepsilon(z^i) &= \min\{s|R_{i,i+s}\le\varepsilon\}\\ &=\min\{s|R_i(x_{i+1},\ldots,x_{i+s})\le\varepsilon \text{ for some } x_{i+1},\ldots,x_{i+s}\in K\}\\ &=\min\left\{s\;\Big|\;\bigcup_{j=i+1}^{i+s} S(x_j,\varepsilon)\supset CK_i(\bar{y}_i+\varepsilon) \text{ for some } x_{i+1},\ldots,x_{i+s}\in K\right\}. \end{aligned}$$

If $x_{i+1},\ldots,x_{i+N_\varepsilon(z^i)}$ are such that $\bigcup_{j=i+1}^{i+N_\varepsilon(z^i)} S(x_j,\varepsilon)\supset CK_i(\bar{y}_i+\varepsilon)$, then $R_i(x_{i+1},\ldots,x_{i+N_\varepsilon(z^i)})\le\varepsilon$, and, due to Lemma 3.3,

$$\sup_{f\in F(z^i)} \varepsilon(x^i,x_{i+1},\ldots,x_{i+N_\varepsilon(z^i)},f)\le\varepsilon. \tag{3.14}$$

Thus, to find the point of the $(i+1)$st informational computation, we have to construct the set $CK_i(\bar{y}_i+\varepsilon)$ and find the minimal s such that the set $CK_i(\bar{y}_i+\varepsilon)$ can be covered with s ρ-balls of radius ε.

If $CK_i(\bar{y}_i+\varepsilon)=\emptyset$, then the problem has already been solved with the prescribed accuracy ε and we can stop the process of computations.

Observe that this problem is simpler than that with a fixed N. In the case being considered, we need not solve the optimization problem of finding R_{iN} at the $(i+1)$st step. For all i's, the role of R_{iN} is played by the constant ε.

4. SPECIFIC COMPUTATIONAL ALGORITHMS

In this section, we concretize the results of the previous section for one-dimensional problems. For higher dimensions, where direct implementation

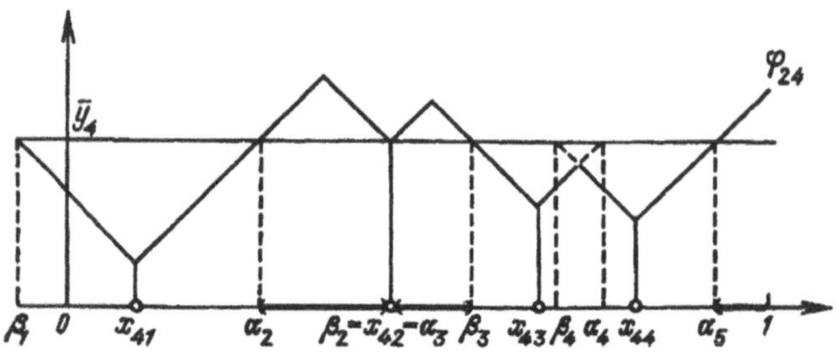

Fig. 9. Construction of a sequentially optimal (by error) algorithm

of a sequentially optimal algorithm is difficult, we show how to use the knowledge of its structure.

4.1. Sequentially optimal (by error) algorithm for the class of one-variable functions satisfying the Lipschitz condition

Assume that the number N of informational computations is fixed, $n=1$, $K=[0,1]$, and $\rho(u,v)=M|u-v|$. Then $S(x,r)=[x-r/M, x+r/M]$ is a segment of the length $2r/M$. In view of Theorem 3.1,

$$x_1 \in \left\{ \frac{1}{2N}, \frac{3}{2N}, \ldots, \frac{2N-1}{2N} \right\}, \tag{4.1}$$

and a sequentially optimal algorithm can select any of the points (4.1) for the first informational computation.

Suppose we have performed i computations at $x_1, \ldots, x_i$; let $y_j = f(x_j)$, $j=1,\ldots,i$, and $z^i=(x^i, y^i)$; let $x_{i1},\ldots,x_{ii}$ be the ordered permutation of the points $x_1,\ldots,x_i$, and let $y_{ij}=f(x_{ij})$. Denote the least and the greatest roots of the equation $y_{ij}+M|x-x_{ij}|=\bar{y}_i$ by β_j and α_{j+1} respectively; i.e.,

$$\beta_j = x_{ij} - \frac{\bar{y}_i - y_{ij}}{M}, \quad \alpha_{j+1} = x_{ij} + \frac{\bar{y}_i - y_{ij}}{M}, \quad j=1,\ldots,i$$

(see Fig.9 illustrating the case $i=4$). Then we have

$$CK_i(\bar{y}_i) = [0,\beta_1) \cup \left[\bigcup_{j=2}^{i} (\alpha_j, \beta_j) \right] \cup (\alpha_{i+1}, 1]$$

(heavy line in Fig.9), and

$$CK_i(\bar{y}_i + r) = \left[\, 0, \beta_1 - \frac{r}{M} \right) \cup \left[\bigcup_{j=2}^{i} \left(\alpha_j + \frac{r}{M}, \beta_j - \frac{r}{M} \right) \cup \left(\alpha_{i+1} + \frac{r}{M}, 1 \right) \right].$$

Here $[0,\beta) \overset{def}{=} \emptyset$ if $\beta \le 0$; $(\alpha,\beta) \overset{def}{=} \emptyset$ if $\alpha \ge \beta$; $(\alpha,1] \overset{def}{=} \emptyset$ if $\alpha \ge 1$. We also denote $l_j = \beta_j - \alpha_j$, $j=1,\ldots,m$.

Apparently, the minimal number of segments of the length $2r/M$ required to cover the set $[0,\beta_1-r/M)$ is equal to $\max\{(\beta_1-r/M)/(2r/M),0\}$. Similarly, for the set $(\alpha_j+r/M,\beta_j-r/M)$ this number is equal to $\max\{(l_j-2r/M)/(2r/M),0\}$, and for the set $(\alpha_{i+1}+r/M,1]$ it is $\max\{(1-\alpha_{i+1}-r/M)/(2r/M),0\}$. Therefore, the inequality $r\geq R_{iN}$ is equivalent to the inequality

$$\left\lceil \max\left\{\frac{\beta_1-r/M}{2r/M},0\right\}\right\rceil+\sum_{j=2}^{i}\left\lceil \max\left\{\frac{l_j/2-r/M}{r/M},0\right\}\right\rceil + \left\lceil \max\left\{\frac{1-\alpha_{i+1}-r/M}{2r/M},0\right\}\right\rceil \leq N-i, \tag{4.2}$$

which yields

$$R_{iN}=\min\left\{r\,\middle|\, r>0,\ \left\lceil \max\left\{\frac{\beta_1-r/M}{2r/M},0\right\}\right\rceil+\sum_{j=2}^{i}\left\lceil \max\left\{\frac{l_j/2-r/M}{r/M},0\right\}\right\rceil + \left\lceil \max\left\{\frac{1-\alpha_{i+1}-r/M}{2r/M},0\right\}\right\rceil \leq N-i\right\} \tag{4.3}$$

The algorithm for finding R_{iN} is based on (4.3) and the following consideration: at least one of the numbers

$$\frac{\beta_1-r/M}{2r/M},\ \frac{l_2-r/M}{r/M},\ldots,\ \frac{l_i/2-r/M}{r/M},\ \frac{1-\alpha_{i+1}-r/M}{2r/M} \tag{4.4}$$

for $r=R_{iN}$ is integer and nonnegative; otherwise there would exist a smaller $r>0$ satisfying (4.2). The algorithm consists of two stages:

1. Find all r's such that the numbers (4.4) take the integer values $N-i,N-i-1,\ldots,1,0$. Enumerate all the obtained r's in the increasing order: $r_1<r_2<\cdots<r_s$, where $s\leq(i+1)(N-i+1)$.

2. Check inequality (4.2) for $r=r_t$ in the order of increase of t. If it holds for the first time for $t=t_0$, then $R_{iN}=r_{t_0}$.

After R_{iN} has been found, Theorem 3.1 gives us possible points of the $(i+1)$st informational computation for a sequentially optimal algorithm:

$$x_{i+1}\in\{(2k_1-1)R_{iN}/M \mid (2k_1-1)R_{iN}/M<\beta_1\} \cup\left[\bigcup_{j=2}^{i}\{\alpha_j+2k_jR_{iN}/M \mid \alpha_j+2k_jR_{iN}/M<\beta_j\}\right] \cup\{1-(2k_{i+1}-1)R_{iN}/M \mid 1-(2k_{i+1}-1)R_{iN}/M>\alpha_{i+1}\}, \tag{4.5}$$

where the numbers $k_1,\ldots,k_{i+1}$ take positive integer values. Any point from the set (4.5) can be chosen by a sequentially optimal algorithm as the point of the $(i+1)$st informational computation.

The set (4.5) may happen to be empty. This would mean that, in the case of the "worst" behaviour of the function, the rest $N-i$ informational computations cannot improve the guaranteed result, and so any point from $K=[0,1]$ can be chosen as x_{i+1}. However, it makes sense to choose

$$x_{i+1}\in\{(2k_1-1)R_{iN}/M \mid (2k_1-1)R_{iN}/M\le\beta_1\}$$
$$\cup\left[\bigcup_{j=2}^{i}\{\alpha_j+2k_jR_{iN}/M \mid \alpha_j+2k_jR_{iN}/M\le\beta_j\}\right]$$
$$\cup\{1-(2k_{i+1}-1)R_{iN}/M \mid 1-(2k_{i+1}-1)R_{iN}/M\le\alpha_{i+1}\}. \tag{4.6}$$

Observe that the methods of covering $CK_i(\bar{y}_i+R_{iN})$ "from left to right" (corresponding to the third line) do not exhaust all the methods satisfying the assumptions of Theorem 3.1, and so (4.5) does not describe the whole set of sequentially optimal algorithms.

We could "symmetrize" the covering methods corresponding to (4.5) and (4.6) in the following way. Let

$$n_1^i=\left\lceil\max\left\{\frac{\beta_1-R_{iN}/M}{2R_{iN}/M},0\right\}\right\rceil,\ n_j^i=\left\lceil\max\left\{\frac{l_j/2-R_{iN}/M}{R_{iN}/M},0\right\}\right\rceil,$$
$$j=2,\dots,i,\quad n_{i+1}^i=\left\lceil\max\left\{\frac{1-\alpha_{i+1}-R_{iN}/M}{2R_{iN}/M},0\right\}\right\rceil. \tag{4.7}$$

It is easily seen that an algorithm with

$$x_{i+1}\in\left\{(2k_1-1)\frac{\beta_1}{2n_1^i+1}\,\middle|\,k_1=1,\dots,n_1^i\right\}$$
$$\cup\left[\bigcup_{j=2}^{i}\left\{\alpha_j+k_j\frac{\beta_j-\alpha_j}{n_j^i+1}\,\middle|\,k_j=1,\dots,n_j^i\right\}\right]$$
$$\cup\left\{\alpha_{i+1}+2k_{i+1}\frac{1-\alpha_{i+1}}{2n_{i+1}^i+1}\,\middle|\,k_{i+1}=1,\dots,n_{i+1}^i\right\} \tag{4.8}$$

is sequentially optimal by error (if $n_j^i=0$, the corresponding set is considered to be empty).

4.2. One-step optimal algorithm for the class of one-variable functions satisfying the Lipschitz condition

In view of (4.1), for $N=1$ we choose

$$x_1=1/2. \tag{4.9}$$

The point x_{i+1} can be selected in accordance with the version (4.7), (4.8) of a sequentially optimal algorithm (in (4.3) we put $N-i=1$). It is easily seen that

$$n_r^i=1,\quad n_j^i=0,\quad j\ne r, \tag{4.10}$$

if the maximal value among $\beta_1, l_1/2, \ldots, l_i/2, 1-\alpha_{i+1}$ is the rth one. If the maximal value is not unique, then we have $n_1^i = \cdots = n_{i+1}^i = 0$ (which corresponds to the case when the set (4.5) is empty), and any point from $K=[0,1]$ can be chosen as x_{i+1}. However, even in this case we will retain (4.10) with r being the ordinal order number of some maximal value (say, the minimum of such numbers). The results of Section 4.1 immediately imply that the algorithm (4.8)–(4.10) is one-step optimal. Note that in the case (4.10) the formula (4.8) can be rewritten as follows:

$$x_{i+1} = \begin{cases} \arg\max_{x\in K} \phi_{2i}(x), & \arg\max_{x\in K} \phi_{2i}(x) \notin \{0,1\}, \\ \beta_1/3, & \arg\max_{x\in K} \phi_{2i}(x) = 0, \\ 1-(1-\alpha_{i+1})/3, & \arg\max_{x\in K} \phi_{2i}(x) = 1, \end{cases}$$

where $\phi_{2i}(x) = \min_{j=1,\ldots,i} \{y_i + M|x-x_j|\}$. This makes it clear that, starting from the third step, the one-step optimal algorithm (4.8)–(4.10) coincides with the algorithm

$$x_i = 0, \quad x_{i+1} = \arg\max_{x\in K} \phi_{2i}(x), \quad i \geq 1, \tag{4.11}$$

which is sometimes called *polygonal method* and has been discussed in many papers, see Danilin [71], Danilin and Piyavskii [67], Lbov and Grunov [76], Piyavskii [67, 72], Shubert [72a, b], Timonov [77], and Vasil'ev [80]. (For $n>1$, one-step optimal algorithm is not described by the formula $x_{i+1} = \arg\max_{x\in K} \phi_{2i}(x)$.)

4.3. Sequentially optimal (counting informational computations) algorithm for the class of one-variable functions satisfying the Lipschitz condition

As before, assume that $n=1$, $K=[0,1]$, $\rho(u,v) = M|u-v|$, but, in contrast with Section 4.1, the problem should be solved with a prescribed accuracy ε using the minimal number of informational computations. We construct a sequentially optimal (counting informational computations) algorithm in accordance with the results of Section 3.3.

The point x_1 is given by (4.1) for $N = N_\varepsilon = \lceil M/(2\varepsilon) \rceil$ (our choice of x_1 could also be based on other methods of covering $[0,1]$ with the minimal number of segments of the length $2\varepsilon/M$).

Suppose we have already performed i informational computations. Then

$$CK_i(\bar{y}_i + \varepsilon) = \left[0, \beta_1 - \frac{\varepsilon}{M}\right) \cup \left[\bigcup_{j=2}^{i} \left(\alpha_j + \frac{\varepsilon}{M}, \beta_j - \frac{\varepsilon}{M}\right)\right] \cup \left(\alpha_{i+1} + \frac{\varepsilon}{M}, 1\right].$$

If this set is empty, the problem has already been solved with the required accuracy and the computational process can be stopped. Otherwise the point x_{i+1} can be chosen either in accordance with (4.5) for $R_{iN} = \varepsilon$, or in accordance with (4.7) and (4.8) for $R_{iN} = \varepsilon$.

Numerical tests show that the smaller the prescribed accuracy ε grows, the more apparent the advantage of the sequentially optimal algorithm over the optimal algorithm becomes (strictly speaking, we compared a modified optimal algorithm and a modified sequentially optimal algorithm, because the both algorithms used adaptive method (4.1), Chapter 2, for estimating the constant M). For the function $f(x) = 3\cos 2x - 2\sin x$, $x \in K = [-1, 1]$, we arrive at this conclusion when analyzing Table 5. In this table, $N_\varepsilon = \lceil M(b-a)/(2\varepsilon) \rceil$ is the number of informational computations required for the optimal algorithm to guarantee the prescribed accuracy ε (here M is the estimate of the Lipschitz constant obtained by (4.1), Chapter 2, on completion of the informational computations); N is the number of informational computations after which the sequentially optimal algorithm guarantees the same accuracy ε; finally, ε_{act} is the actual accuracy of the solution given by the sequentially optimal algorithm.

Table 5

ε	0.1	0.01	0.001
N_ε	15	82	269
N	66	980	9803
N/N_ε	0.227	0.084	0.027
ε_{act}	0.7×10^{-5}	0.2×10^{-7}	0.4×10^{-8}

We see that, for global optimization, the advantage of the sequentially optimal algorithm over the optimal algorithm is more demonstrative than for integration or function recovery. This is not surprising since in integration and recovery problems we have, in a sense, to investigate the function's behaviour all over the domain, while in global optimization we can confine our search to the subdomain that looks "promising" in the light of the information we already have.

One cannot but notice that the actual accuracy of the solution is several orders of magnitude better than the prescribed accuracy. This is characteristic of many problems. It can be explained by the fact that, if the function "behaves well" in a neighbourhood of the extremum, then the estimate of the Lipschitz constant for this neighbourhood turns out to be too large. This should be taken into account when solving real-life problems.

All numerical tests also show that, for ε small enough, the sequentially optimal algorithm performs better than the one-step optimal algorithm (we used initial segments of Fourier series with random coefficients as test functions).

4.4. On the possibility of using the universal scheme for constructing sequentially optimal algorithms

The scheme of constructing sequentially optimal optimization algorithms for functions of one variable can be applied not only to the class F_ρ where $\rho(u,v)=M|u-v|$ and the information consists of function values, but also to other computation models. As well as for numerical integration, these models should satisfy condition (3.37), Chapter 2. We once again outline the key points of the scheme for constructing sequentially optimal optimization algorithms.

At the $(i+1)$st step, we find the "promising" domain $\{x \mid \phi_{2i}(x) > \bar{y}_i\}$, where, as before, ϕ_{2i} is the upper envelope for the functions from the subclass $F(z^i) \subset F$ determined by the information z^i. Assume that the closure of this domain consists of the intervals

$$[0,\beta_1],[\alpha_2,\beta_2],\ldots,[\alpha_i,\beta_i],[\alpha_{i+1},1]. \tag{4.12}$$

Then we find the optimal allocation of computational resources among the intervals (4.12) and, with this allocation, construct optimal algorithms for the functional classes corresponding to these intervals. After that, the point of the first informational computation of any of the obtained algorithms can be chosen as x_{i+1}. Thus, the problem of constructing a sequentially optimal algorithm can be reduced to a series of problems of constructing optimal algorithms and integer problems of resource allocation.

The problem of constructing a one-step optimal algorithm is considerably simpler. In many cases, solving this problem leads to an efficient computational algorithm, while the problem of constructing a sequentially optimal algorithm looks intractable. One-step optimal algorithm for minimizing convex functions (Chernous'ko [70b]) is a good example. The same setting leads to an efficient algorithm for minimizing functions whose second derivative is piecewise continuous and bounded on a segment in the case where informational computations are evaluations of the function and its first derivative at some points within the segment, see Chuyan [86].

4.5. Construction of an optimal (counting informational computations) algorithm in two dimensions

We continue to work with the class $F=F_\rho$. Assume now that $n=2$, $K=\{(u^1,u^2) \mid 0 \le u^i \le c^i,\ i=1,2,\}$, $\rho(u,v)=M\max\{|u^1-v^1|,|u^2-v^2|\}$, and we wish to construct a sequentially optimal (counting informational computations) algorithm guaranteeing a prescribed accuracy ε. Put $r=\varepsilon/M$, $l=2r$. The set $CK_i(\bar{y}_i+\varepsilon)$ is obtained by excluding coordinate squares with side not less than l (ρ-balls with radius not less than ε) from K. For the $(i+1)$st informational computation, a sequentially optimal algorithm can choose any center of any covering of this set with the minimal number of coordinate squares with side l.

Put $x_1=(r,r)$, $x_2=(3r,r)$. Suppose we have already performed the computations at $x_1,\dots,x_i$, and let $y_j=f(x_j)$, $j=1,\dots,i$. Select positive integer s such that the intersection of the line $u^2=c^2-sl$ with $CK_i(\bar{y}_i+\varepsilon)$ is not empty and the intersection of the line $u^2=c^2-(s+1)l$ with $CK_i(\bar{y}_i+\varepsilon)$ is empty. Cover the one-dimensional set

$$\{u=(u^1,u^2) \mid u\in CK_i(\bar{y}_i+\varepsilon),\ u^2=c^2-sl\}$$

with the minimal number of segments of length l. Consider any of the coordinate squares for which these segments are the "upper" sides. We could choose the center of this square as x_{i+1}, but we would rather proceed in the following way. If the "left" side of the square lies entirely within the set $K_i(\bar{y}_i+\varepsilon)$, we move the square "to the right" as far as possible so that its "left" side is still in $K_i(\bar{y}_i+\varepsilon)$; after that, if its "lower" side lies entirely within $K_i(\bar{y}_i+\varepsilon)$, we move the square "up" as far as possible so that its "lower" side is still in $K_i(\bar{y}_i+\varepsilon)$. Then we take the center of the obtained square as x_{i+1}.

To understand why the above algorithm is sequentially optimal, it is sufficient to observe that the problem of covering the set $CK_i(\bar{y}_i+\varepsilon)$ with the minimal number of coordinate squares with side l can be reduced to the problems of covering the strips

$$\{u=(u^1,u^2) \mid u\in CK_i(\bar{y}_i+\varepsilon),\ c^2-l\le u^2\le c^2\},$$
$$\{u=(u^1,u^2) \mid u\in CK_i(\bar{y}_i+\varepsilon),\ c^2-2l\le u^2\le c^2-l\},\dots$$

with the minimal number of such squares. Each of these problems is equivalent to a simple problem of covering the corresponding one-dimensional set

$$\{u=(u^1,u^2) \mid u\in CK_i(\bar{y}_i+\varepsilon),\ u^2=c^2\},$$
$$\{u=(u^1,u^2) \mid u\in CK_i(\bar{y}_i+\varepsilon),\ u^2=c^2-l\},\dots$$

with the minimal number of segments of length l.

4.6. Multidimensional case

The considerations of the previous section lead us to the following conclusion: even for $n=2$, though we still can give a constructive description of a sequentially optimal algorithm, its usage is not advisable because of its high combinatory complexity and inadequate computer storage requirements. For $n>2$, even in the simplest case

$$K=\{u \mid 0\le u^i\le 1,\ i=1,\dots,n\},\quad \rho(u,v)=M\max_{i=1,\dots,n}|u^i-v^i|, \qquad (4.13)$$

serious difficulties come up in the process of solving the problem of optimal covering for $CK_i(\bar{y}_i+\varepsilon)$.

However, as we mentioned in Section 4.2 of Chapter 1, knowing the structure of sequentially optimal algorithms often enables us to construct efficient computational algorithms that approximate sequentially optimal algorithms in some sense and have acceptable combinatory complexity.

We now describe one of the possible approaches. First, we fix a lattice $\{u_1, \dots, u_N\}$ such that its exhaustion guarantees, according to our estimate, the required accuracy of the solution (sometimes it is impossible to guarantee the prescribed accuracy since the quasi-metric ρ determining the functional class may not be known exactly and may be updated in the process of computations). For example, in the case (4.13), a reasonable choice will be the cubic lattice with $N = m^n$ nodes

$$\left\{ \left(\frac{j_1}{2m}, \dots, \frac{j_n}{2m} \right) \middle| \, j_k = 1, 3, \dots, 2m-1; \; k = 1, \dots, n \right\}.$$

If the quasi-metric ρ is not known exactly, it makes sense to use the points of LP_τ-sequence (see Sobol [69, 79, 82], Sobol and Statnikov [81]) as $u_1, \dots, u_N$; in this case it is not necessary to fix the number of points N in advance.

After that, the solution process amounts to exhaustion of the nodes $\{u_1, \dots, u_N\}$. If the number of nodes is comparatively small and so exhaustive search is possible, then there will be no problem (if we know ρ, the described algorithm is an approximation to the optimal algorithm). Otherwise we can use the following method based on the same ideas as sequentially optimal algorithms.

The first step should satisfy the condition

$$x_1 \in \{u_1, \dots, u_N\}. \tag{4.14}$$

Suppose we have performed i steps, $i \geq 1$. Then

$$x_{i+1} \in \{u_j \mid 1 \leq j \leq N, \; \phi_{2i}(u_j) \geq \bar{y}_i\} = \{u_1, \dots, u_N\} \cap CK_i(\bar{y}_i). \tag{4.15}$$

In other words, we must exclude from the lattice $\{u_1, \dots, u_N\}$ those points at which the function is certain not to take its maximal value (the quasi-metric ρ is assumed to be known exactly, say, we may know that ρ has the form (4.13) and our estimate of M is exact), and then choose one of the rest points. Concrete methods of selecting points from the sets (4.14) and (4.15) can vary (as well as for sequentially optimal algorithms); the same goes for methods of program implementation of the algorithm (4.14), (4.15). Different methods of program implementation can be based on different ways of organizing data structures (by data structures we mean lists of points u_j, function values, points at which function values have already been computed, excluded points, etc.).

The above algorithm can also be regarded as a method for selection of starting points for local optimization. Then the algorithm should be supplemented with a block containing one or more local maximization methods. As well as the one-dimensional optimization algorithms from Sections 4.1-4.3, the above algorithm can be modified for real-life problems using other heuristic tricks, such as adaptive estimation of the Lipschitz constant (see Section 4.2 of Chapter 2), interchanging different strategies of global and local search, applying more realistic stopping criteria (in this connection, see Section 4.3 for discussion of relations between error estimates and actual errors of the solution of test problems), and so on.

5. CASE OF APPROXIMATE INFORMATION

In this section, we consider the case where informational computations are approximate, i.e., the function f is evaluated with some errors. As before, we assume that $f \in F = F_\rho$, the set K is compact, and the quasi-metric ρ is continuous on $K \times K$.

Computational errors are often considered to be random variables with a known (usually normal) or not completely known distribution. Stochastic approximation is an example of this approach.

We will take a different approach. As well as in Section 5 of Chapter 2, we will assume the errors to be indeterminate rather than stochastic variables that can take any values from a certain set. But, in contrast with Section 5 of Chapter 2, there will be no optimization with respect to $\delta_1, \ldots, \delta_N$ (maximal errors of informational computations at $x_1, \ldots, x_N$). We just assume that none of them exceeds some given δ. Our aim is to construct a sequentially optimal algorithm (rather than an optimal nonadaptive algorithm we constructed in Section 5 of Chapter 2).

5.1. Statement of the problem

We now proceed to strict formulations. Assume that the number of informational computations N is fixed and any informational computation at an arbitrary point $x \in K$ gives, instead of the exact value $y = f(x)$, two numbers $y^- = f^-(x)$ and $y^+ = f^+(x)$ such that

$$y^- \leq y \leq y^+, \quad y^+ - y^- \leq 2\delta. \tag{5.1}$$

This setting naturally arises in many situations, for instance, in the situation where every value of f is obtained through numerical solution of some problem, say, $f(x) = \min_{u \in U} g(x, u)$ or $f(x) = \int_U g(x, u) du$, with a prescribed accuracy δ. The guaranteed *a posteriori* accuracy of evaluating $f(x)$ may appear to be less than δ, which is reflected in assumptions (5.1).

On completion of the informational computations, we have the vectors $x^N = (x_1, \ldots, x_N)$, $y_-^N \stackrel{def}{=} (y_1^-, \ldots, y_N^-)$, and $y_+^N \stackrel{def}{=} (y_1^+, \ldots, y_N^+)$. Denote

$$z_-^N = (x^N, y_-^N), \quad z_+^N = (x^N, y_+^N),$$

$$F(x^N, y_-^N, y_+^N) = \{f \in F \mid y_j^- \le f(x_j) \le y_j^+,\ j=1,\dots,N\},$$

$$\phi_1(x; z_-^N) = \phi_1(x; x^N, y_-^N) = \max_{j=1,\dots,N} \{y_j^- - \rho(x, x_j))\},$$

$$\phi_2(x; z_+^N) = \phi_2(x; x^N, y_+^N) = \min_{j=1,\dots,N} \{y_j^+ + \rho(x, x_j))\}$$

(necessity of listing z_-^N and z_+^N explicitly among the arguments of the functions ϕ_1 and ϕ_2 respectively will be clear from the construction below). It is easy to see that

$$F(x^N, y_-^N, y_+^N) = \{f \in F \mid \phi_1(x; x^N, y_-^N) \le f(x) \le \phi_2(x; x^N, y_+^N)\}, \qquad (5.2)$$

i.e., ϕ_1 and ϕ_2 are the sharp lower and upper envelopes for the functions from the class $F(x^N, y_-^N, y_+^N)$ respectively. However, we do not necessarily have $\phi_1(x_j; x^N, y_-^N) = y_j^-$ and $\phi_2(x_j; x^N, y_+^N) = y_j^+$, $j=1,\dots,N$ (cf. Lemmas 1.1 and 1.2 of Chapter 2).

For the problem being considered, we have to redefine (though in full accordance with the notions introduced before) the criterion for estimating efficiency of an algorithm and the notion of sequentially optimal algorithm since the previous definitions were formulated under the assumption that f was evaluated exactly.

As we wish our results to be guaranteed, it looks logical to approximate $\max_{x\in K} f(x)$ with the least possible value of the maximum of functions from $F(x^N, y_-^N, y_+^N)$:

$$\bar{y}_j \overset{def}{=} \max_{x\in K} \phi_1(x; z_-^N) = \max_{i=1,\dots,N} y_i^- .$$

Then we have $f(x_{i_0}) \ge \phi_1(x_{i_0}; z_-^N) = \bar{y}_i$, where $i_0 \in \{1,\dots,N\}$ is determined by the condition $y_{i_0}^- = \bar{y}_i$.

Like in the case of exact information, the criterion will be the error of estimating the maximum:

$$\max_{x\in K} f(x) - \max_{i=1,\dots,N} f^-(x_i) = \max_{x\in K} f(x) - \bar{y}_N . \qquad (5.3)$$

After N computations with the results x^N, y_-^N, y_+^N, the worst-case error is

$$\max_{f\in F(x^N, y_-^N, y_+^N)} \left[\max_{x\in K} f(x) - \bar{y}_N\right] = \max_{x\in K} \phi_2(x' z_+^N) - \bar{y}_N \qquad (5.4)$$

(this equation holds due to (5.2)).

A situation (x, y_-^N, y_+^N) is called *realizable* iff

$$0 \le y_j^+ - y_j^- \le 2\delta, \quad j=1,\dots,N, \quad F(x^N, y_-^N, y_+^N) \ne \emptyset.$$

In accordance with Section 6 of Chapter 1, a sequentially optimal (by error) algorithm in any situation that may arise in the solution process selects the point x_{i+1} of the next informational computation in the way that minimizes, in the case of the "worst" subsequent behaviour of f, the error of estimating its maximum (using the criterion (5.3)).

The strict definition is analogous to Definition 2 from Section 6.3 of Chapter 1. Namely, in the setting under consideration, an algorithm is called ***sequentially optimal (by error)*** iff its choice of the point x_{i+1} of the $(i+1)$st informational computation, $i=0,1,\ldots,N-1$, delivers the outer minimum in

$$\min_{x_{i+1}} \max_{y_{i+1}^-, y_{i+1}^+} \ldots \min_{x_N} \max_{y_N^-, y_N^+} \left[\max_{x\in K} \phi_2(x; z_+^N) - \bar{y}_N \right], \tag{5.5}$$

where x_j, y_j^-, y_j^+, $j=i+1,\ldots,N$, are such that the situation (x^N, y_-^N, y_+^N) is realizable.

5.2. Sequentially optimal (by error) algorithm

We show that, like in the case of exact information, construction of a sequentially optimal algorithm amounts to solving a series of optimal covering problems. Define

$$R_0^*(x_1,\ldots,c_N) = \max_{x\in K} \min_{j=1,\ldots,N} \rho(x, x_j),$$

$$R_i^*(x_{i+1},\ldots,c_N) = \max_{x\in K} \min \Big\{ \min_{j=1,\ldots,N} [\rho(x,x_j) - \bar{y}_i + y_j^+ - 2\delta],$$

$$\min_{j=i_1,\ldots,N} \rho(x, x_j) \Big\}, \quad i \geq 1.$$

The structure of a sequentially optimal algorithm is clear from the following theorem (where the situation (x^N, y_-^N, y_+^N) is assumed to be realizable).

Theorem 5.1 (Sukharev [75]). *For* $i=0,1,\ldots,N-1$ *we have*

$$\min_{x_{i+1}} \max_{y_{i+1}^-, y_{i+1}^+} \ldots \min_{x_N} \max_{y_N^-, y_N^+} \left[\max_{x\in K} \phi_2(x; z_+^N) - \bar{y}_N \right]$$
$$= \min_{x_{i+1},\ldots,x_N \in K} R_i^*(x_{i+1},\ldots,x_N) + 2\delta.$$

Proof. We can rewrite (5.5) as follows:

$$\max_{y_N^-, y_N^+} \left[\max_{x\in K} \phi_2(x; z_+^N) - \bar{y}_N \right] = \max_{y_N^-} \left[\max_{x\in K} \phi_2(x; x^N, y_+^{N-1}, \bar{y}_N + 2\delta) - \bar{y}_N \right]. \tag{5.6}$$

The requirement that the situation (x^N, y_-^N, y_+^N) be realizable puts on y_N^- the following restrictions:

$$\phi_1(x_N; z_-^{N-1}) - 2\delta \leq y_N^- \leq \phi_2(x_N; z_+^{N-1}). \tag{5.7}$$

First, we compute (5.6) without considering restrictions (5.7). The outer maximum on the right-hand side of (5.6) is attained at $y_N^- = \bar{y}_{N-1}$. Indeed, the function inside this minimum is apparently nondecreasing in y_N^- for $y_N^- \leq \bar{y}_{N-1}$. If $y_N^- = \bar{y}_{N-1} + \alpha$, $\alpha \geq 0$, then

$$\begin{aligned}
&\max_{x\in K}\phi_2(x; x^N, y_+^{N-1}, y_N^- + 2\delta) - \bar{y}_N \\
&= \max_{x\in K}\phi_2(x; x^N, y_+^{N-1}, \bar{y}_{N-1} + \alpha + 2\delta) - \max\{y_1^-, \ldots, y_{N-1}^-, \bar{y}_{N-1} + \alpha\} \\
&= \max_{x\in K}\min\{\phi_2(x; z_+^{N-1}), \bar{y}_{N-1} + \alpha + 2\beta + \rho(x, x_N)\} - \bar{y}_{N-1} - \alpha \\
&\leq \max_{x\in K}\min\{\phi_2(x; z_+^{N-1}), \bar{y}_{N-1} + 2\delta + \rho(x, x_N)\} - \bar{y}_{N-1} \\
&= \max_{x\in K}\phi_2(x; x^{N-1}, y_+^{N-1}, \bar{y}_{N-1} + 2\delta) - \bar{y}_{N-1},
\end{aligned}$$

thus the function inside the maximum with respect to y_N^- on the right-hand side of (5.6) for $y_N^- \geq \bar{y}_{N-1}$ does not exceed the value of this function at $y_N^- = \bar{y}_{N-1}$.

Now we show that considering restrictions (5.7) does not change the obtained maximal value. Obviously, $y_N^- = \bar{y}_{N-1}$ satisfies the left inequality in (5.7), and violation of the right inequality for $y_N^- = \bar{y}_{N-1}$ (i.e., the situation $\bar{y}_{N-1} > \phi_2(x_N; z_+^{N-1})$) does not prevent the maximum in (5.6) under restrictions (5.7) from coinciding with the value that the function inside the maximum with respect to y_N^- takes at $y_N^- = \bar{y}_{N-1}$. To demonstrate this, it suffices to show that, for any $x \in K$, the function $\phi_2(x; x^N, y_+^{N-1}, y_N^- + 2\delta)$ (and, therefore, the function $\max_{x\in K}\phi_2(x; x^N, y_+^{N-1}, y_N^- + 2\delta)$) is constant with respect to y_N^- for $y_N^- \geq \phi_2(x_N; z_+^{N-1}) - 2\delta$.

For the proof, we rewrite the latter inequality in the form

$$y_N^- \geq \min_{j=1,\ldots,N-1}\{y_j^+ + \rho(x_N, x_j)\} - 2\delta = y_{j_0}^+ + \rho(x_N, x_{j_0}) - 2\delta,$$

where $1 \leq j_0 \leq N-1$. Hence

$$\begin{aligned}
y_N^- + 2\delta + \rho(x, x_N) &\geq y_{j_0}^+ + \rho(x_N, x_{j_0}) + \rho(x, x_N) \\
&\geq y_{j_0}^+ + \rho(x, x_{j_0}) \\
&\geq \min_{j=1,\ldots,N-1}\{y_j^+ + \rho(x, x_j)\} = \phi_2(x; z_+^{N-1}).
\end{aligned}$$

Thus, for $y_N^- \geq \phi_2(x_N; z_+^{N-1}) - 2\delta$ we have

$$\begin{aligned}
\phi_2(x; x^N, y_+^{N-1}, y_N^- + 2\delta) &= \min\{\phi_2(x; z_+^{N-1}), y_N^- + 2\delta + \rho(x, x_N)\} \\
&= \phi_2(x; z_+^{N-1}),
\end{aligned}$$

and so the function $\phi_2(x; x^N, y_+^{N-1}, y_N^- + 2\delta)$ is constant with respect to y_N^-.

In the same way we show that the maximum with respect to y_{N-1}^- and y_{N-1}^+ in (5.5) computed under the corresponding restrictions coincides with the value of the function inside the maximum at $y_{N-1}^- = \bar{y}_{N-2}$, $y_{N-1}^+ = \bar{y}_{N-2} + 2\delta$, and so on. Thus, we have proved the first equation in the following chain:

$$\min_{x_{i+1}} \max_{y^-_{i+1}, y^+_{i+1}} \ldots \min_{x_N} \max_{y^-_N, y^+_N} \left[\max_{x \in K} \phi_2(x; z^N_+) - \bar{y}_N\right]$$
$$= \min_{x_{i+1}, \ldots, x_N} \max_{x \in K} \phi_2(x; x^N, y^i_+, \bar{y}_i + 2\delta, \ldots, \bar{y}_i + 2\delta) - \bar{y}_i$$
$$= \min_{x_{i+1}, \ldots, x_N} \max_{x \in K} \min\{y^+_1 + \rho(x, x_1), \ldots, y^+_i + \rho(x, x_i),$$
$$\bar{y}_i + 2\delta + \rho(x, x_{i+1}), \ldots, \bar{y}_i + 2\delta + \rho(x, x_N)\} - \bar{y}_i$$
$$= \min_{x_{i+1}, \ldots, x_N} R^*(x_{i+1}, \ldots, x_N) + 2\delta. \tag{5.8}$$

In (5.8) it is assumed that $i \geq 1$. Considering now (5.8) for $i=1$, we easily prove the theorem for $i=0$, which completes the proof. □

Observe that

$$R^*_i(x_{i+1}, \ldots, x_N)$$
$$= \min\left\{ r \;\middle|\; \bigcup_{j=1}^{i} S(x_j, \bar{y}_i - y^+_j + 2\delta + r) \cup \left[\bigcup_{j=i+1}^{N} S(x_j, r)\right] \supset K \right\}$$

(see (3.4)). Thus, Theorem 5.1. enables us to reduce the problem of constructing a sequentially optimal algorithm to a series of special type optimal covering problems in the case of approximate information, too. Comparing the formulas for $R^*_i(x_{i+1}, \ldots, x_N)$ and $R_i(x_{i+1}, \ldots, x_N)$ (see (3.4)), we find no significant difference between these problems and the corresponding optimal covering problems for the case of exact information. In particular, if $n=1$ and $\rho(u,v) = M|u-v|$, then, after some obvious modifications of the algorithm from Section 4.1, we obtain a sequentially optimal (by error) algorithm for the case of approximate information.

Note that, if after i informational computations in a situation (x^i, y^i_-, y^i_+) the guaranteed accuracy (worst-case error) does not exceed 2δ, then the rest informational computations do not guarantee any further reduction of the error. Indeed, suppose that in a situation (x^i, y^i_-, y^i_+) the worst-case error is not greater than 2δ. This means that

$$\max_{x \in K} \phi_2(x; z^i) - \bar{y}_i = \max_{x \in K} \min_{j=1,\ldots,i} [y^+_j + \rho(x, x_j)] - \bar{y}_i \leq 2\delta,$$

which yields

$$\max_{x \in K} \min_{j=1,\ldots,i} [\rho(x, x_j) - \bar{y}_i + y^+_j - 2\delta] \leq 0.$$

Then for arbitrary $x_{i+1}, \ldots, x_N \in K$ we have

$$R^*_i(x_{i+1}, \ldots, x_N) = \max_{x \in K} \min_{j=1,\ldots,i} [\rho(x, x_j) - \bar{y}_i + y^+_j - 2\delta],$$

and so the best accuracy $\min_{x_{i+1}, \ldots, x_N} R^*_i(x_{i+1}, \ldots, x_N) + 2\delta$ that can be guaranteed in the situation (x^i, y^i_-, y^i_+) by the rest $N-i$ informational computations coincides with the accuracy guaranteed after the first i informational computations.

6. ONE-STEP OPTIMAL STOCHASTIC ALGORITHM

Let $K=[0,1]$, $\rho(u,v)=M|u-v|$, $F=F_\rho$, i.e.,

$$F=\{f \mid |f(u)-f(v)|\leq M|u-v|,\ u,v\in[0,1]\}. \tag{6.1}$$

The aim of this section is to construct a one-step optimal stochastic algorithm in the sense of the definition (7.15), (7.16), Chapter 1, under the assumption that informational computations are computations of exact values of the function f at points of the segment K, i.e., $X_i=K$ and $\Sigma_i=\Sigma$, $i\leq 1$, where Σ is the set of all probability measures on the σ-algebra of Borel subsets of K.

6.1. Reduction of the problem of constructing a one-step optimal stochastic algorithm to the problem of solving a collection of antagonistic games on the unit square

First we recall some notations introduced before and introduce into consideration some new notations. For $i\geq 1$ we denote[1)]

$$\varepsilon(x^i,f)=\max_{x\in K} f(x)-\max_{j=1,\dots,i} f(x_j),$$

$$\bar{y}_i=\max_{j=1,\dots,i} y_j,\quad \phi_{2i}(x)=\min_{j=1,\dots,i}\{y_j+M|x-x_j|\},$$

$$h_a(x)=-M|x-a|,\quad h_a^{i+1}(x)=\max\{\phi_{2i}(a)-M|x-a|,\bar{y}_i\},$$

$$g_1(a,x)=h_a(a)-h_a(x)=M|x-a|,$$

$$\begin{aligned} g_{i+1}(a,x)&=\varepsilon(x^i,x,h_a^{i+1})\\ &=\max_{\bar{x}\in K} h_a^{i+1}(\bar{x})-\max\{h_a^{i+1}(x_1),\dots,h_a^{i+1}(x_i),h_a^{i+1}(x)\}\\ &=h_a^{i+1}(a)-h_a^{i+1}(x). \end{aligned}$$

Lemma 6.1. *Let the functional class F be defined by* (6.1), *and let σ_j be an optimal mixed strategy of the minimizing player in the game on the unit square with the payoff function g_j. Then the algorithm determined by the set of probability measures $\{\sigma_j\}_{j\geq 1}$ is one-step optimal on the class F among all adaptive stochastic algorithms, that is, σ_1 satisfies* (7.5), *Chapter* 1, *and σ_{i+1}, $i\geq 1$, satisfies* (7.16), *Chapter* 1.

---

1) For simplicity of notation, we do not indicate explicitly dependence of h_a^{i+1} and g_{i+1} (as well as of ϕ_{2i} which was introduced before) on z^i.

Proof. Let $f\in F(z^i)$, $i\geq 1$, and let $\max_{x\in K} f(x)=f(a)$. Clearly, for an arbitrary $x\in K$ we have $f(x)\geq f(a)-M|x-a|$, which yields

$$\begin{aligned}\varepsilon(x^i,x,f)&=f(a)-\max\{f(x),\bar{y}_i\}\leq f(a)-\max\{f(a)-M|x-a|,\bar{y}_i\}\\ &\leq \phi_{2i}(a)-\max\{\phi_{2i}(a)-M|x-a|,\bar{y}_i\}=h_a^{i+1}(a)-h_a^{i+1}(x)\\ &=\varepsilon(x^i,x,h_a^{i+1})=g_{i+1}(a,x).\end{aligned} \tag{6.2}$$

Observe that the function $k_a^{i+1}(x)\stackrel{def}{=}\min\{h_a^{i+1}(x),\phi_{2i}(x)\}$ satisfies the equation $\varepsilon(x^i,x,h_a^{i+1})=\varepsilon(x^i,x,k_a^{i+1})$ and $k_a^{i+1}\in F(z^i)$. Since $x\in K$ is arbitrary, in view of (6.2) we have, for any $\sigma\in\Sigma$,

$$\int_K \varepsilon(x_i,x,f)\sigma\{dx\}\leq\int_K \varepsilon(x^i,x,h_a^{i+1})\sigma\{dx\}$$

$$=\int_K \varepsilon(x^i,x,k_a^{i+1})\sigma\{dx\}\leq\max_{a\in K}\int_K \varepsilon(x^i,x,k_a^{i+1})\sigma\{dx\},$$

and, since $f\in F(z^i)$ is arbitrary,

$$\max_{f\in F(z^i)}\int_K \varepsilon(x^i,x,f)\sigma\{dx\}\leq\max_{a\in K}\int_K \varepsilon(x^i,x,k_a^i)\sigma\{dx\}.$$

(Apparently, for the functional class (6.1) all the maximums here and below are attained.) On the other hand, since $k_a^{i+1}\in F(z^i)$ for all $a\in K$, the opposite inequality also holds. In view of the equation $\varepsilon(x^i,x,k_a^{i+1})=g_{i+1}(a,x)$, this yields

$$\max_{f\in F(z^i)}\int_K \varepsilon(x^i,x,f)\sigma\{dx\}=\max_{a\in K}\int_K g_{i+1}(a,x)\sigma\{dx\}. \tag{6.3}$$

Let σ_{i+1} be an optimal mixed strategy of the minimizing player in the game on the unit square with the payoff function g_{i+1}. This means (see, e.g., Karlin [59]) that

$$\max_{a\in K}\int_K g_{i+1}(a,x)\sigma_{i+1}\{dx\}=\min_{\sigma\in\Sigma}\max_{a\in K}\int_K g_{i+1}(a,x)\sigma\{dx\}.$$

Since (6.3) holds for any $\sigma\in\Sigma$, the latter equation implies equation (7.16) of Chapter 1. Lemma 2.1 yields equation (7.15) of Chapter 1, which completes the proof. □

6.2. Solution of antagonistic games

Solution of the game with the payoff function g_1 is no problem. Denote by I_c the probability measure concentrated at a point c.

Lemma 6.2. *The strategies $\sigma_1=I_{1/2}$ and $\tau_1=I_0/2+I_1/2$ are optimal mixed strategies for the minimizing and maximizing players respectively, and* *the value of the game on the unit square with the payoff function* g_1 *is equal to* $M/2$. □

The proof of Lemma 6.2 is trivial and we omit it. Note that the game with the payoff function g_1 is a special case of the game that was dealt with in the proof of Theorem 2.2.

We now proceed to the solution of the game on the unit square with the payoff function g_{i+1} for $i \geq 1$. Let the set $\{x \mid \phi_{2i}(x) > \bar{y}_i\}$ consist of n non-overlapping intervals, $n \leq i+1$. Denote the endpoints of the jth interval by b_j and d_j. Then

$$P_i \stackrel{def}{=} \{x \mid \phi_{2i}(x) \geq \bar{y}_i\} = \bigcup_{j=1}^{n} [b_j, d_j].$$

Put $\alpha_j = b_j - d_j$ if $b_j \neq 0$ and $d_j \neq 1$, and put $\alpha_j = 2(d_j - b_j)$ if $b_j = 0$ or $d_j = 1$. Assume that the intervals are enumerated in such a way that

$$\alpha_1 \geq \alpha_2 \geq \cdots \geq \alpha_n > 0. \tag{6.4}$$

Put $c_j = \arg \max_{x \in [b_j, d_j]} \phi_{2i}(x)$. Clearly, we have $c_j = 0$ if $b_j = 0$; $c_j = 1$, if $d_j = 1$; and $c_j = (b_j + d_j)/2$ otherwise. Fig.10(a) illustrates the case $i = 5$, $n = 4$.

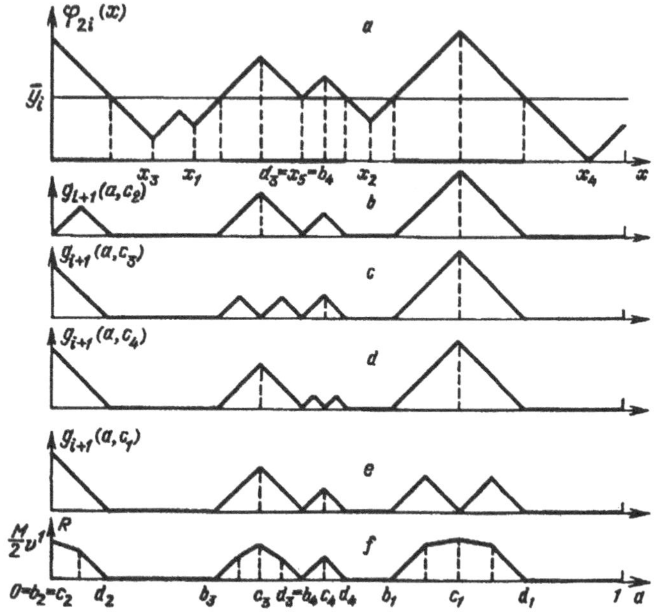

Fig. 10. On optimality of the strategy σ_{i+1} for the minimizing player in the game with the payoff function g_{i+1} in the case where $q_1^1 \leq 1/2$

To obtain the solution of the game, we can use the following scheme of argument. First, from domination considerations we conclude that it is sufficient to seek spectrum points of optimal strategies for the both players in the set $P(z^i)$. Then we show that these spectra contain at most one point from every segment $[b_j, d_j]$, $j>1$, namely c_j. For $j=1$, the optimal strategies' spectra can contain (besides c_1) the points $\xi_1=b_1+(d_1-b_1)/3$ and $\eta_1=b_1+2(d_1-b_1)/3$ from $[b_1, d_1]$, provided that $b_1\neq 0$ and $d_1\neq 1$. If $b_1=0$ or $d_1=1$, then the optimal strategies' spectra can contain (besides c_1) the point $\zeta_1=(b_1+d_1)/2\in[b_1, d_1]$. Thus, the optimal strategies' spectra contain at most $n+2$ points each. This enables us to reduce solution of the game on the unit square to solution of matrix games.

The proof of the above statement about the points of the optimal strategies' spectra lying in $[b_j, d_j]$, $j>1$, is not difficult. To find the additional spectrum points from $[b_1, d_1]$, we have to solve the game with the payoff function g_{i+1} on the square $[b_1, d_1]\times[b_1, d_1]$. If $b_1\neq 0$ and $d_1\neq 1$, then on this square g_{i+1} is a butterfly-shaped payoff function, see Karlin [59]. In this situation, the mixed strategy $I_{\xi_1}/2+I_{\eta_1}/2$ is optimal for the both players. Instead of proving these and other statements (which would be necessary for the natural process of finding the solution of the game), we give a short formal scheme with all the required proofs.

Consider a game with the matrix $A^1=(a^1_{jk})\overset{def}{=}(g_{i+1}(c_j, c_k))$, $j,k=1,\dots,n$. Using the formula $g_{i+1}(a,x)=h^{i+1}_a(a)-h^{i+1}_a(x)$, we get

$$A^1=\frac{M}{2}\begin{pmatrix} 0 & \alpha_1 & \alpha_1 & \dots & \alpha_1 \\ \alpha_2 & 0 & \alpha_2 & \dots & \alpha_2 \\ \alpha_3 & \alpha_3 & 0 & \dots & \alpha_3 \\ \vdots & \vdots & \vdots & \ddots & \vdots \\ \alpha_n & \alpha_n & \alpha_n & \dots & 0 \end{pmatrix}^{1)}.$$

In order to solve the game with the matrix A^1 under the above assumption (6.4), we put $v^1_j=(j-1)/(\alpha_1^{-1}+\alpha_2^{-1}+\dots+\alpha_j^{-1})$, $j=1,\dots,n$.

Lemma 6.3. *Let $v^1_1<v^1_2<\dots<v^1_{j_1}$, and let $v^1_{j_1}\le v^1_{j_1+1}$ or $j_1=n$. Then $Mv^1/2\overset{def}{=}Mv^1_{j_1}/2$ is the value of the game with the matrix A^1, the strategy*

$(q^1_1, q^1_2,\dots,q^1_{j_1}, q^1_{j_1+1},\dots,q^1_n)=(1-v^1/\alpha_1, 1-v^1/\alpha_2,\dots,1-v^1/\alpha_{j_1}, 0,\dots,0)$ *is optimal for the minimizing player, and the strategy*

$$(p^1_1, p^1_2,\dots,p^1_{j_1}, p^1_{j_1+1},\dots,p^1_n)$$
$$=\left(\frac{1}{\alpha_1}\cdot\frac{v^1}{j_1-1}, \frac{1}{\alpha_2}\cdot\frac{v^1}{j_1-1},\dots,\frac{1}{\alpha_{j_1}}\cdot\frac{v^1}{j_1-1}, 0,\dots,0\right)$$

---

1) For simplicity of notation, we do not indicate explicitly dependence of the matrix A^1, the matrices A^2 and A^3 introduced below, and their elements either on z^i or on the number of step.

is optimal for the maximizing player.

Proof. In what follows, we use the symbol $\Leftrightarrow$ to indicate equivalency of inequalities. It can be verified directly that

$$v_j^1 < v_{j+1}^1 \Leftrightarrow v_j^1 < \alpha_{j+1} \Leftrightarrow v_{j+1}^1 < \alpha_{j+1}, \quad j=1,\dots,n-1, \tag{6.5}$$

$$\sum_{j=1}^{n} p_j^1 = 1, \quad p_j^1 \geq 0, \ j=1,\dots,n, \quad \sum_{k=1}^{n} q_j^1 = 1, \tag{6.6}$$

$$\sum_{j=1}^{n} a_{jk}^1 p_j^1 = \frac{M}{2} v^1, \quad k \leq f_1, \tag{6.7}$$

$$\sum_{j=1}^{n} a_{jk}^1 p_j^1 = \frac{f_1}{f_1 - 1} \cdot \frac{M}{2} v^1 > \frac{M}{2} v^1, \quad k \geq j_1 + 1, \tag{6.8}$$

$$\sum_{k=1}^{n} a_{jk}^1 q_k^1 = \frac{M}{2} v^1, \quad j \leq j_1. \tag{6.9}$$

Making use of (6.4), (6.5) and the condition $v_{j_1}^1 \geq v_{j_1+1}^1$, we obtain

$$\sum_{k=1}^{n} a_{jk}^1 q_k^1 = \frac{M}{2} \alpha_j \leq \frac{M}{2} \alpha_{j_1+1} \leq \frac{M}{2} v^1, \quad j \geq j_1 + 1. \tag{6.10}$$

Finally, we write the following chain of inequalities which are equivalent due to (6.5) (the first inequality in this chain holds by the assumption of the lemma):

$$v_{j-1}^1 < v_{j_1}^1 \Leftrightarrow v_{j_1-1}^1 < \alpha_{j_1} \Leftrightarrow v^1 < \alpha_{j_1} \Leftrightarrow q_{j_1}^1 > 0.$$

In view of (6.4), this yields $q_k^1 > 0$, $k=1,\dots,j_1$. These inequalities in combination with (6.6)-(6.10) prove the lemma. □

Lemma 6.4. *If* $q_1^1 \leq 1/2$, *then*

$$\sigma_{i+1} = \sum_{i=1}^{n} q_k^1 I_{c_k} \quad \textit{and} \quad \tau_{i+1} = \sum_{j=1}^{n} p_j^1 I_{c_j}$$

are optimal mixed strategies of the minimizing and maximizing players respectively and the value of the game on the unit square with the payoff function g_{i+1} *is equal to* $Mv^1/2$.

Proof. Observe that $q_k^1 \leq 1/2$, $k=1,\dots,n$, since, due to (6.4) and the assumption of the lemma, we have $1/2 \geq q_1^1 \geq q_2^1 \geq \cdots \geq q_n^1$. Therefore, the function

$$R \stackrel{def}{=} \int_K g_{i+1}(a,x)\sigma_{i+1}\{dx\} = \sum_{i=1}^{n} g_{i+1}(a, c_k) q_k^1,$$

regarded as a function of the argument a, attains its maximum on $[b_j, d_j]$ at the point c_j, $j=1,\dots,n$. This is clear from Fig.10(b-f), which depicts the graphs of the functions $g_{i+1}(a, c_j)$, $j=1,\dots,n$, and also the graph of the function R (all the functions are regarded as functions of a) in the situation corresponding to Fig.10(a). Thus, we have

$$\max_{a\in K} R = \max_{j=1,\dots,n} \sum_{k=1}^{n} g_{i+1}(c_j, c_k) q_k^1 = \max_{j=1,\dots,n} \sum_{k=1}^{n} a_{jk}^1 q_k^1 = \frac{M}{2} v^1. \tag{6.11}$$

Consider now the function of x

$$R' \stackrel{def}{=} \int_K g_{i+1}(a, x) \tau_{i+1}\{da\} = \sum_{j=1}^{n} g_{i+1}(c_j, x) p_j^1.$$

Clearly (see Fig.11), this function attains its minimum on $[b_k, d_k]$ at the point c_k, $k=1,\dots,n$. Hence

$$\min_{x\in K} R' = \min_{k=1,\dots,n} \sum_{j=1}^{n} g_{i+1}(c_j, c_k) p_j^1 = \min_{k=1,\dots,n} \sum_{j=1}^{n} a_{jk}^1 p_j^1 = \frac{M}{2} v^1. \tag{6.12}$$

Equations (6.11) and (6.12) prove the lemma. □

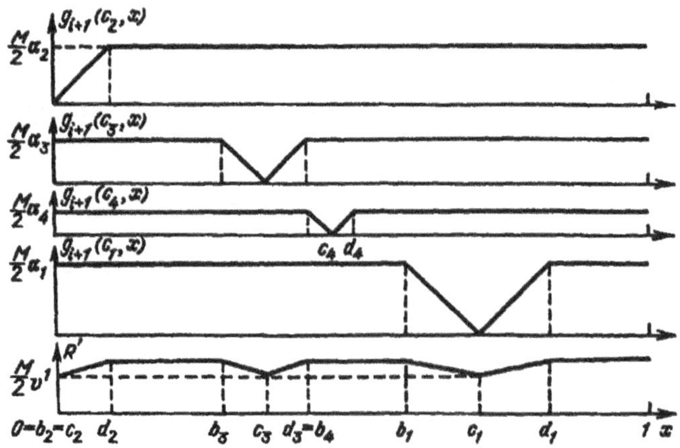

Fig. 11. On optimality of the strategy τ_{i+1} for the maximizing player in the game with the payoff function g_{i+1} in the case where $q_1^1 \leq 1/2$

Consider now a game with the matrix

$$A^2 = (a_{jk}^2)_{j,k=1,\dots,n} \stackrel{def}{=} \frac{M}{2} \begin{pmatrix} \alpha_1/3 & 2\alpha_1/3 & 2\alpha_1/3 & \dots & 2\alpha_1/3 \\ \alpha_2 & 0 & \alpha_2 & \dots & \alpha_2 \\ \alpha_3 & \alpha_3 & 0 & \dots & \alpha_3 \\ \vdots & \vdots & \vdots & \ddots & \vdots \\ \alpha_n & \alpha_n & \alpha_n & \dots & 0 \end{pmatrix}.$$

Put $v_j^2 = j/(3\alpha^{-1} + \alpha_2^{-1} + \cdots + \alpha_j^{-1})$, $j = 1, \ldots, n$.

Lemma 6.5. *Let $v_1^2 < v_2^2 < \cdots < v_{j_2}^2$, and let $v_{j_2}^2 \geq v_{j_2+1}^2$ or $j_2 = n$. Let, in addition, $q_1^1 > 1/2$. Then $Mv^2/2 \stackrel{def}{=} Mv_{j_2}^2/2$ is the value of the game with the matrix A^2, the strategy*

$$(q_1^2, q_2^2, \ldots, q_{j_2}^2, q_{j_2+1}^2, \ldots, q_n^2)$$
$$= (2 - 3v^2/\alpha_1, 1 - v^2/\alpha_2, \ldots, 1 - v^2/a_{j_2}, 0, \ldots, 0)$$

is optimal for the minimizing player, $q_1^2 > 1/2$, and the strategy

$$(p_1^2, p_2^2, \ldots, p_{j_2}^2, p_{j_2+1}^2, \ldots, p_n^2)$$
$$= \left(\frac{3}{\alpha_1} \cdot \frac{v^2}{j_2}, \frac{1}{\alpha_2} \cdot \frac{v^2}{j_2}, \ldots, \frac{1}{\alpha_{j_2}} \cdot \frac{v^2}{j_2}, 0, \ldots, 0\right)$$

is optimal for the maximizing player.

Proof. It can be verified directly that

$$v_j^2 < v_{j+1}^2 \Leftrightarrow v_j^2 < \alpha_{j+1} \Leftrightarrow v_{j+1}^2 < \alpha_{j+1}, \quad j = 1, \ldots, n-1, \tag{6.13}$$

$$\sum_{j=1}^n p_j^2 = 1, \quad p_j^2 \geq 0, \ j = 1, \ldots, n,$$

$$\sum_{k=1}^n q_k^2 = 1, \quad \sum_{j=1}^n a_{jk}^2 p_j^2 = \frac{M}{2} v^2, \quad k \leq j_2,$$

$$\sum_{j=1}^n a_{jk}^2 p_j^2 = \frac{j_2 + 1}{j_2} \cdot \frac{M}{2} v^2 > \frac{M}{2} v^2, \quad k \geq j_2 + 1,$$

$$\sum_{k=1}^n a_{jk}^2 q_k^2 = \frac{M}{2} v^2, \quad j \leq j_2.$$

Making use of (6.4), (6.13) and the condition $v_{j_2}^2 \geq v_{j_2+1}^2$, we obtain

$$\sum_{k=1}^n a_{jk}^2 q_k^2 = \frac{M}{2} \alpha_j \leq \frac{M}{2} \alpha_{j_2+1} \leq \frac{M}{2} v^2, \quad j \geq j_2 + 1.$$

Then, in the same way as in the proof of Lemma 6.3, we make sure that $q_k^2 > 0$, $k = 2, \ldots, j_2$. To complete the proof, it suffices to show that $q_1^2 > 1/2$.

From the assumption $q_1^1 > 1/2$ we easily derive an equivalent inequality

$$\alpha_2^{-1} + \cdots + \alpha_{j_1}^{-1} > (2j_1 - 3)\alpha_1^{-1}. \tag{6.14}$$

Let $j_1 > j_2$. Then $v_{j_2}^1 < v_{j_2+1}^1$ and $v_{j_2+1}^2 < v_{j_2}^2$. By virtue of (6.5) and (6.13), $v_{j_2}^1 < \alpha_{j_2+1} \leq v_{j_2}^2$. We have

$$v_{j_2}^1 < v_{j_2}^2 \Leftrightarrow \alpha_2^{-1} + \cdots + \alpha_{j_2}^{-1} > (2j_2 - 3)\alpha_1^{-1} \Leftrightarrow q_1^2 > 1/2.$$

If $j_1 = j_2$, then (6.14) gives the required inequality $q_1^2 > 1/2$.

Finally, assume that $j_1<j_2$. Then $v^2_{j_1}<v^2_{j_1+1}$ and $v^1_{j_1+1}\le v^1_{j_1}$. By virtue of (6.4) and (6.13), $v^2_{j_1}<\alpha_{j_1+1}\le v^1_{j_1}$. We have

$$v^2_{j_1}<v^1_{j_1}\Leftrightarrow \alpha_2^{-1}+\cdots+\alpha_{j_1}^{-1}<(2j_1-3)\alpha_1^{-1},$$

which contradicts (6.14). The obtained contradiction proves the inequality $q^2_1>1/2$ and thus completes the proof of the lemma. □

Recall our notations $\xi_1=b_1+(d_1-b_1)/3$ and $\eta_1=b_1+2(d_1-b_1)/3$. Put, in addition, $\bar{q}^2_1=\sum_{k=2}^{n} q^2_k$.

Lemma 6.6. *If $q^1_1>1/2$, $b_1\neq 0$, and $d_1\neq 1$, then*

$$\sigma_{i+1}=\bar{q}^2_1 I_{c_1}+(q^2_1-\bar{q}^2_1)\left(\frac{1}{2}I_{\xi_1}+\frac{1}{2}I_{\eta_1}\right)+\sum_{k=2}^{n} q^2_k I_{c_k}$$

and

$$\tau_{i+1}=p^2_1\left(\frac{1}{2}I_{\xi_1}+\frac{1}{2}I_{\eta_1}\right)+\sum_{j=2}^{n} p^2_j I_{c_j}$$

are optimal mixed strategies of the minimizing and maximizing players respectively and the value of the game on the unit square with the payoff function g_{i+1} is equal to $Mv^2/2$.

Proof. Observe that if $q^1_1>1/2$, then, by Lemma 6.5, we have $q^2_1>1/2$ and, therefore, $\bar{q}^2_1<1/2$ and $q^2_1-\bar{q}^2_1>0$. Thus, σ_{i+1} and, obviously, τ_{i+1} are mixed strategies. We have

$$R=Q+(q^2_1-\bar{q}^2_1)G,$$

where

$$Q=\bar{q}^2_1 g_{i+1}(a,c_1)+\sum_{k=2}^{n} q^2_k g_{i+1}(a,c_k),$$

$$G=[g_{i+1}(a,\xi_1)+g_{i+1}(a,\eta_1)]/2.$$

If $a\in[b_1,d_1]$, then $g_{i+1}(a,c_1)=g_{i+1}(a,\xi_1)=g_{i+1}(a,\eta_1)$; hence, $R=\sum_{k=1}^{n} q^2_k g_{i+1}(a,c_k)$ and the maximum with respect to a on the segment $[b_j,d_j]$, $j\geq 2$, is attained at c_j (this can be proved in the same way as in Lemma 6.4: it is sufficient to observe that $q^2_j<\sum_{k\neq j} q^2_k$, $j\geq 2$). Therefore,

$$\max_{a\in[b_j,d_j]} R=\int_K g_{i+1}(c_j,x)\sigma_{i+1}\{dx\}=\sum_{k=1}^{n} a^2_{jk}q^2_k,\quad j\geq 2. \tag{6.15}$$

Fig.12 shows that

$$\operatorname{Arg}\max_{a\in[b_1,d_1]} R=[b_1+(d_1-b_1)/4,b_1+3(d_1-b_1)/4],$$

which yields

$$\max_{a\in[b_1,d_1]} R = \int_K b_{i+1}(c_1,x)\sigma_{i+1}\{dx\} = \frac{M}{2}\left[\bar{q}_1^2\cdot 0 + (q_1^2-\bar{q}_1^2)\frac{\alpha_1}{3} + \sum_{k=2}^n q_k^2\alpha_1\right]$$

$$= \frac{M}{2}\left(q_1^2\frac{\alpha_1}{3} + \sum_{k=2}^n q_k^2\frac{2\alpha_1}{3}\right) = \sum_{k=1}^n a_{1k}^2 q_k^2. \tag{6.16}$$

Noticing that $R=0$ if $a\notin\bigcup_{j=1}^n[b_j,d_j]$, we derive from (6.15) and (6.16)

$$\max_{a\in K} R = \max_{j=1,\ldots,n}\sum_{k=1}^n a_{jk}^2 q_k^2 = \frac{M}{2}v^2. \tag{6.17}$$

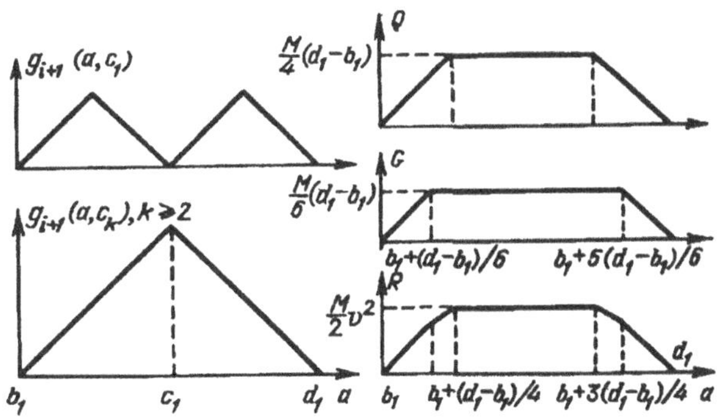

Fig. 12. On optimality of the strategy σ_{i+1} for the minimizing player in the game with the payoff function g_{i+1} in the case where $q_1^1>1/2$, $b_1\neq 0$, $d_1\neq 1$

Now we prove that the function

$$R' = p_1^2 G' + \sum_{j=2}^n p_j^2 g_{i+1}(c_j,x),$$

where $G' = [g_{i+1}(\xi_1,x)+g_{i+1}(\eta_1,x)]/2$, attains its minimum with respect to x on the segment $[b_k,d_k]$ at c_k. For $k=1$, this is clear from Fig.13: we have

$$\operatorname{Arg}\min_{a\in[b_1,d_1]} R' = [b_1+(d_1-b_1)/3,\, b_1+2(d_1-b_1)/3];$$

for $k\geq 2$, this can be proved in the same way as in Lemma 6.4. Now we easily obtain

$$\min_{x\in K} R' = \min_{k=1,\ldots,n}\sum_{j=1}^n a_{jk}^2 p_j^2 = \frac{M}{2}v^2. \tag{6.18}$$

Equations (6.17) and (6.18) prove the lemma. □

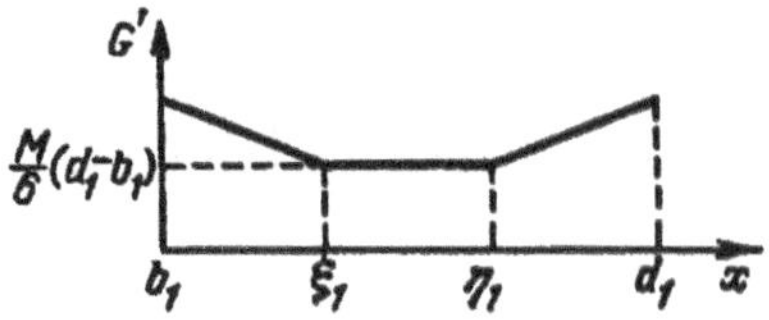

Fig. 13. On optimality of the strategy τ_{i+1} for the maximizing player in the game with the payoff function g_{i+1} in the case where $q_1^1 > 1/2$, $b_1 \neq 0$, $d_1 \neq 1$

We now consider another game (the last one) with the matrix

$$A^3 = (a_{jk}^3)_{j,k=1,\dots,n} \overset{def}{=} \frac{M}{2}\begin{pmatrix} \alpha_1/4 & 3\alpha_1/4 & 3\alpha_1/4 & \dots & 3\alpha_1/4 \\ \alpha_2 & 0 & \alpha_2 & \dots & \alpha_2 \\ \alpha_3 & \alpha_3 & 0 & \dots & \alpha_3 \\ \vdots & \vdots & \vdots & \ddots & \vdots \\ \alpha_n & \alpha_n & \alpha_n & \dots & 0 \end{pmatrix}.$$

Put $v_j^3 = (j - 1/2)(2\alpha^{-1} + \alpha_2^{-1} + \dots + \alpha_j^{-1})$, $j = 1, \dots, n$.

Lemma 6.7. *Let $v_1^3 < v_2^3 < \dots < v_{j_3}^3$, and let $v_{j_3}^3 \geq v_{j_3+1}^3$ or $j_3 = n$. Let, in addition, $q_1^1 > 1/2$. Then $Mv^3/2 \overset{def}{=} Mv_{j_3}^3/2$ is the value of the game with the matrix A^3, the strategy*

$$(q_1^3, q_2^3, \dots, q_{j_3}^3, q_{j_3+1}^3, \dots, q_n^3)$$
$$= (3/2 - 2v^2/\alpha_1, 1 - v^2/\alpha_2, \dots, 1 - v^2/a_{j_3}, 0, \dots, 0)$$

is optimal for the minimizing player, $q_1^2 > 1/2$, and the strategy

$$(p_1^3, p_2^3, \dots, p_{j_3}^3, p_{j_3+1}^3, \dots, p_n^3)$$
$$= \left(\frac{2}{\alpha_1} \cdot \frac{v^3}{j_3 - 1/2}, \frac{1}{\alpha_2} \cdot \frac{v^3}{j_3 - 1/2}, dots, \frac{1}{\alpha_{j_3}} \cdot \frac{v^3}{j_3 - 1/2}, 0, \dots, 0\right)$$

is optimal for the maximizing player. □

The proof is similar to the proof of Lemma 6.5.

Recall the notation $\zeta_1 = (b_1 + d_1)/2$ introduced before for the case where $b_1 = 0$ or $d_1 = 1$. Put, in addition, $\bar{q}_1^3 = \sum_{k=2}^{n} q_k^3$.

Lemma 6.8. *If $q_1^1 > 1/2$, $b_1 = 0$ and $d_1 = 1$, then*

$$\sigma_{i+1} = \bar{q}_1^3 I_{c_1} + (q_1^3 - \bar{q}_1^3)\left(\frac{1}{2}I_{\zeta_1} + \frac{1}{2}I_{c_1}\right) + \sum_{k=2}^{n} q_k^3 I_{c_k}$$

and

$$\tau_{i+1} = p_1^3\left(\frac{1}{2}I_{\zeta_1} + \frac{1}{2}I_{c_1}\right) + \sum_{j=2}^{n} p_j^2 I_{c_j}$$

are optimal mixed strategies of the minimizing and maximizing players respectively and the value of the game on the unit square with the payoff function g_{i+1} is equal to $Mv^3/2$.

Proof. Observe that if $q_1^1 > 1/2$, then, by Lemma 6.7, we have $q_1^3 > 1/2$ and, therefore, $\bar{q}_1^3 < 1/2$ and $q_1^3 - \bar{q}_1^3 > 0$. Thus, σ_{i+1} and, obviously, τ_{i+1} are mixed strategies. We have

$$R = Q' + (q_1^3 - \bar{q}_1^3)G'',$$

where

$$Q' = \bar{q}_1^3 g_{i+1}(a, c_1) + \sum_{k=2}^{n} q_k^3 g_{i+1}(a, c_k),$$

$$G'' = [g_{i+1}(a, \zeta_1) + g_{i+1}(a, c_1)]/2.$$

If $a \notin [b_1, d_1]$, then $g_{i+1}(a, c_1) = g_{i+1}(a, \zeta_1)$; hence, $R = \sum_{k=1}^{n} q_k^3 g_{i+1}(a, c_k)$ and the maximum with respect to a on the segment $[b_j, d_j]$, $j \geq 2$, is attained at c_j (this can be proved in the same way as in Lemma 6.4: it is sufficient to observe that $q_j^3 < \sum_{k \neq j} q_k^3$, $j \geq 2$). Therefore,

$$\max_{a \in [b_j, d_j]} R = \int_K g_{i+1}(c_j, x)\sigma_{i+1}\{dx\} = \sum_{k=1}^{n} a_{jk}^3 q_k^3, \quad j \geq 2. \tag{6.19}$$

Fig.14 shows that

$$\operatorname{Arg} \max_{a \in [b_1, d_1]} R = [\zeta_1, c_1],$$

which yields

$$\max_{a \in [b_1, d_1]} R = \int_K b_{i+1}(c_1, x)\sigma_{i+1}\{dx\} = \frac{M}{2}\left[\bar{q}_1^3 \cdot 0 + (q_1^3 - \bar{q}_1^3)\frac{\alpha_1}{4} + \sum_{k=2}^{n} q_k^3 \alpha_1\right]$$

$$= \frac{M}{2}\left(q_1^3 \frac{\alpha_1}{4} + \sum_{k=2}^{n} q_k^3 \frac{3\alpha_1}{4}\right) = \sum_{k=1}^{n} a_{1k}^3 q_k^3. \tag{6.20}$$

Noticing that $R = 0$ if $a \notin \bigcup_{j=1}^{n} [b_j, d_j]$, we derive from (6.19) and (6.20)

$$\max_{a \in K} R = \max_{j=1,\ldots,n} \sum_{k=1}^{n} a_{jk}^3 q_k^3 = \frac{M}{2} v^3. \tag{6.21}$$

Now we prove that the function

$$R' = p_1' G''' + \sum_{j=2}^{n} p_j^3 g_{i+1}(c_j, x),$$

where $G''' = [g_{i+1}(\zeta_1, x) + g_{i+1}(c_1, x)]/2$, attains its minimum with respect to x on the segment $[b_k, d_k]$ at c_k. For $k = 1$, this is clear from Fig.15: we

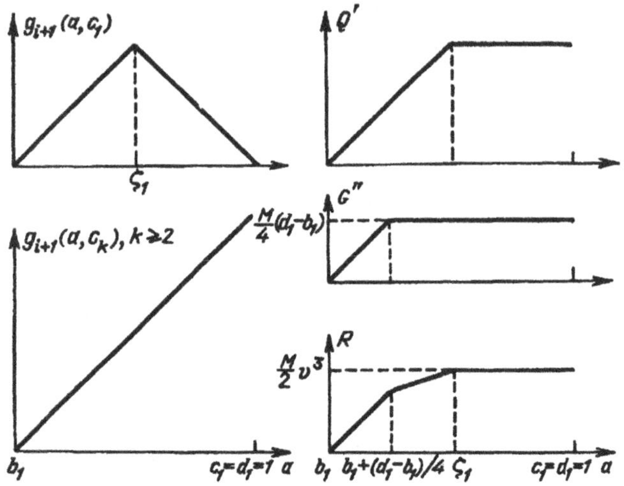

Fig. 14. On optimality of the strategy σ_{i+1} for the minimizing player in the game with the payoff function g_{i+1} in the case where $q_1^1 > 1/2$ and $b_1 = 0$ or $d_1 = 1$

have

$$\operatorname{Arg} \min_{a \in [b_1, d_1]} R' = [\zeta_1, c_1];$$

for $k \geq 2$, this can be proved in the same way as in Lemma 6.4. Now we easily obtain

$$\min_{x \in K} R' = \min_{k=1,\dots,n} \sum_{j=1}^{n} a_{jk}^3 p_j^3 = \frac{M}{2} v^3. \tag{6.22}$$

Equations (6.21) and (6.22) prove the lemma. □

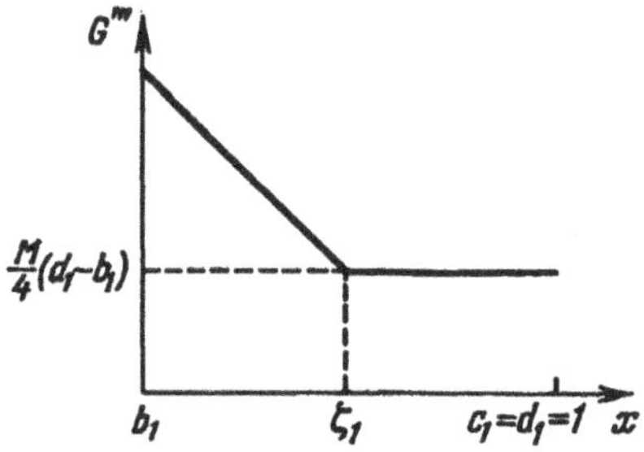

Fig. 15. On optimality of the strategy τ_{i+1} for the maximizing player in the game with the payoff function g_{i+1} in the case where $q_1^1 > 1/2$ and $b_1 \neq 0$ or $d_1 \neq 1$

6.3. Optimization algorithm

In conclusion, we describe a stochastic algorithm for maximization of a function f satisfying the Lipschitz condition with a given constant M on $[0,1]$, i.e., belonging to the functional class (6.1).

At the first step, we put $x_1 = 1/2$ and compute $y_1 = f(x_1)$. Suppose we have already performed i steps and computed $y_j = f(x_j)$, $j=1,\dots,i$, $i \geq 1$.

At the $(i+1)$st step, we proceed as follows. We compute n, b_j, d_j, α_j, and c_j for $j=1,\dots,n$ according to the formulas from Section 6.2. After that, we compute $v^1, q_1^1, \dots, q_n^1$.

If $q_1^1 \leq 1/2$, we put x_{i+1} equal to a realization of the random variable taking the values c_k with the probabilities q_k^1, $k=1,\dots,n$. Then we compute $y_{i+1} = f(x_{i+1})$.

If $q_1^1 > 1/2$, $b_1 \neq 0$ and $d_1 \neq 0$, we compute $\xi_1, \eta_1, q_1^2, \dots, q_n^2, \bar{q}_1^2$ and put x_{i+1} equal to a realization of the random variables taking the values c_j with the probabilities q_j^2, $j=2,\dots,n$, the value c_1 with the probability $\bar{q}_1^2$, and the values ξ_1 and η_1 with the equal probabilities $(q_1^2 - \bar{q}_1^2)/2$. Then we compute $y_{i+1} = f(x_{i+1})$.

If $q_1^1 > 1/2$ and, in addition, $b_1 = 0$ or $d_1 = 1$, we compute $\zeta_1, q_1^3, \dots, q_n^3, \bar{q}_1^3$ and put x_{i+1} equal to a realization of the random variables taking the values c_j with the probabilities q_j^3, $j=2,\dots,n$, the value c_1 with the probability $(q_1^3 + \bar{q}_1^3)/2$, and the value ζ_1 with the probability $(q_1^3 - \bar{q}_1^3)/2$. Then we compute $y_{i+1} = f(x_{i+1})$.

The stopping criterion can be chosen in various ways, for instance, we can choose the criterion (4.20) from Chapter 1.

Denote the above algorithm by A.

Theorem 6.1 (Sukharev [81c]). *The algorithm A is one-step optimal on the functional class* (6.1) *among all adaptive stochastic algorithms for global maximization.* □

The proof of Theorem 6.1 is given by Lemmas 6.1-6.8.

Note that before the $(i+1)$st step we can estimate the average accuracy guaranteed after N steps in the following way: *a priori* guaranteed accuracy is equal to $Mv^1/2$ if $q_1^1 \leq 1/2$; to $Mv^2/2$ if $q_1^1 \leq 1/2$, $b_1 \neq 0$ and $d_1 \neq 1$; to $Mv^3/2$ if $q_1^1 \leq 1/2$ and, in addition, $b_1 = 0$ or $d_1 = 1$.

Numerical comparison of the one-step optimal stochastic algorithm (with the stopping criterion (4.20), Chapter 1) and the sequentially optimal (counting informational computations) deterministic algorithm from Section 4.3 for the functions of the form $\sum_{k=1}^{9}\left(a_k \cos\frac{k\pi x}{100} + b_k \sin\frac{k\pi x}{100}\right)$, $x \in [0,100]$, where a_k and b_k are independent uniformly distributed random variables, has shown a certain advantage of the stochastic algorithm over the deterministic one.

CHAPTER 5

SOME SPECIAL CLASSES OF EXTREMAL PROBLEMS

In this chapter, we study optimal methods for solving nonlinear equations and systems of equations and for maximizing a minimum function with constrained variables, and also optimal methods for solving multi-criterion problems. All these problems essentially fit into the computation model of Chapter 1 and can be represented as extremal problems of some special type.

1. SOLUTION OF EQUATIONS AND SYSTEMS OF EQUATIONS

Methods for solving nonlinear equations of the form $f(x)=0$ and systems of equations of the form $f_i(x)=0$, $i=1,\ldots,n$, have been discussed in detail in textbooks, reference books and monographs, see Bakhvalov [73], Berezin and Zhidkov [66], Ivanov [86], Ortega and Pheinboldt [70], Ostrovski [63], Traub [82], Traub and Woźniakowski [80], Zaguskin [60], etc. Many papers deal with problems of complexity, efficiency and optimality for different algorithmic classes, especially for the class of iterative algorithms. Investigation of iterative algorithms' complexity was initiated by the monograph by Traub [82]; a good survey with annotated bibliography is given in Traub and Woźniakowski [80]. An important role in solving equations belongs to the problems of stability of the solution depending on the initial data and to the problems of regularization, see Goncharskii and Leonov [73], Ivanov, Vasin and Tanana [78], Morozov [74], Tanana [81], Tikhonov [63a, b, 65], Tikhonov and Arsenin [79]. Among the books and papers dealing with solution of equations and systems of equations and most closely connected with the general computation model of Chapter 1 we can mention Bellman and Dreyfus [62], Booth [67, 69], Chernous'ko [68], Eichhorn [68], Gal [77], Gross and Johnson [59], Hyafil [77], Ivanov [75], Kiefer [57], Kruger [75, 76], Maistrovskii [72], Micchelli and Miranker [75], Sikorski [82, 84], Sonnevend [77], Todd [76], Traub, Wasilkowski and Woźniakowski [83], Traub and Woźniakowski [80], Vasil'ev, P.P. [83a, b, 84, 85].

1.1. Sequentially optimal algorithm for seeking the root of a function with values of different signs at the endpoints of the segment

Consider the problem of finding a root of a function f satisfying the Lipschitz condition with a given constant M on $K=[a,b]$ and taking values of different signs at the endpoints a and b. To be more definite, let

$$f\in F_1 \stackrel{def}{=} \{f \mid |f(u)-f(v)|\le M|u-v|,\ u,v\in[a,b],\ f(a)\le 0,\ f(b)\ge 0\}.$$

Apparently, every function from the class F_1 has at least one root on the segment $[a,b]$. Assume that informational computations are evaluations of f, and let $B=K$ and $S(f)=\{x_*\in K \mid f(x_*)=0\}$ (thus we slightly generalize the setting from Section 1 of Chapter 1, allowing subsets of B alongside single points of B as possible values of $S(f)$). With $\gamma(S(f),\beta)$ being the distance from a point β approximating the unknown root to the set $S(f)$, standard argument (see, e.g., Traub and Woźniakowski [80]) shows that the *bisection algorithm* is sequentially optimal (the algorithm that at every step divides the interval containing a root into two equal parts):

$$x_{i+1}=\frac{a_i+b_i}{2}, \quad i=0,1,\ldots, \tag{1.1}$$

where

$$a_0-a, \quad b_0=b,$$

$$a_{i+1}=\begin{cases} a_i, & y_{i+1}>0,\\ x_{i+1}-y_{i+1}/M, & y_{i+1}<0,\end{cases}$$

$$b_{i+1}=\begin{cases} x_{i+1}-y_{i+1}/M, & y_{i+1}>0,\\ b_i, & y_{i+1}<0;\end{cases}$$

if $y_i=0$, the computational process stops – a root has been found.

We now introduce into consideration another criterion of efficiency, which is sometimes called *residual criterion*. Let F be the class of functions that have at least one root in K. If $\beta\in K$ approximates a root of f, then $|f(\beta)|$ is called *residual of the solution*. The aim of "the computer" is to minimize the residual, and so the problem of solving the equation is reduced to an extremal problem of a special type – the problem of minimizing a function with a given (zero) minimum.

For estimating efficiency of computational algorithms, we use the criterion

$$\gamma(S(f),\beta)=|f(\beta)|; \tag{1.2}$$

then

$$\varepsilon_N(z^N)=\inf_{\beta\in K}\max_{f\in F(z^N)}|f(\beta)|, \tag{1.3}$$

and the terminal operation is the function $\tilde{\beta}_*$ defined by formulas (4.3) and (4.4) from Chapter 1. Thus, the root of f is approximated by the point $\beta_*(z^N)$ delivering the infimum in (1.3). Note that, in this setting, the function γ defined by (1.2) is not a metric in B, and so $\tilde{\beta}_*(z^N)$cannot be regarded as the Chebyshev center of the uncertainty set (cf. Section 4.5 of Chapter 1). Nevertheless, we will still call $\tilde{\beta}_*$ the central terminal operation.

For an arbitrary function $f \in F_1(z^i)$, $i \geq 1$, alongside the inequality $\phi_{1i}(x) \leq f(x) \leq \phi_{2i}(x)$, $x \in [a,b]$, where

$$\phi_{1i}(x) = \max_{j=1,\ldots,i} \{y_j - M|x - x_j|\},$$

$$\phi_{2i}(x) = \min_{j=1,\ldots,i} \{y_j + M|x - x_j|\},$$

the inequality $M(x-b) \leq f(x) \leq M(x-a)$, $x \in [a,b]$, is also valid (otherwise the conditions $f(a) \leq 0$, $f(b) \geq 0$ would be violated). Therefore, $\phi_{3i}(x) \leq f(x) \leq \phi_{4i}(x)$, $x \in [a,b]$, where

$$\phi_{3i}(x) = \max\{\phi_{1i}(x), M(x-b)\},$$

$$\phi_{4i}(x) = \min\{\phi_{2i}(x), M(x-a)\}.$$

Observe that, for any realizable situation z^i, any point $(x,y) \in \{(x,y) \mid a \leq x \leq b,\ \phi_{3i}(x) \leq f(x) \leq \phi_{4i}(x)\}$ belongs to the graph of at least one function from $F_1(z^i)$. Hence, we have

$$\max_{f \in F_1(z^N)} |f(\beta)| = \max\{|\phi_{3N}(\beta)|, |\phi_{4N}(\beta)|\},$$

and (1.3) takes the form

$$\varepsilon_N(z^N) = \min_{\beta \in [a,b]} \max\{|\phi_{3N}(\beta)|, |\phi_{4N}(\beta)|\}. \tag{1.4}$$

The bisection algorithm appears to be sequentially optimal (by error) on the class F_1 in the case (1.2), (1.3), too.

Theorem 1.1 (Sukharev [76a]). *The algorithm* (1.1) *is sequentially optimal (by error) on the class F_1 for any given number N of informational computations. The result guaranteed by this algorithm is equal to $M(b-a)\gamma_N$, where $\gamma_N = 1/(2^{N+2}-2)$. The central terminal operation $\tilde{\beta}_*$ is determined by the following condition: if the minimal number among $|y_1|, \ldots, |y_N|, M(b_N - a_N)/2$ is the jth one, then $\tilde{\beta}_*(Z^N)$ is the jth number from the set $\{x_1, \ldots, x_N, (a_N + b_N)/2\}$.*

Proof. We show that the algorithm (1.1) is optimal by error in the classes $\tilde{X}^N$ of all adaptive algorithms and the best guaranteed result is equal to $M(a-b)\gamma_N$.

First we verify that the result guaranteed by the algorithm (1.1) is not worse than $M(b-a)\gamma_N$. Let $f \in F_1$, let the points $x_1, \ldots, x_N$ be obtained according to (1.1), and let $y_j = f(x_j)$, $j = 1, \ldots, N$.

If $|y_j| \leq M(b-a)$ for at least one number $j \in \{1, \ldots, N\}$, then $|\phi_{3N}(x_j)| = |\phi_{4N}(x_j)| = |y_j| \leq M(b-a)\gamma_N$, which, due to (1.4), yields $\varepsilon_N(z^N) \leq M(b-a)\gamma_N$.

Assume now that

$$|y_j| > M(b-a)\gamma_N, \quad j=1,\dots,N. \tag{1.5}$$

We show that in this case

$$b_j - a_j < (b-a)\left(\frac{1}{2^j} - \frac{2^j-1}{2^{j-1}}\gamma_N\right), \quad j=1,\dots,N. \tag{1.6}$$

Indeed, (1.1) and (1.5) imply

$$b_1 - a_1 = \frac{b-a}{2} - \frac{|y_1|}{M} < \frac{b-a}{2} - (b-a)\gamma_N = (b-a)\left(\frac{1}{2} - \gamma_N\right).$$

Suppose that (1.6) holds for $j=i-1$, $i \le N$. Then

$$\begin{aligned} b_i - a_i &= \frac{b_{i-1}-a_{i-1}}{2} - \frac{|y_i|}{M} \\ &< \frac{b-a}{2}\left(\frac{1}{2^{i-1}} - \frac{2^{i-1}-1}{2^{i-2}}\gamma_N\right) - (b-a)\gamma_N \\ &= (b-a)\left(\frac{1}{2^i} - \frac{2^i-1}{2^{i-1}}\gamma_N\right), \end{aligned}$$

which gives induction proof of inequality (1.6). In particular, (1.6) yields

$$M\frac{b_N - a_N}{2} < M(b-a)\left(\frac{1}{2^{N+1}} - \frac{2^N-1}{2^N}\gamma_N\right) = M(b-a)\gamma_N.$$

From here, taking into account (1.4), we get

$$\begin{aligned} \varepsilon_N(z^N) &\le \max\left\{\left|\phi_{3N}\left(\frac{a_N+b_N}{2}\right)\right|, \left|\phi_{4N}\left(\frac{a_N+b_N}{2}\right)\right|\right\} \\ &= M\frac{b_N-a_N}{2} < M(b-a)\gamma_N \end{aligned}$$

(see Fig.16). Thus, we have verified that the result guaranteed by the algorithm (1.1) is not worse than $M(b-a)\gamma_N$. Moreover, it is easy to see (Fig.16) that the minimum in (1.4) is attained at one of the points $x_1,\dots,x_N,(a_N+b_N)/2$ and equals to

$$\varepsilon_N(z^N) = \min\left\{|y_1|,\dots,|y_N|, M\frac{b_N-a_N}{2}\right\}. \tag{1.7}$$

Therefore, approximation $\tilde{\beta}_*(z^N)$ to the root should be set equal to the jth of the points $x_1,\dots,x_N,(a_N+b_N)/2$ if the minimal number among $|y_1|,\dots,|y_n|, M(b_N-a_N)/2$ is the jth one.

Now we make sure that it is impossible to guarantee a better result than $M(b-a)\gamma_N$. That is, we show that for any adaptive algorithm $\tilde{x}^N \in \tilde{X}^N$ there exists a function $f \in F_1$ such that $\varepsilon_N(x^N, y^N) \ge M(b-a)\gamma_N$, where the vector x^N is obtained by formulas (3.2) from Chapter 1 for the algorithm $\tilde{x}^N$ being applied to the function f, and $y_j = f(x_j)$, $j=1,\dots,N$. For the proof, it is sufficient to consider a function f for which $|y_j| = M(b-a)\gamma_N$, $j=1,\dots,N$, and the signs of successively computed numbers $y_1,\dots,y_N$ satisfy the following rule. If the signs of f at two points from the set

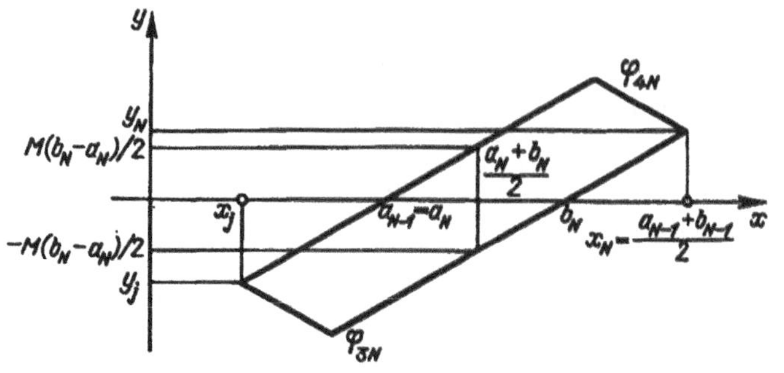

Fig. 16. Computation of $\varepsilon_N(z^N)$ for the class F_1

$a, x_1, \ldots, x_{j-1}, b$, the one closest to x_j from the left and the one closest to x_j from the right, coincide, then y_j has the same sign. If these signs are different, it means that x_j lies inside a segment localizing a root (denote this segment by $[a_{j-1}, b_{j-1}]$); then the sign of y_j is such that the length of the next segment $[a_j, b_j]$ localizing the root is determined by the condition $b_j - a_j = \max\{b_{j-1} - x_j, x_j - a_{j-1}\} - (b-a)\gamma_N$. It is easy to see that such a function $f \in F_1$ exists and for this function $\varepsilon_N(x^N, y^N) = M(b-a)\gamma_N$. This proves optimality of the algorithm (1.1) in the class $\tilde{X}^N$ of all adaptive algorithms. We now prove that this algorithm is sequentially optimal by error.

Suppose we have already evaluated the function f at points $x_1, \ldots, x_i$, $i < N$, and obtained values $y_1, \ldots, y_i$ respectively; assume that $[a_i, b_i]$ is a segment localizing a root, and we have to perform the rest $N - i$ evaluations of f. Then the above argument yields that the best guaranteed result in the situation z^i is equal to

$$\min\{|y_1|, \ldots, |y_i|, M(b-a)\gamma_{N-1}\}. \tag{1.8}$$

Clearly, the result (1.8) can be guaranteed by the algorithm (1.1). Hence, the algorithm (1.1) is sequentially optimal. □

Remark 1. In the situation where $\min\{|y_1|, \ldots, |y_i|\} \leq M(b-a)\gamma_{N-i}$, the rest $N - i$ informational computations cannot guarantee any improvement of the result, and so the choice of x_{i+1} is arbitrary. We can use this freedom to improve the guaranteed result by selecting x_{i+1} at random, i.e., by applying stochastic algorithms. For instance, if $x_{i+1} \in [a_i, b_i]$ is a uniformly distributed random variable, then the result guaranteed on the average is better than the result that can be guaranteed by deterministic algorithms. □

Remark 2. The point chosen by the algorithm (1.1) for the next informational computation does not depend on the total number of informational computations N. Therefore, the algorithm (1.1) is sequentially optimal (counting informational computations) for any prescribed accuracy ε. The computational process should stop after N steps if

$$\varepsilon_N(z^N)=\min\{|y_1|,\ldots,|y_N|,M(b_N-a_N)/2\}\le\varepsilon. \quad \Box$$

1.2. Optimal error algorithms for the functional class F_2

Denote by F_2 the class of functions satisfying the Lipschitz condition with a given constant M on $[a,b]$ and having at least one root on this segment. Clearly, $F_1\supset F_1$.

We construct optimal error algorithms under the assumption that efficiency is estimated with the residual criterion.

Suppose we have already performed N computations $y_j=f(x_j)$, $j=1,\ldots,N$. The following three cases may occur:

Case 1. Some of the numbers $y_1,\ldots,y_N$ have different signs.

Case 2. All the numbers $y_1,\ldots,y_N$ are of the same sign, and the set

$$[a,b]\setminus\bigcup_{j=1}^{N}(x_j-|y_j|/M,x+|y_j|/M)$$

consists of at least two nonintersecting segments (functions from $F_2(z^N)$ can take zero values only at the points from this set; it is not empty since every function from F_2 has at least one root).

Case 3. All the numbers $y_1,\ldots,y_N$ are of the same sign, and

$$[a,b]\setminus\bigcup_{j=1}^{N}(x_j-|y_j|/M,x+|y_j|/M)=[a_N,b_N].$$

In Cases 1 and 2 every point

$$(x,y)\in\{(x,y)\mid a\le x\le b,\ \phi_{1N}(x)\le y\le\phi_{2N}(x)\}$$

belongs to the graph of at least one function $f\in F_2(z^N)$, therefore,

$$\varepsilon_N(z^N)=\min_{\beta\in[a,b]}\max\{|\phi_{1N}(\beta)|,|\phi_{2N}(\beta)|\}. \tag{1.9}$$

Moreover, in Case 2

$$\varepsilon_N(z^N)=\min_{j=1,\ldots,N}|y_j|. \tag{1.10}$$

In Case 3 every function $f\in F_2(z^N)$ has a root in $[a_n,b_n]$ and, therefore,

$$\min\{-M(x-a_N),M(x-b_N)\}\le f\le\max\{M(x-a_N),-M(x-b_N)\}.$$

Combining these inequalities with the inequality $\phi_{1N}(x)\le f(x)\le\phi_{2N}(x)$, we get

$$\phi_{5N}(x)\le f(x)\le\phi_{6N}(x),$$

where

$$\phi_{5N}(x)=\max\{\phi_{1N}(x),\min[-M(x-a_N),M(x-b_N)]\},$$
$$\phi_{6N}(x)=\min\{\phi_{1N}(x),\max[M(x-a_N),-M(x-b_N)]\}.$$

Hence, in Case 3 we have

$$\varepsilon_N(z^N)=\min_{\beta\in B}\max\{|\phi_{5N}(\beta),|\phi_{6N}(\beta)|\}$$
$$=\min\left\{|y_1|,\ldots,|y_N|,M\frac{b_N-a_N}{2}\right\}, \quad (1.11)$$

see Fig.17.

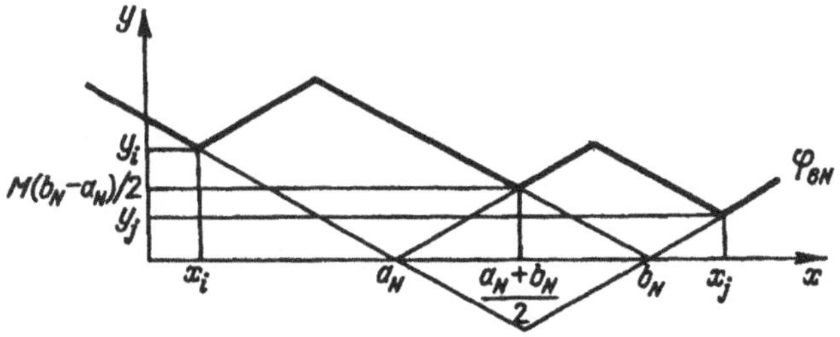

Fig. 17. Computation of $\varepsilon_N(z^N)$ for the class F_2 in Case 3

Finding the central terminal operation $\tilde{\beta}_*$ is no problem. After N informational computations, approximation $\tilde{\beta}_*(z^N)$ to the root of the function in Cases 1 and 3 should be the jth of the points $x_1,\ldots,x_N$, $(a_N+b_N)/2$, provided that the minimal number among $|y_1|,\ldots,|y_N|,(b_N-a_N)/2$ is the jth one. In Case 2, it should be the jth of the points $x_1,\ldots,x_N$, provided that the minimal number among $|y_1|,\ldots,|y_N|$ is the jth one. In Case 1, by $[a_N,b_N]$ we mean the segment that has the minimal length among all the segments localizing a root (if there are several such segments). If this condition still does not determine the segment $[a_N,b_N]$ uniquely, we will choose the leftmost of all the possible segments.

Assume that the terminal operation we use is the central one: $\tilde{\beta}=\tilde{\beta}_*$. Then the following theorem is valid:

Theorem 1.2 (Sukharev [76a]). *The nonadaptive algorithms*

$$x^N(i)=(x_1^0,\ldots,x_{i-1}^0,x_{i+1}^0,\ldots,x_{N+1}^0),\quad i=1,\ldots,N+1,$$

where

$$x_j^0=a+\frac{2j-1}{2(N+1)}(b-a),\quad j=1,\ldots,N+1,$$

are optimal by error on F_2 in the class $\tilde{X}^N$ of all adaptive algorithms and guarantee the result

$$v\stackrel{def}{=}M(b-a)/[2(N+1)].$$

Proof. First, we show that the algorithm $x^N(i)$ does guarantee the result v, $i=1,\dots,N+1$. Let $y_i^0=|f(x_j^0)|>v$, $j=1,\dots,i-1,i+1,\dots,N+1$, $f\in F_2$. Then either we have Case 1 and $\operatorname{sgn} y_{i-1}^0\cdot\operatorname{sgn} y_{i+1}^0=-1$, or we have Case 3. In the both cases

$$[a_N,b_N]=[x_{i-1}^0+|y_{i-1}^0|/M,x_{i+1}^0-|y_{i+1}^0|/M],$$

and, in view of (1.7),

$$\varepsilon_N(z^N)\le M\frac{b_N-a_N}{2}\le\frac{M}{2}\left[\frac{2(b-a)}{N+1}-\frac{2v}{M}\right]=v,$$

where $z^N=(x^N(i),y_1^0,\dots,y_{i-1}^0,y_{i+1}^0,\dots,y_{N+1}^0)$.

To complete the proof, it is sufficient to verify that no algorithm $\tilde{x}^N\in\tilde{X}^N$ guarantees a better result than v. We demonstrate that, for any algorithm $\tilde{x}^N\in\tilde{X}^N$, there exists a function $f\in F_2$ such that $\varepsilon_N(x^N,y^N)=v$, where the vector x^N is obtained by formulas (3.2), Chapter 1, for the algorithm $\tilde{x}^N$ being applied to f, and $y_j=f(x_j)$, $j=1,\dots,N$.

First we show that, for any vector $(x_1,\dots,x_N)$ such that $a\le x_1\le\dots\le x_N\le b$, either at least two of the following $N+1$ inequalities hold:

$$x_1-a\ge\frac{b-a}{2(N+1)},\quad x_{j+1}-x_j\ge\frac{b-a}{N+1},\ j=1,\dots,N-1,$$
$$b-x_N\ge\frac{b-a}{2(N+1)},\tag{1.12}$$

or at least one of the following three inequalities holds:

$$x_1-a>\frac{3(b-a)}{2(N+1)},\tag{1.13}$$

$$x_{k+1}-x_k>\frac{2(b-a)}{N+1},\quad k\in\{1,\dots,N-1\},\tag{1.14}$$

$$b-x_N>\frac{3(b-a)}{2(N+1)}.\tag{1.15}$$

Suppose that at most one of inequalities (1.12) is valid. This may occur in the following three cases:

$$x_{j+1}-x_j<\frac{b-a}{N+1},\ j=1,\dots,N-1,\quad b-x_N<\frac{b-a}{2(N+1)};\tag{1.16}$$

$$x_1-a<\frac{b-a}{2(N+1)},\quad x_{j+1}-x_j<\frac{b-a}{N+1},\ j=1,\dots,k-1,k+1,\dots,N-1,$$

$$k\in\{1,\dots,N-1\},\quad b-x_N<\frac{b-a}{2(N+1)};\tag{1.17}$$

$$x_1-a<\frac{b-a}{2(N+1)},\quad x_{j+1}-x_j<\frac{b-a}{N+1},\ j=1,\dots,N.\tag{1.18}$$

In the case (1.16), summing all the N inequalities gives

$$(b-a)-(x_1-a)<\frac{(2N-1)(b-a)}{2(N+1)},$$

which yields (1.13). Similarly, (1.17) implies (1.14), and (1.18) implies (1.15).

Consider now the vector $x^N=(x_1,\dots,x_N)$ obtained by formulas (3.2), Chapter 1, for the algorithm $\tilde{x}^N$ in the situation $y_1=\dots=y_N=v$. Apparently, the situation $y_1=\dots=y_j=v$ is realizable for any $x_1,\dots,x_j$, and so the values $\tilde{x}_{j+1}(z^j)$ are well-defined for $j=2,\dots,N-1$. Without loss of generality we can assume that $a\le x_1\le\dots\le x_N\le b$. Consider the function $f(x)=\max\limits_{j=1,\dots,n}\{v-M|x-x_j|\}$. Due to Lemma 2.7 of Chapter 1, it satisfies the Lipschits condition with the constant M. It has roots in those of the $N+1$ intervals $[0,x_1),(x_1,x_2),\dots,(x_{N-1},x_n,b]$ for which the corresponding inequalities from (1.12) hold. The above argument proves that there is at least one such interval. Therefore, $f\in F_2$. It is easy to see that $y_j=f(x_j)=v$, $j=1,\dots,N$, hence, applying the algorithm $\tilde{x}^N$ to f gives the vector x^N defined above.

If at least two inequalities in (1.12) are valid, it means that we have Case 2 and $\varepsilon_N(x^N,y^N)=v$ due to (1.10). If at most one of inequalities (1.12) is valid, we have Case 3. In this case, if, in addition, (1.13) holds, then $b_N-a_N=x_1-a-v/M>(b-a)/(N+1)$; if (1.14) holds, then $b_N-a_N=x_{k+1}-x_k-2v/M>(b-a)/(N+1)$; finally, if (1.15) holds, then $b_N-a_N=b-x_N-v/M>(b-a)/(N+1)$. Hence, in Case 3 we also have $\varepsilon_N(x^N,y^N)=\min\{v,M(b_N-a_N)/2\}=v$, see (1.11). Thus, in the both possible cases $\varepsilon_N(x^N,y^N)=v$, which completes the proof. □

Note that, according to the above theorem, an N-step optimal algorithm computes values of the function at the N centers of an optimal covering of $[a,b]$ with $N+1$ segments of equal length.

1.3. Sequentially optimal (by error) algorithm for the functional class F_2

We briefly describe construction of a sequentially optimal (by error) algorithm under the assumption that, as before, efficiency is estimated with the residual criterion (1.2) (all justifications similar to those from the proof of Theorem 1.2 are omitted).

A sequentially optimal algorithm may perform the first informational computation at any of the points that can be chosen by optimal algorithms, that is,

$$x_1\in\left\{a+\frac{2j-1}{2(N+1)}(b-a)\;\middle|\;j=1,\dots,N+1\right\}.$$

Suppose we have already performed i informational computations, $i \geq 1$, and $\operatorname{sgn} y_1 = \cdots = \operatorname{sgn} y_i$. Denote $m_i = \min\{|y_1|, \ldots, |y_i|\}$, and denote by r_{iN} the radius of optimal covering of the set

$$[a, b] \setminus \bigcup_{j=1}^{i} (x_j - |y_j|/M, x_j + |y_j|/M) \tag{1.19}$$

with $N - i + 1$ segments of equal length. In a situation z^i the best guaranteed result is equal to $\min\{Mr_{iN}, m_i\}$. If $Mr_{iN} < m_i$, then x_{i+1} can be any center of an optimal covering of the set (1.19). If $Mr_{iN} \geq m_i$, then the rest $N - i$ informational computations cannot guarantee any improvement of the result, and so the choice of x_{i+1} is arbitrary.

On the other hand, if stochastic algorithms are allowed, then, like for the class F_1 in the situation where $\min\{|y_1|, \ldots, |y_i|\} \leq M(b_i - a_i)\gamma_{N-i}$ (cf. Remark 1 after Theorem 1.1), the result guaranteed by deterministic algorithms can be improved on the average by random choice of x_{i+1}.

The algorithm for computation of r_{iN} is described in Sukharev [75, 76a]; it is analogous to the algorithm for computation of R_{iN} (see Section 4.1 of Chapter 4).

Assume now that we have performed i informational computations and $\operatorname{sgn} y_1 = \cdots = \operatorname{sgn} y_{i-1} \neq \operatorname{sgn} y_i$. To be more definite, let $y_1, \ldots, y_{i-1} > 0$ and $y_i < 0$. Denote by x_{i_1} and x_{i_2} the points from the set $\{x_1, \ldots, x_{i-1}\}$ closest to x_i from the left and from the right respectively (one of the two points x_{i_1} and x_{i_2} may not exist since all the points $x_1, \ldots, x_{i-1}$ may lie on the same side of x_i). The shorter of the segments $[x_{i_1} + y_{i_1}/M, x_i + y_i/M]$ and $[x_i - y_i/M, x_{i_2} - y_{i_2}/M]$ is chosen as $[a_i, b_i]$. Then x_{i+1} and the points of the rest informational computations are selected in accordance with the algorithm (1.1).

1.4. Solution of systems of equations

Consider a system of equations

$$f_i(x) = 0, \quad i = 1, \ldots, m, \quad x \in K \subset \mathbb{R}^n, \tag{1.20}$$

assuming that $f_i \in F_\rho$, $i = 1, \ldots, m$ (where ρ is some quasi-metric), and that the system has at least one solution. The problem of finding a solution of the system (1.20) is equivalent to the problem of minimizing the residual function $\max_{i=1,\ldots,m} |f_i(x)|$, $x \in K$, or, which is the same, to the problem of maximizing $f(x) = -\max_{i=1,\ldots,m} |f_i(x)|$ on K. Due to Lemma 2.7 of Chapter 1, the function f belongs to the class F_ρ, and so the problem can be solved approximately with the help of the algorithms developed in Section 3 of Chapter 4. However, we should take into account that now we know the maximal value of f, which is zero. This enables us to modify the algorithms from Section 3 of Chapter 4 so that they become simpler and at the same time more efficient.

Assume that the accuracy ε of maximization is prescribed, that is, we have to find a point x_ε such that $f(x_\varepsilon) \geq -\varepsilon$. Clearly, this can be done with the help of some modification of the algorithm described in Section 3.3 of Chapter 4. The required modification amounts to replacing all the sets $CK_i(\bar{y}_i + \varepsilon)$ by the sets $CK_i(-\varepsilon)$.

The algorithm from Section 3.3 of Chapter 4 and its modification can still be applied if the system (1.20) is not assumed to be consistent. In this case, if after i steps the stopping criterion for the nonmodified algorithm $CK_i(\bar{y}_i + \varepsilon) = \emptyset$ is satisfied and at the same time $\bar{y}_i < -\varepsilon$, or, for the modified algorithm, $CK_i(-\varepsilon) = \emptyset$ and $\bar{y}_i < -\varepsilon$, this means that the system (1.20) is inconsistent.

If $n > 2$, even in the comparatively simple case (4.13), Chapter 4, direct implementation of these algorithms is impossible for the reasons that were mentioned in Section 4.6 of Chapter 4. In this situation, the solution can be obtained with the help of the algorithm (4.14), (4.15), Chapter 4. The modification of this algorithm analogous to the above one makes the algorithm simpler and more efficient. The modified algorithm selects the point of the $(i+1)$st informational computation according to the rule

$$x_{i+1} \in \{u_j \mid 1 \leq j \leq N,\ \phi_{2i}(u_j) \geq -\varepsilon\} \tag{1.21}$$

(cf. (4.15), Chapter 4).

It is advisable to supplement this algorithm with a block for local refinement of the solution. For instance, in Vasil'ev and Sukharev [83] the local refinement block works in the following way: the solution obtained by the algorithm (1.21) is taken as a starting point for one of the Newton-type methods suggested by Brown [67, 69].

2. MAXIMIZATION OF A MINIMUM FUNCTION WITH COUPLED VARIABLES

Application of the best guaranteed result principle in operations research and game theory requires solution of various minimax and maximin problems, see Germeier [71, 76]. The simplest one is the maximin problem with uncoupled variables (see Dem'yanov and Malozemov [72], Fedorov [79]):

$$\max_{x \in X} \min_{v \in V} Q(x, v). \tag{2.1}$$

In this section, we deal with a more general than (2.1) maximin problem with coupled variables:

$$f(x) \stackrel{def}{=} \min_{v \in B(x)} Q(x, v) \to \sup_{x \in K}, \tag{2.2}$$

where

$$B(x)=\{v\in V \mid \phi_j(x,v)\geq 0,\ j=1,\dots,m\},\quad x\in X,$$

$K=\{x\in X \mid B(x)\neq\emptyset\}$, and V is a metric space. Problems of the type (2.2) often arise in the theory of hierarchical systems and games with information exchange, see Germeier [76].

2.1. When does the minimum function belong to a functional class determined by a quasi-metric?

It is easy to give examples (see, e.g., Fedorov [79]) of the minimum function (2.2) being discontinuous even for smooth $Q,\phi_1,\dots,\phi_m$. In this case maximization of such a function is difficult. Our aim is to establish conditions on $Q,\phi_1,\dots,\phi_m$ guaranteeing some properties of the minimum function that would enable us to construct algorithms for its maximization. For instance, if

$$|f(x_1)-f(x_2)|\leq\rho(x_1,x_2),\quad x_1,x_2\in K, \tag{2.3}$$

and ρ is a quasi-metric, then the problem (2.2) can be solved using the algorithms from Chapter 4. In this connection, we are interested in conditions under which (2.3) holds and the function ρ in (2.3) is a quasi-metric.

We start with establishing a few important properties of the minimum function f defined by (2.2).

Assume that the following conditions are satisfied:

$$|Q(x_1,v)-Q(x_2,v)|\leq\alpha(x_1,x_2),\quad x_1,x_2\in X,\ v\in V, \tag{2.4}$$

where α is a quasi-metric;

$$|\phi_j(x_1,v)-\phi_j(x_2,v)|\leq\beta(x_1,x_2),\quad x_1,x_2\in X,\ v\in V,\ j=1,\dots,m, \tag{2.5}$$

where β is a quasi-metric;

$$|Q(x,v_1)-Q(x,v_2)|\leq\gamma(r(v_1,v_2)),\quad x\in X,\ v_1,v_2\in V, \tag{2.6}$$

where $\gamma(0)=0$, the function γ is defined and nondecreasing on $[0,\infty)$, and r is the distance in V;

$$\min_{j=1,\dots,m}\phi_j(x,v)\leq-\omega(r(v,B(x)))\quad\text{for } v\notin B(x), \tag{2.7}$$

where $\omega(0)=0$, the function ω is defined, continuous and nondecreasing on $[0,\infty)$, $\lim_{r\to\infty}\omega(r)=\infty$, and $r(v,A)=\inf_{\bar v\in A} r(v,\bar v)$ is the distance from the point v to the set A.

Theorem 2.1 (Sukharev and Fedorov [81]). *If assumptions* (2.4)-(2.7) *are valid, then the function f of the form* (2.2) *satisfies condition* (2.3) *with*

$$\rho(x_1,x_2)=\alpha(x_1,x_2)+\gamma[\omega^{-1}(\beta(x_1,x_2))], \tag{2.8}$$

where ω^{-1} is the inverse to ω.

Proof. In view of (2.5) and (2.7), for arbitrary $x_1, x_2 \in X$ and $v \notin B(x_2)$ we have

$$\min_{1\le j\le m} \phi_j(x_1, v) \le \min_{1\le j\le m} \phi_j(x_2, v) + \beta(x_1, x_2) \le -\omega(r(B(x_2))) + \beta(x_1, x_2).$$

If $r(v, B(x_2)) > \omega^{-1}(\beta(x_1, x_2))$, then $\min_{1\le j\le m} \phi_j(x_1, v) < 0$, and so $v \notin B(x_1)$.

Thus, the set $B(x_1)$ lies in a $\omega(\beta(x_1, x_2))$-neighbourhood of the set $B(x_2)$, that is, for any point $v_1 \in B(x_1)$ there is a point $v_2 \in B(x_2)$ such that $r(v_1, v_2) \le \omega^{-1}(\beta(x_1, x_2))$. Therefore,

$$Q(x_2, v_2) - Q(x_1, v_1) = [Q(x_2, v_2) - Q(x_1, v_2)] + [Q(x_1, v_2)) - Q(x_1, v_1)]$$
$$\le \alpha(x_1, x_2) + \gamma[\omega^{-1}(\beta(x_1, x_2))] = \rho(x_1, x_2),$$

which yields that $f(x_2) \le Q(x_1, v_1) + \rho(x_1, x_2)$ for an arbitrary $v_1 \in B(x_1)$, hence $f(x_2) \le f(x_1) + \rho(x_1, x_2)$. The inequality $f(x_1) \le f(x_2) + \rho(x_1, x_2)$ can be established in the same way, which completes the proof. □

Assume in addition that V is a linear normed space with a norm $\|\cdot\|$,

$$r(v_1, v_2) = \|v_1 - v_2\|, \quad v_1, v_2 \in V, \tag{2.9}$$

the function γ has the property

$$\gamma(r_1 + f_2) \le \gamma(r_1) + \gamma(r_2), \quad r_1, r_2 \ge 0, \tag{2.10}$$

and the function ω is continuous and has the property

$$\omega(r_1 + r_2) \ge \omega(r_1) + \omega(r_2), \quad r_1, r_2 \ge 0. \tag{2.11}$$

Observe that, for a continuous nondecreasing on $[0, \infty)$ function ω, assumption (2.11) implies

$$\omega^{-1}(\omega_1 + \omega_2) \le \omega^{-1}(\omega_1) + \omega^{-1}(\omega_2), \quad \omega_1, \omega_2 \ge 0. \tag{2.12}$$

Theorem 2.2 (Sukharev and Fedorov [81]). *Let assumptions* (2.4)-(2.7) *and* (2.9)-(2.11) *be valid. Then the function* ρ *defined by* (2.8) *is a quasi-metric and the upper bound* (2.3) *is sharp for functions of the form* (2.2).

Proof. The fact that ρ has all the properties of a quasi-metric follows directly from our assumptions and inequality (2.12) which is implied by these assumptions.

We now check that the upper bound (2.3) is sharp. For arbitrary $a \in K$ and $v_0 \in V$, put

$$Q(x, v) = -\alpha(a, x) - \gamma(r(v, v_0)), \tag{2.13}$$

$$\phi_j(x, v) = \beta(a, x) - \omega(r(v, v_0)), \quad j = 1, \dots, m, \tag{2.14}$$

where α, β, γ, and ω satisfy all the assumptions of the theorem. Conditions (2.4)-(2.6) are easy to verify. We check if (2.7) is valid. Let $v \notin B(x) = \{v \mid r(v, v_0) \le \omega^{-1}(\beta(a, x))\}$. Then (see Fig.18)

$$r(v, v_0) = r(v, B(x)) + \omega^{-1}(\beta(a, x)).$$

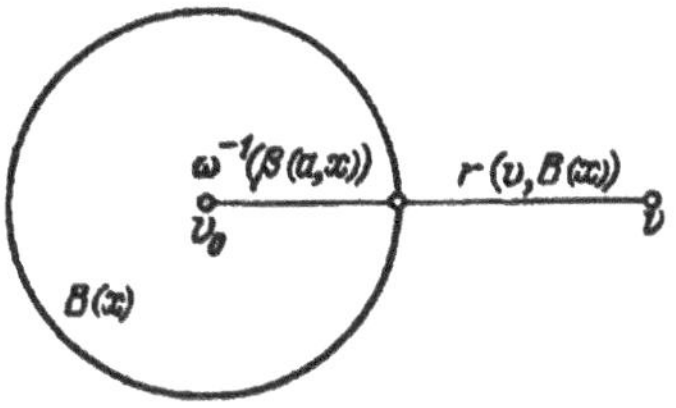

Fig. 18. To the proof of Theorem 2.2

In view of (2.11) this yields

$$\omega(r(v,v_0)) = \omega[r(v,B(x)) + \omega^{-1}(\beta(a,x))] \geq \omega(r(v,B(x))) + \beta(a,x),$$

which is equivalent to inequality (2.7).

Compute now the function (2.2) in the case (2.13), (2.14):

$$\min_{v\in B(x)} Q(x,v) = -\alpha(a,x) - \max_{v\in\{v \mid r(v,v_0)\leq\omega^{-1}(\beta(a,x))\}} \gamma(r(v,v_0))$$
$$= -\alpha(a,x) - \gamma[\omega^{-1}(\beta(a,x))] = -\rho(a,x). \quad (2.15)$$

Putting $a = x_1$ in (2.13) and (2.14) and taking (2.15) into account, we obtain for $f(x) = \min_{v\in B(x)} Q(x,v)$

$$|f(x_1) - f(x_2)| = \rho(x_1, x_2),$$

which completes the proof. □

2.2. Optimal algorithms for maximization of the minimum function

We now formulate the main result on optimal methods for solving the problem (2.2). Suppose that assumptions (2.4)-(2.7) and (2.9)-(2.11) are satisfied. Denote by $\bar{F}$ the subclass of the functional class F_ρ containing all functions of the form (2.2). Let, as before, X^N and $\tilde{X}^N$ denote the classes of nonadaptive and adaptive algorithms with N informational computations respectively.

Theorem 2.3 (Sukharev and Fedorov [81]). *A global optimization algorithm optimal by error in X^N on the class F_ρ is also optimal by error both in X^N and $\tilde{X}^N$ on the subclass $\bar{F}$; the best guaranteed results on F_ρ and $\bar{F}$ coincide.*

Proof. In the proof of Theorem 2.2 we established that all the functions

$$h_a(x) = -\rho(a,x), \quad a \in K, \quad (2.16)$$

belong to the subclass $\bar{F}$. Thus, the fact that an algorithm $x_0^N \in X^N$ optimal on F_ρ is also optimal in X^N on $\bar{F}$, as well as coincidence of the best results guaranteed in X^N on F_ρ and $\bar{F}$, follow from Remark 1 after Theorem 2.1 of Chapter 4.

Now it is easy to see that the class $\bar{F}$ contains all constants and satisfies conditions (5.14) and (5.15), Chapter 1. For instance, (5.15) holds because, for any $x \in K$,

$$f(x) = \min_{v \in B(x)} Q(x,v) \Rightarrow \max\{f(x),0\} = \min_{v \in B(x)} \max\{Q(x,v),0\}.$$

Therefore, due to Theorem 5.3 of Chapter 1., the algorithm x_0^N is optimal by error in $\tilde{X}^N$ on the subclass $\bar{F}$. □

The obtained results in principle solve the problem of optimal maximization of functions f of the form (2.2). The corresponding algorithms have been derived in Chapter 4. However, the problems of constructing optimal coverings of the set K that arise in implementation of optimal algorithms, as well as the problems of optimal computation of function values, are still open. There are some cases though in which a complete solution to the problem has been obtained, for instance, the case $B(x) = \{v \in V \mid \psi_j^-(x) \le v \le \psi_j^+(x),\ j = 1, \dots, l\}$. An example of solving a specific problem of this type is given in Sukharev and Fedorov [81].

2.3. On the assumptions of Theorems 2.1-2.3

In this section, we make some remarks on the assumptions of Theorems 2.1-2.3 and mention some special cases of these assumptions that are of practical interest.

Assumptions (2.4) and (2.5) are valid for arbitrary bounded functions; assumption (2.6) is valid, for instance, for uniformly continuous functions with modulus of continuity γ, provided that (2.9) and (2.10) hold. The only considerably restrictive assumption is (2.7). It is valid, for instance, for $\omega(r) = Rr$ with $R > 0$ (regularity condition) in the following cases (see Fedorov [79]):

Case 1. The functions $\phi_j(x,v)$ are continuous on the product of compact sets $X \times \bar{V}$, where $\bar{V} \supset V$ and V is a finite set, $j = 1, \dots, m$.

Case 2. The functions $\phi_j(x,v)$ are linear with respect to v on a polyhedral set $V \supset \mathbb{R}^n$, i.e., $\phi_j(x,v) = \{a_j, v\} + b_j(x)$, $j = 1, \dots, m$.

Case 3. The functions $\phi_j(x,v)$ are concave with respect to v on a convex compact set $V \subset \mathbb{R}^n$, continuous with respect to (x,v), and satisfy the Slater condition, i.e., for every $x \in X$ there is $\bar{v} \in V$ such that $\phi_j(x,\bar{v}) > 0$, $j = 1, \dots, m$.

Let X be a normed linear space with a norm $\|\cdot\|$, and let $\sigma \le 1$. Put $\alpha(x_1,x_2) = L\|x_1 - x_2\|^\sigma$, $\beta(x_1,x_2) = P\|x_1 - x_2\|$, $\gamma(r) = Mr^\sigma$, and $\omega = Rr$.

Then $\rho(x_1, x_2) = (L + M(P/R)^\sigma)\|x_1 - x_2|^\sigma$, and thus assumption (2.3) is the Hölder condition with the constant $L + M(P/R)^\sigma$ and the exponent σ. For $\sigma = 1$ it becomes the Lipschitz condition.

In conclusion we note that the result of Theorem 2.1 can be extended to the problem of finding a multiple maximin with constrained variables, see Sukharev and Fedorov [85].

3. OPTIMIZATION WITH SEVERAL CRITERIA

In decision theory we often have to deal with the situation where the decision maker is interested in increasing all the components of a vector criterion $f(x) = (f_1(x), \ldots, f_k(x))$ for x belonging to some set $K \in \mathbb{R}^n$. There have been developed a lot of principles of making decisions in this type of situations, see Berezovskii, Borzenko and Kemper [81], Germeier [71, 76], Gvishiani and Emel'yanov [78], Keeney and Raiffa [76], Larichev and Nikiforov [86], Moiseyev [79], Podinovskii and Gavrilov [75], Podinovskii and Nogin [82]. In this section, we consider one of the most traditional principles connected to selecting the set of unimprovable, or Pareto-optimal, vectors from the set $Y = \{y = f(x) \mid x \in K\}$.

3.1. Statement of the problem

A vector $y^* = f(x^*)$ is called *unimprovable* on the set K iff there is no vector $y \in Y$ such that $y \prec y^*$, i.e., there is no $y \in Y$ such that $y_i \geq y_i^*$, $i = 1, \ldots, k$, and the inequality is strict for at least one i. We also say that y^* *is not dominated* by vectors from the set Y or that y^* is *Pareto-optimal* in Y. The point $x^* \in K$ is called an *effective point*. The set of all the vectors unimprovable on K is called the *Pareto set* and denoted by Y^*.

Solution of many problems (or the first stage of the solution) amounts to finding the set of effective points (at the second stage we may have to maximize some additional criterion on the set of effective points). The problem of computer aided design of complex technological systems is a typical example (see Krasnoshchekov [84], Krasnoshchekov, Morozov and Fedorov [79a, b], Krasnoshchekov and Petrov [83], Krasnoshchekov, Petrov and Fedorov [86]).

Let $\varepsilon_1 > 0, \ldots, \varepsilon_k > 0$ and $\varepsilon = (\varepsilon_1, \ldots, \varepsilon_k)$. Our aim is to choose appropriate points $x_1, \ldots, x_N \in K$, compute the values of the vector function f at these points, and, on the basis of this information, single out a subset Y_ε^* of the set $Y_\varepsilon \stackrel{def}{=} \{f(x_1), \ldots, f(x_N)\}$ such that Y_ε^* is a *Pareto ε-approximation* to the set Y^*.

A set Y' is called a *Pareto ε-approximation* to Y'' iff the vectors from Y' are pairwise incomparable (i.e., there are no two vectors $u, v \in Y'$ such that $u \prec v$) and

$$\sup_{y'' \in Y''} \inf_{y' \in Y'} \max_{i=1,\ldots,k} \frac{y_i'' - y_i'}{\varepsilon_i} \leq 1;$$

in other words, under the assumption that the infimum is attained, for any $y'' \in Y''$ there is a vector $y' \in Y'$ such that $y_i' \geq y_i'' - \varepsilon_i$ for all $i = 1, \ldots, k$.

Moreover, the number of points N should be the minimal integer number such that we can guarantee obtaining a solution.

This setting is close to that for the general computation model from Chapter 1 in the case of fixed accuracy ε. They differ, in particular, in that ε now is a vector.

Assume that, for any $x, x' \in K$,

$$|f_i(x) - f_i(x')| \leq M_i \rho(x, x'), \quad i = 1, \ldots, k, \tag{3.1}$$

where $M_1, \ldots, M_k$ are given constants and ρ is a quasi-metric. In addition, assume for simplicity that

$$\varepsilon_1 / M_1 = \cdots = \varepsilon_k / M_k \stackrel{def}{=} \delta. \tag{3.2}$$

Consider the problem of construction of optimal algorithms for solving problems with several criteria. Like in the general computation model from Chapter 1, all such algorithms consist of two stages.

At the fist stage, we perform informational computations (in the case being considered, informational computations are evaluations of the vector function f at points from the set K) and thus gather information about the problem. At the second stage, we construct an approximation to the solution on the basis of the obtained information (in the case being considered, it amounts to selecting a subset Y_ε^* of the set Y_ε). In practice, we do not always proceed to the second stage after the first one has been completed. If adaptive algorithms are applied, intermediate results of the second stage can be used at the first stage for better selection of the points of informational computations.

In our setting, the subset Y_ε^* which is selected from Y_ε at the second stage of every algorithm will always consist of all the vectors unimprovable on $K_\varepsilon \stackrel{def}{=} \{x_1, \ldots, x_N\}$. In other words, Y_ε^* will consist of all the vectors from Y_ε that are not dominated by vectors from Y_ε. In principle, effective organization of the second stage of computations is not difficult, and so, in what follows, by *algorithm* we mean just a method of choosing the points $x_1, \ldots, x_N$.

Denote by F the class of vector functions (problems) f satisfying (3.1). An algorithm is called *optimal* in a given algorithmic class iff it solves every problem from F using at most N steps (i.e., at most N informational computations), and in this algorithmic class there is no algorithm that solves every problem from F using less than N steps. By solving the problem we mean constructing a Pareto ε-approximation Y_ε^* to the set Y^*.

3.2. Optimal algorithms for optimization with several criteria

The following two theorems (and their proofs) are similar to the corresponding statements about ordinary optimization problems (see Chapter 4). In these theorems, δ is assumed to be given by (3.2); we also assume that there exists a finite δ-net in K, i.e., a finite set $A \subset K$ such that $\bigcup_{a \in A} \{x \mid \rho(x,a) \leq \delta\} \supset K$.

Theorem 3.1 (Sukharev [82b]). *An algorithm choosing a δ-net with the minimal number of elements as K_ε is optimal in the class of all nonadaptive algorithms.*

Proof. Denote by N_δ the minimal number of elements that is enough to form a δ-net in K. Clearly, vectors from the set Y_ε^* of the vectors unimprovable on K_ε are pairwise incomparable. Let $y'' = f(x'') \in Y^*$ and $x'' \in K$. There exists an element $x \in K_\varepsilon$ such that $\rho(x, x'') \leq \delta$. According to (3.1), we have $|y_i - y_i''| \leq \varepsilon_i$, where $y = f(x)$, $i = 1, \dots, k$. If $y \notin Y_\varepsilon^*$, there exists a vector $y' \in Y_\varepsilon^*$ such that $y' \prec y$. For this vector we have $y_i' > y_i \geq y_i'' - \varepsilon_i$, $i = 1, \dots, k$. Thus we have shown that the algorithm under consideration solves any problem from F using N_δ steps.

Suppose there exists a nonadaptive algorithm $(x_1^0, \dots, x_\nu^0)$ that solves any probelm from F using $\nu < N_\delta$ steps. By the definition of N_δ, there is a point $b \in K$ such that $\rho(x_i^0, b) \geq \delta + \gamma$, $i = 1, \dots, \nu$, where $\gamma > 0$. Put

$$f_j(x) = \max\{-M_j \rho(x,b) + M_j \delta + M_j \gamma, 0\}, \quad j = 1, \dots, k.$$

Due to Lemmas 2.1 and 2.7 of Chapter 1, $f \in F$. It is easy to see that

$$Y^* = \{(\varepsilon_1 + M_1\gamma, \dots, \varepsilon_k + M_k\gamma)\},$$

$$f(x_i^0) = 0, \quad i = 1, \dots, \nu, \quad Y_\varepsilon^* = \{(0, \dots, 0)\}.$$

Apparently, Y_ε^* is not an ε-approximation to Y^*. This contradiction completes the proof. □

Theorem 3.2 (Sukharev [82b]). *There is no adaptive algorithm that solves every problem from F using less than N_δ steps. Thus, every optimal nonadaptive algorithm is also optimal in the class of all adaptive algorithms.*

Proof. Suppose there is an adaptive algorithm $\tilde{x}_0^\nu$ that solves any problem from F using less than N_δ steps. Denote by $x_1^0, \dots, x_\nu^0$ $(\nu < N_\delta)$ the points of informational computations of the algorithm $\tilde{x}_0^\nu$ when applied to the vector function with identically zero components. Clearly, if we apply $\tilde{x}_0^\nu$ to the vector function f constructed in the proof of Theorem 3.1, we will obtain the same function values at these points. Thus, in this case the algorithm $\tilde{x}_0^\nu$ fails to solve the problem. This contradiction completes the proof. □

We have shown that adaption does not help to reduce the number of informational computations for the "worst" problem from F. However, there are adaptive algorithms that can solve "nice" problems from F much faster than nonadaptive algorithms.

To construct a sequentially optimal (counting informational computations) algorithm, we have to solve complex auxiliary problems, which makes the algorithm impractical. So, instead of this, we will construct an optimal algorithm which is not sequentially optimal but, nevertheless, in "nice" cases requires considerably fewer informational computations than the optimal nonadaptive algorithm. After some modification, this algorithm admits a rather simple program implementation. Due to the definition of optimality, in "bad" cases the numbers of informational computations required by both the adaptive and nonadaptive algorithms are the same.

At the first step, we choose an arbitrary point from a δ-net consisting of N_δ elements for the first informational computation. Then we choose any other point of the same net for the second informational computation.

Suppose we have performed i steps, $i \geq 2$, $x^i = (x_1, \dots, x_i)$, $f(x_m) = y_m = (y_m 1, \dots, y_{mk})$, $m = 1, \dots, i$, $y^i = (y_1, \dots, y_i)$, and $z^i = (x^i, y^i)$. Put

$$\phi_{2ij}(x) = \min_{m=1,\dots,i} \{y_{mj} + M_j \rho(x, x_m)\}, \quad j = 1, \dots, k.$$

Due to Lemmas 1.1 and 1.2 of Chapter 2, the vector function $\phi_{2i} \overset{def}{=} (\phi_{2i1}, \dots, \phi_{2ik})$ (cf. notation (1.3), Chapter 2, for scalar setting) is a sharp upper envelope for vector functions from F taking the values $y_1, \dots, y_i$ at the points $x_1, \dots, x_i$ respectively.

Set

$$C_i = K \setminus \bigcup_{m=1}^{i} \{x \mid \phi_{2i}(x) \leq y_m + \varepsilon\} = K \setminus \bigcup_{m=1}^{i} \bigcup_{j=1}^{k} \{x \mid \phi_{2ij}(x) \leq y_{mj} + \varepsilon_j\}$$

$$= K \setminus \bigcup_{m=1}^{i} \bigcap_{j=1}^{k} \bigcup_{p=1}^{i} \left\{ x \;\middle|\; \rho(x, x_p) \leq \frac{y_{mj} - y_{pj}}{M_j} + \delta \right\}. \tag{3.3}$$

If $C_i = \emptyset$, the computational process stops. If $C_i \neq \emptyset$, denote by $N_\delta(z^i)$ the minimal number of elements from K (but not necessarily from C_i) that form a δ-net in C_i. At the $(i+1)$st step, we choose an arbitrary point from a δ-net in C_i consisting of $N_\delta(z^i)$ elements.

Denote the described algorithm by α.

Theorem 3.3 (Sukharev [82b]). *The algorithm α is optimal.*

Proof. Let $\{x^i_{i+1},\ldots,x^i_{i+N_\delta(z^i)}\}$ be a δ-net in C_i, and let $y''=f(x'')\in Y^*$, $x''\in K$. If $x''\in C_i$, then there is a point $x\in\{x^i_{i+1},\ldots,x^i_{i+N_\delta(z^i)}\}$ such that $\rho(x,x'')\leq\delta$, and, therefore, $|y_j-y''_j|\leq\varepsilon_j$, where $y=f(x)$, $j=1,\ldots,k$. If $x''\notin C_i$, then, by the definition of C_i, there is a point $x\in\{x_1,\ldots,x_i\}$ such that $\phi_{2i}(x'')\leq f(x)+\varepsilon$. In this case we have $y''\leq y+\varepsilon$, where $y=f(x)$, since ϕ_{2i} is an upper envelope. The obtained inequalities prove that performing the rest informational computations at the points $x^i_{i+1},\ldots,x^i_{i+N_\delta(z^i)}$ guarantees solution of the problem.

In view of (3.3), we have

$$C_i\subset C_{i-1}\setminus\{x \mid \rho(x,x_i)\leq\delta\}. \tag{3.4}$$

This yields

$$N_\delta(z^i)\leq N_\delta(z^{i-1})-1 \tag{3.5}$$

since, by construction, x_i is one of the points of a δ-net in C_{i-1} consisting of $N_\delta(z^{i-1})$ elements, and the rest $N_\delta(z^{i-1})-1$ points of this net form, due to (3.4), a δ-net in C_i. Inequality (3.5) holds for any $i\geq 2$; moreover, we have $N_\delta(z^1)\leq N_\delta-1$, i.e.,

$$N_\delta(z^i)\leq N_\delta(z^{i-1})-1\leq N_\delta(z^{i-2})-2\leq\cdots\leq N_\delta-i.$$

Thus, for any i, our choice of x_{i+1} guarantees that the solution will be found after at most $N_\delta-i$ steps. This means that the total number of steps will not exceed N_δ, and, therefore, the algorithm α is optimal. □

Note that the algorithm α is an analogue of the algorithm from Section 3.3 of Chapter 4.

3.3. Program implementation of the optimal algorithm

Direct program implementation of the algorithm α for arbitrary K and ρ is practically impossible because it requires constructing a δ-net with the minimal number of elements at every step, which, in turn, leads to a complicated geometrical problem of constructing a optimal covering. To make the problem more tractable, we introduce some simplifying assumptions.

Let K be an n-dimensional coordinate parallelepiped with edges $2\delta m_1,\ldots,2\delta m_n$, where $m_1,\ldots,m_n$ are integer, and let $\rho(u,v)=\max_{i=1,\ldots,n}|u^i-v^i|$. Let us partition the edges of K into $m_1,\ldots,m_n$ equal parts respectively and draw through the partitioning points hyperplanes parallel to the coordinate hyperplanes. Thus, K is partioned into $N=m_1\cdot\ldots\cdot m_n$ coordinate cubes with edge 2δ. Denote the centers of these cubes by $u_1,\ldots,u_N$. Clearly, the set $U\stackrel{def}{=}\{u_1,\ldots,u_N\}$ is a δ-net in K with the minimal number of elements, and so $N=N_\delta$.

The main simplification of the original algorithm α is based on the same idea as the algorithm (4.14), (4.15) from Chapter 4. The idea is that, after i steps, we construct a δ-net in K in such a way that this δ-net contains points from the set U only. It is easy to see that

$$U_i \stackrel{def}{=} U \setminus \bigcup_{m=1}^{i} \{u \mid \phi_{2i}(u) \leq y_m\} \tag{3.6}$$

is such a δ-net. Apparently, it is enough for the union in (3.6) to include only the sets corresponding to the vectors y_m that are not dominated by the vectors y_j, $j=1,\ldots,m-1,m+1,\ldots,i$. With these assumptions, program implementation of the multi-criterion optimization algorithm becomes rather simple, though the number of elements of the set U_i may appear to be greater than $N_\delta(z^i)$.

To implement the algorithm, we organize an N-element storage array. The jth element of the array corresponds to the jth point u_j (for this, we need a subroutine establishing one-to-one correspondence between the numbers and coordinates of the points). After i steps of the algorithm, the jth element of the array contains the following records: k components of the vector function f for the elements from List 1 defined below; k components of the vector function ϕ_{2i} for the elements from List 2 defined below; all the necessary pointers.

At the first step, we choose randomly with equal probabilities $j_1 \in \{1,\ldots,N\}$, put $x_1 = u_{j1}$, compute $y_1 = f(x_1)$, and organize two linked lists. List 1 contains the element corresponding to the point x_1, and List 2 contains all the rest elements. For the elements from List 2, we compute the values of ϕ_{2i}.

Suppose we have performed i steps and computed vectors $y_1,\ldots,y_i$. List 1 now contains the elements corresponding to those already computed vectors that are not dominated by vectors from the set $\{y_1,\ldots,y_i\}$. The elements corresponding to dominated vectors are among the excluded elements. List 2 consists of the elements that have not been excluded.

At the $(i+1)$st step, we choose randomly with equal probabilities an element from List 2. Suppose the jth element of the array has been chosen. Then we successively compare the vector $\phi_2(u_j)$ with the vectors y_m from List 1.

If $\phi_{2i}(u_j) \leq y_m$ for some vector y_m from List 1, then we choose the element adjacent to the jth one in List 2, exclude the jth element, and successively compare the vector ϕ_{2i} corresponding to the newly chosen element with the vectors y_m from List 1.

Otherwise we put $x_{i+1} - u_j$, compute y_{i+1}, and compute

$$\phi_{2,i+1,p}(u) = \min\{\phi_{2ip}(u_i), y_{i+1,p} + M_p\rho(x_{i+1}, u_l)\}, \quad p=1,\ldots,k,$$

for all the points u_l corresponding to the elements from List 2. Then we successively compare the vector y_{i+1} with the vectors y_m from List 1. If $y_{i+1} \leq y_m$ for some vector y_m, then we exclude the jth array element (which corresponds to the vector y_{i+1}). Otherwise we exclude from List 1 all the vectors dominated by y_{i+1} (if any) and include the jth element into List 1.

The computational process stops when List 2 is empty. The solution is given by the vectors y_m from List 1.

In conclusion, we make some comments on possible modifications of the original setting and the algorithm α.

In practice, the constants $M_1, \dots, M_k$ are often unknown or not known exactly. In this case, the algorithm should start with a block for estimating the constants. Such a block can be constructed using recomendations from Section 4.2 of Chapter 2. Clearly, algorithms with a block for estimating the constants no longer guarantee solution and no longer enjoy the optimality property. However, it is easy to show that in this setting we can not guarantee that the solution will be obtained after N steps for any N.

In some situations, we have to approximate a certain subset of the set Y^* of unimprovable vectors rather than the entire set Y^*, say, we may have to approximate $Y^*(w) = \{y \mid y \in Y^*,\ y_i \geq w_i,\ i = 1, \dots, k\}$. In this case, having selected the jth element of the array from List 2 at the $(i+1)$st step, we start with comparing the vectors $\phi_{2i}(u_j)$ and $w - \varepsilon$. If $\phi_{2is}(u_j) < w_s - \varepsilon_s$ for at least one s, we exclude the jth element. Some other obvious modifications of the algorithm are also required.

Note that, although the algorithm described in this section has some obvious shortcomings in comparison with the optimal nonadaptive algorithm, i.e., exhaustion of the set U (it requires more storage, we have to organize linked lists, compute vectors ϕ_{2i} and compare them with vectors f, etc.), it can save us a lot of computer time in case of high complexity of informational computations.

We conclude with an observation that the problem of approximating the Pareto set is numerically unstable with respect to errors of computation of the partial criteria $f_1, \dots, f_k$ and can be regularized using the methods developed by Fedorov [79], Moiseyev [79], Molodtsov [78, 80], Popov [80], Tikhonov and Arsenin [79].

BIBLIOGRAPHY

Afanas'ev, A.Yu. [73], *The search for the minimum of a function with a bounded second derivative*, [Russian], Zh. Vychisl. Mat. Mat. Fiz. **14** no. 4 (1974), 1018-1021; USSR Comput. Math. Math. Phys. **14** no. 4 (1974), 191-195, [English transl.].

Afanas'ev, A.Yu., and Novikov, V.A. [77], *On the search for the minimum of a function with a bounded third derivative*, [Russian], Zh. Vychisl. Mat. Mat. Fiz. **17** no. 4 (1977), 1031-1034; USSR Comput. Math. Math. Phys. **17** no. 4 (1977), [English transl.].

Aho, A., Hopcroft, J., and Ullman, J. [75], *The Design and Analysis of Computer Algorithms*, Addison-Wesley, Reading (USA), 1975.

Aird, T.J., and Rice, J.R. [77], *Systematic search in high dimensional sets*, SIAM J. Numer. Anal. **14** no. 2 (1977), 296-312.

Akimova, I.Ya. [84], *Application of Voronoi diagrams to combinatorial problems: Survey*, [Russian], Izv. AN SSSR, Ser. Tekhn. Kibern. no. 2 (1984), 102-109.

Alperovich, E.Ye., Batishchev, D.I., and Strongin, R.G. [73], *Theoretical and applied aspects of testing search algorithms*, [Russian], Issues of Cybernetics. Problems of Random Search, Nauchn. Soviet AN SSSR po Kompl. Probl. "Kibernetika", Moscow, USSR, 1973, pp. 53-56.

Anderssen, R.S. [72], *Global optimization*, Optimization., Univ. of Queensland Press, St. Luchia, Australia, 1972, pp. 26-48.

Anderssen, R.S., and Bloomfield, P. [75], *Properties of the random search in global optimization*, J. Optimization Theory and Appl. **16** no. 5/6 (1975), 383-398.

Anuchina, N.N., Babenko, K.I., Godunov, S.K., et al. [79], *Theoretical basis of constructing numerical algorithms for problems of mathematical physics* (K.I. Babenko, ed.), [Russian], Nauka, Moscow, USSR, 1979.

Archetti, F., and Betro, B. [78a], *Some remarks on dimensionality reduction techniques in global optimization problems*, Report **A-49**, Università di Pisa, Dipartimento di Ricerca Operativa e Scienze Statistiche, Pisa, Italy, 1978.

[78b], *On the effectiveness of uniform random sampling in global optimization problems*, Report **A-51**, Università di Pisa, Dipartimento di Ricerca Operativa e Scienze Statistiche, Pisa, Italy, 1978.

[78c], *A stopping criterion for global optimization algorithms*, Report **A-56**, Università di Pisa, Dipartimento di Ricerca Operativa e Scienze Statistiche, Pisa, Italy, 1978.

[78d], *A probabilistic algorithm for global optimization*, Calcolo **16** no. 3 (1978), 335-343.

[80], *Stochastic models and optimization*, Bolletino della Unione Matematica Italiana **5** no. 17-A (1980), 225-301.

Archetti, F., and Szegö, G.P. [80], *Global optimization algorithms*, Nonlinear optimization, theory and algorithms, Birkhauser, Boston, 1980, pp. 429-469.

Avriel, M., and Wilde, D.J. [66], *Optimality proof for the symmetric Fibonacci search technique*, Fibonacci Quart **4** no. 3 (1966), 265-269.

[68], *Golden block search for the maximum of unimodal functions*, Manage. Sci. **14** no. 5 (1968), 307-319.

Baba, N. [81], *Convergence of a random optimization method for constrained optimization problems*, J. Optimization Theory and Appl. **33** no. 4 (1981), 451-461.

[83], *A hybrid algorithm for finding a global minimum*, Int. J. Control **37** no. 5 (1983), 929-942.

Babenko, V.F. [76a], *Asymptotically sharp bounds for the remainder for the best cubature formulas for several classes of functions*, [Russian], Mat. Zametki **19** no. 3 (1976), 313-322; Math. Notes **19** no. 3 (1976), 187-193, [English transl.].

[76b], *Sharp asymptotics for the remainders for the optimal weighted cubature formulas for several classes of functions*, [Russian], Mat. Zametki **20** no. 4 (1976), 589-595; Math. Notes **20** no. 4 (1976), [English transl.].

[77], *On the optimal error bound for cubature formulae on certain classes of continuous functions*, Analysis Matematica **3** no. 1 (1977), 3-9.

Babich, M.D., and Shevchuk, L.B. [82], *On an algorithm for approximate solution of systems of nonlinear equations*, [Russian], Kibernetika no. 2 (1982), 74-79.

Babii, A.N. [78], *An algorithm for finding the global extremum of a function of several variables with a prescribed accuracy*, [Russian], Kibernetika no. 5 (1978), 52-56.

Bakhvalov, N.S. [59], *On approximate computation of multiple integrals*, [Russian], Vestn. Mosk. Un-ta, Ser. Mat., Mekhan., Astron., Fiz., Khim. no. 4 (1959), 3-18.

[61], *An estimate of the mean remainder term in quadrature formulae*, [Russian], Zh. Vychisl. Mat. Mat. Fiz. **1** no. 1 (1961), 64-77; USSR Comput. Math. Math. Phys. **1** no. 1 (1961), 68-82, [English transl.].

[62], *On determination of the initial step and estimation of the principal term of the error in numerical integration with automatic step selection*, [Russian], Numerical Methods and Programming, Izd-vo Mosk. Un-ta, Moscow, USSR, 1962, pp. 69-79.

[64], *On optimal bounds for the convergence of quadrature processes and Monte Carlo type integration methods on functional classes*, [Russian], Numerical Methods for Solving Differential and Integral Equations and Quadrature Formulas, Nauka, Moscow, USSR, 1964, pp. 5-63.

[65], *On computation of multiple integrals with automatic step selection*, [Russian], Numerical Methods and Programming, Izd-vo Mosk. Un-ta, Moscow, USSR, 1965, pp. 237-240.

[66], *On algorithms for selection of the step of integration*, [Russian], Numerical Methods and Programming, Izd-vo Mosk. Un-ta, Moscow, USSR, 1966, pp. 3-8.

[70], *Properties of optimal methods for solution of problems of mathematical physics*, [Russian], Zh. Vychisl. Mat. Mat. Fiz. **10** no. 3 (1970), 555-568; USSR Comput. Math. Math. Phys. **10** no. 3 (1970), 1-19, [English transl.].

[71], *On the optimality of linear methods for operator approximation in convex classes of functions*, [Russian], Zh. Vychisl. Mat. Mat. Fiz. **11** no. 4 (1971), 1014-1018; USSR Comput. Math. Math. Phys. **11** no. 4 (1971), 244-249, [English transl.].

[72], *Lower bounds for the asymptotic characteristics of classes of functions with a dominating mixed derivative*, [Russian], Mat. Zametki **12** no. 6 (1972), 655-664; Math. Notes **12** no. 6 (1972), [English transl.].

[73], *Numerical Methods*, [Russian], Nauka, Moscow, 1973.

Basso, P. [82], *Iterative methods for the localization of the global maximum*, SIAM J. Numer. Anal. **19** no. 4 (1982), 781-792.

Batukhtin, V.D., and Maiboroda, L.A. [84], *Extremum seeking methods for discontinuous functions*, [Russian], Izv. AN SSSR, Ser. Tekhn. Kibern. no. 4 (1984), 192-202.

Beamer, J.H, and Wilde, D.J. [69], *Minimax optimization of unimodal functions of one variable*, Manage. Sci. **15** no. 9 (1969), 528-538.

[70], *Minimax optimization of unimodal functions by variable block search*, Manage. Sci. **16** no. 9 (1970), 529-541.

[71], *Minimax optimization of a unimodal function by variable block derivative search with time delay*, J. Comb. Th. **10** no. 2 (1971), 160-173.

Belaya, N.I. [78], *An algorithm for construction of an optimal by error derivative of a function in the class $C_{2,L,N}$*, [Russian], Izv. Vuzov, Ser. Matematika no. 8 (1978), 31-43.

Belaya, N.I., and Ivanov, V.V. [85], *Reproduction algorithms that are of optimal accuracy for derivatives of certain classes of functions*, [Russian], Zh. Vychisl. Mat. Mat. Fiz. **25** no. 3 (1985), 456-461; USSR Comput. Math. Math. Phys. **25** no. 3 (1985), 87-91, [English transl.].

Bellman, R. [57], *Dynamic Programming*, Princeton, New Jersey, 1957.

Bellman, R., and Dreyfus, S. [62], *Applied Dynamic Programming*, Princeton, New Jersey, 1962.

Berezin, I.S., and Zhidkov, N.P. [66], *Methods of Computations*, [Russian], vol. 1, Nauka, Moscow, USSR, 1966.

Berezovskii, A.I., Borzenko, V.I., and Kemper, L.M. [81], *Binary Relations in Multicriterion Optimization*, [Russian], Nauka, Moscow, USSR, 1981.

Berezovskii, A.I., Danilenko, L.S., Dulskaya, V.A., and Ivanov, V.V. [79], *Approximation of functions,*, Rep. 79-48, [Russian], Institute of Cybernetics, Kiev, 1979.

Berezovskii, A.I., and Ivanov, V.V. [77], *On optimal by error uniform spline approximation*, [Russian], Izv. Vuzov, Ser. Matematika no. 10 (1977), 14-24.

Betro, B. [84], *Bayesian testing of nonparametric hypotheses and its application to global optimization*, J. Optimization Theory and Appl. **42** no. 1 (1984), 31-50.

Boender, C.G.E, Rinnooy Kan, A.H.G., Timmer, G.T., and Stougie, L. [82], *A stochastic method for global optimization*, Math. Programming **22** no. 2 (1982), 125-140.

Boikov, I.V. [83a], *Optimal by Error Algorithms for Approximate Computation of Singular Integrals*, [Russian], Izd-vo Sarat. Un-ta, Saratov, USSR, 1983.

[83b], *Optimal Computational Methods for Automatic Regulation Problems*, [Russian], Izd-vo Penz. Politekhn. In-ta, Penza, USSR, 1983.

Bojanov, B.D. [73], *Optimal rate of integration and ε-entropy of a class of analytic functions*, [Russian], Mat. Zametki **14** no. 1 (1973), 3-10; Math. Notes **14** no. 1 (1973), 551-556, [English transl.].

[75], *Best methods of interpolation for certain classes of differentiable functions*, [Russian], Mat. Zametki **17** no. 4 (1975), 511-524; Math. Notes **17** no. 4 (1975), 301-309, [English transl.].

[86], *Comparison theorems in optimal recovery*, Optimal Algorithms, BAS, Sofia, 1986, pp. 15-50.

Bojanov, B.D., and Chernogorov, V.G. [77], *An optimal interpolation formula*, J. Approx. Theor **20** no. 3 (1977), 264-274.

Booth, R.S. [67], *Location of zeros of derivatives*, SIAM J. Appl. Math. **15** no. 6 (1967), 1496-1501.

[69], *Location of zeros of derivatives, II*, SIAM J. Appl. Math. **17** no. 2 (1969), 409-415.

Boult, T. [86], *Some examples and applications of information-based complexity*, Optimal Algorithms, BAS, Sofia, 1986, pp. 51-64.

Brooks, S.H. [52], *A discussion of random methods for seeking maxima*, Operations Res. **6** no. 2 (1952), 244-251.

Brown, K.M. [67], *Solution of simultaneous nonlinear equations*, Comm. Assoc. Comput. Mach. **10** no. 11 (1967), 728-729.

[69], *A quadratically convergent Newton-like method based upon Gaussian elimination*, SIAM J. Numer. Analysis **6** no. 4 (1969), 560-569.

Brusov, V.S., and Piyavskii, S.A. [71], *A computational algorithm for optimality covering a plane region*, [Russian], Zh. Vychisl. Mat. Mat. Fiz. **11** no. 2 (1971), 304-312; USSR Comput. Math. Math. Phys. **11** no. 2 (1971), 17-27, [English transl.].

Chernous'ko, F.L. [68], *An optimal algorithm for finding the roots of an approximately computed function*, [Russian], Zh. Vychisl. Mat. Mat. Fiz. **8** no. 4 (1968), 705-724; USSR Comput. Math. Math. Phys. **8** no. 4 (1968), 1-23, [English transl.].

[70a], *Optimal search for the extremum of a unimodal function*, [Russian], Zh. Vychisl. Mat. Mat. Fiz. **10** no. 4 (1970), 922-933; USSR Comput. Math. Math. Phys. **10** no. 4 (1970), 15-21, [English transl.].

[70b], *Optimal search for the minimum of a convex function*, [Russian], Zh. Vychisl. Mat. Mat. Fiz. **10** no. 6 (1970), 1355-1366; USSR Comput. Math. Math. Phys. **10** no. 6 (1970), 20-33, [English transl.].

Chernous'ko, F.L., and Melikyan, A.A. [78], *Game Problems of Search and Control*, [Russian], Nauka, Moscow, USSR, 1978.

Chichinadze, V.K. [83], *Solution of Nonconvex Nonlinear Optimization Problems*, [Russian], Nauka, Moscow, USSR, 1983.

Chuyan, O.R. [84], *On optimal algorithms for seeking the extremum of a differentiable function*, [Russian], Vestn. Mosk. Un-ta, Ser. Vychisl. Matem. i Kibern., no. 3 (1984), 28-34.

[86], *An optimal single-step algorithm for maximizing doubly differentiable functions*, [Russian], Zh. Vychisl. Mat. Mat. Fiz. **26** no. 3 (1986), 383-397; USSR Comput. Math. Math. Phys. **26** no. 2 (1986), 37-47, [English transl.].

Clarke, F.H. [83], *Optimization and nonsmooth analysis*, Wiley, New York, 1983.

Collatz, L. [64], *Funktionalanalysis und Numerische Mathematik*, Springer-Verlag, Berlin, 1964.

Converse, A.O. [67], *The use of uncertainty in a simultaneous search*, Operations Res. **15** no. 6 (1967), 1088-1095.

Coxeter, H. [62], *The classification of zonohedra by means of projective diagrams*, Journal de mathématiques pures et appliquées **41** no. 2 (1962), 137-156.

Danilin, Yu.M. [71], *Estimation of the efficiency of an absolute–minimum–finding algorithm*, [Russian], Zh. Vychisl. Mat. Mat. Fiz. **11** no. 4 (1971), 1026-1031; USSR Comput. Math. Math. Phys. **11** no. 4 (1971), 261-267, [English transl.].

Danilin, Yu.M., and Piyavskii, S.A. [67], *An algorithm for finding the absolute minimum*, [Russian], Seminar. Theory of Optimal Solutions, vol. 2, Izd-vo IK AN USSR, Kiev, USSR, 1967, pp. 25-37, [Russian].

De Boor, C. [71a], *On writing an automatic integration algorithm*, Mathematical software, Academic Press, New York, 1971, pp. 417-449.

[71b], *CARDE: An algorithm for numerical quadrature*, Mathematical software, Academic Press, New York, 1971, pp. 417-449.

De Boor, C., and Rice, J.R. [79], *An adaptive algorithm for multivariate approximation*, J. Approx. Theory **25** no. 4 (1979), 337-360.

Dem'yanov, V.F., Malozemov, V.N. [72], *Introduction to Minimax*, [Russian], Nauka, Moscow, USSR, 1972.

Dem'yanov, V.F., Vasil'ev, L.V. [81], *Non-differentiable Optimization*, [Russian], Nauka, Moscow, USSR, 1981.

Devroye, L. [78], *Progressive global random search of continuous functions*, Math. Programming **15** no. 3 (1978), 330-342.

Dixon, L.C.W. [80], *Reflections on nondifferentiable optimization. Part 1, Ball gradient*, J. Optimization Theory and Appl. **32** no. 2 (1980), 123-133.

Dixon, L.C.W., and Gaviano, M. [80], *Reflections on nondifferentiable optimization. Part 2. Convergence*, J. Optimization Theory and Appl. **32** no. 3 (1980), 259-276.

Dixon, L.C.W., Szegö, G.P. [75], *Towards global optimization, I* (L.C.W. Dixon, G.P. Szegö, eds.), North-Holland, Amsterdam, 1975.

[78], *Towards global optimization, II* (L.C.W. Dixon, G.P. Szegö, eds.), North-Holland, Amsterdam, 1978.

Dixon, V.A. [74], *Numerical quadrature: A survey of the available algorithms*, Software for numerical mathematics, Academic Press, London, New York, 1974, pp. 105-137.

Eichhorn, B.H. [68], *On sequential search*, Selected statistical papers. V.1, Math. Centrum, Amsterdam, 1968, pp. 81-95.

Einarsson, B. [74], *Testing and evaluation of some subroutines for numerical quadrature*, Software for numerical mathematics, Academic Press, London, New York, 1974, pp. 149-157.

Evtushenko, Yu.G. [71], *Numerical Methods for finding global extrema (case of a non-uniform mesh)*, [Russian], Zh. Vychisl. Mat. Mat. Fiz. **11** no. 6 (1971), 1390-1403; USSR Comput. Math. Math. Phys. **11** no. 6 (1971), 38-54, [English transl.].

[74], *Methods for finding the global extremum*, [Russian], Operations Research, VTs AN SSSR, Moscow, USSR, 1974, pp. 39-68.

[82], *Methods for Solving Extremal Problems and their Application to Optimization Systems*, [Russian], Nauka, Moscow, USSR, 1982.

Fedorov, V.V. [71], *Theory of Optimal Experimental Disign*, [Russian], Nauka, Moscow, USSR, 1971.

[79], *Numerical Methods of Maximin*, [Russian], Nauka, Moscow, USSR, 1979.

Fine, T. [66], *Optimum search for the location of the maximum of a unimodal function*, IEEE Trans. on Information Theory **12** no. 2 (1966), 103-111.

Finney, D.J. [60], *An Introduction to the Theory of Experimental Design*, The University of Chicago Press, 1960.

Fisher, S.D., and Micchelli, C.A. [84], *Optimal sampling of holomorphic functions*, Amer. J. Math. **106** no. 3 (1984), 593-609.

Fosdick, L.C. [79], *Performance evaluation of numerical software* (L.G. Fosdick, ed.), North-Holland, Amsterdam, 1979.

Gabdulkhayev, B.G. [80], *Optimal Approximations to the Solutions of Linear Problems*, [Russian], Izd-vo Kaz. Un-ta, Kazan', USSR, 1980.

Gaffney, P.W. [77], *The range of possible values of $f(x)$*, Computer Science and Systems Division Rep., Harwell, AERE, Oxfordshire, 1977.

[78], *To compute the optimal interpolation formula*, Math. Comp. **32** no. 143 (1978), 763-777.

Gaffney, P.W., and Powell, M.J.D. [76], *Optimal interpolation*, Numerical analysis, Lecture Notes in Math., v.506, Springer-Verlag, Berlin, New York, 1976, pp. 90-100.

Gal, S. [71], *Sequential minimax search for a maximum when prior information is available*, SIAM J. Appl. Math. **21** no. 4 (1971), 590-595.

[72], *Multidimensional minimax search for a maximum*, SIAM J. Appl. Math. **23** no. 4 (1972), 513-526.

[77], *Optimal sequential and parallel search for finding a root*, J. Comb. Th. Ser. A **23** no. 1 (1977), 1-14.

Gal, S., and Micchelli, C.A. [80], *Optimal sequential and nonsequential procedures for evaluating a functional*, Applicable Analysis **10** no. 2 (1980), 105-120.

Gametskii, A.F. [63], *On the optimality of Voronoi's principal lattice of the first type among the lattices of the first type of any number of dimensions*, [Russian], Dokl. Akad. Nauk SSSR **151** (1963), 482-484.

Ganshin, G.S. [76], *Computation of the maximum of a function*, [Russian], Zh. Vychisl. Mat. Mat. Fiz. **16** no. 1 (1976), 30-39; USSR Comput. Math. Math. Phys. **16** no. 1 (1976), [English transl.].

[77], *Optimal passive algorithms for evaluating the maximum of a function in an interval*, [Russian], Zh. Vychisl. Mat. Mat. Fiz. **17** no. 3 (1977), 562-571; USSR Comput. Math. Math. Phys. **17** no. 3 (1977), 8-17, [English transl.].

[83], *Computation of the maximum of a function of several variables*, [Russian], Kibernetika no. 2 (1983), 61-63.

Gaponenko, Yu.L. [81], *A method for finding the global extremum of a nonlinear functional*, [Russian], Kibernetika no. 2 (1983), 58-61.

Garey, M., Johnson, D.S. [79], *Computers and Intractability*, Freeman, San Francisco, 1979.

Gaviano, M, [75], *Some general results on the convergence of random search algorithms in minimization problems*, Towards global optimization, North-Holland, Amsterdam, 1975, pp. 149-157.

Germeier, Yu.B. [67], *Methodological and Mathematical Foundations of Operations Research and Game Theory*, [Russian], Izd-vo Mosk. Un-ta, Moscow, USSR, 1967.

[71], *Introduction to Operations Research Theory*, [Russian], Nauka, Moscow, USSR, 1971.

[76], *Games with Non-analytic Interests*, [Russian], Nauka, Moscow, USSR, 1976.

Germeier, Yu.B., Krylov, I.A. [72], *Maximin search by the method of discrepancies*, [Russian], Zh. Vychisl. Mat. Mat. Fiz. **12** no. 4 (1972), 871-881; USSR Comput. Math. Math. Phys. **12** no. 4 (1972), 36-48, [English transl.].

Girlin, S.K. [78], *On optimal by error interpolation and minimization of functions from the class* $C_{2,L_1,L_2,\ldots,L_m,N}$, [Russian], Izv. Vuzov, Ser. Matematika, no. 10 (1978), 95-98.

Girlin, S.K., Ganshin, G.S. [83], *Optimal algorithms for finding the minimum of a function*, [Russian], Izv. Vuzov, Ser. Matematika, no. 10 (1983), 75-77.

Glinkin, I.A. [81a], *On optimal integration of monotonic functions*, [Russian], Mathematical Methods in Operations Research, Izd-vo Mosk. Un-ta, Moscow, USSR, 1981, pp. 37-46.

[81b], *On optimal search for the extremum of several functions*, [Russian], Mathematical Methods in Operations Research, Izd-vo Mosk. Un-ta, Moscow, USSR, 1981, pp. 46-54.

[83], *Optimal algorithms for integrating convex functions*, [Russian], Zh. Vychisl. Mat. Mat. Fiz. **23** no. 2 (1983), 267-277; USSR Comput. Math. Math. Phys. **23** no. 2 (1983), 6-12, [English transl.].

[84], *On the best quadrature formula for the class of convex functions*, [Russian], Mat. Zametki **35** no. 5 (1984), 697-707; Math. Notes **35** no. 5 (1984), [English transl.].

[85], *On search for the global extremum of a function that can be represented by an integral over a domain of changable configuration*, [Russian], Issues of Cybernetics. Models and Methods of Global Optimization, Nauchn. Soviet AN SSSR po Kompl. Probl. "Kibernetika", Moscow, USSR, 1985, pp. 91-97.

Glinkin, I.A., Sukharev, A.G. [80], *Sequentially optimal integration algorithm*, GosFAP, No. П004578, (Annotation in: *Algorithms and Programs*, no. 6 (1980), П004578 [Russian]).

[84], *Evaluation of multidimensional integrals using Peano-curve type involutions*, [Russian], Zh. Vychisl. Mat. Mat. Fiz. **24** no. 8 (1984), 1259-1263; USSR Comput. Math. Math. Phys. **24** no. 8 (1984), 182-185, [English transl.].

[85], *Performance of some numerical integration algorithms and their application to solution of extremal problems*, [Russian], Issues of Cybernetics. Models and Methods of Global Optimization, Nauchn. Soviet AN SSSR po Kompl. Probl. "Kibernetika", Moscow, USSR, 1985, pp. 23-37.

Glushkov, V.M. [75], *On interactive method of solving optimization problems*, [Russian], Kibernetika no. 4 (1975), 2-6.

[80], *On system optimization*, [Russian], Kibernetika no. 5 (1980), 89-90.

Glushkov, V.M., Ivanov, V.V., Mikhalevich, V.S., Sergienko, I.V., and Stognii, A.A. [77], *Reserves of optimization of computations*, [Russian], Rep., Institute of Cybernetics, Kiev, 1977.

Glushkov, V.M., Ivanov, V.V., and Yanenko, V.M. [83], *Modeling of Developing Systems*, [Russian], Nauka, Moscow, USSR, 1983.

Goldstein, A.A. [77], *Optimization of Lipschitz continuous functions*, Math. Programming **13** no. 1 (1977), 14-22.

Golomb, M. [77], *Interpolation operators as optimal recovery schemes for classes of analytic functions*, Optimal estimation in approximation theory, Plenum Press, New York, London, 1977, pp. 93-138.

Goncharskii, A.V., Leonov, A.S., and Yagola, A.O. [73], *A generalized discrepancy principle*, [Russian], Zh. Vychisl. Mat. Mat. Fiz. **13** no. 2 (1973), 294-302; USSR Comput. Math. Math. Phys. **13** no. 2 (1973), 25-37, [English transl.].

Gorodetskii, S.N., Neimark, Yu.I., and Fadeyev, V.A. [83], *Non-parametrical adaptive stochastic models in global optimization search problems*, [Russian], Mathematical Statistics and Its Applications, vol. 9, Izd-vo TGU, Tomsk, USSR, 1983, pp. 52-58.

Grebennikov, A.I. [78], *The optimal approximation of non-linear operators*, [Russian], Zh. Vychisl. Mat. Mat. Fiz. **18** no. 3 (1978), 762-766; USSR Comput. Math. Math. Phys. **18** no. 3 (1978), 236-242, [English transl.].

[83], *Spline Method and Solution of Ill-posed Problems of Approximation Theory*, [Russian], Izd-vo Mosk. Un-ta, Moscow, USSR, 1983.

Grishagin, V.A., and Strongin, R.G. [84], *Optimization of multi-extremum functions with monotonically unimodal constraints*, [Russian], Izv. AN SSSR, Ser. Tekhn. Kibern. no. 4 (1984), 203-208.

Gromenko, V.M., and Gurin, L.S. [74], *An optimal algorithm of finding the extremum for polynomials with bounded modulus*, [Russian], Avtomatika i Vychisl. Tekhn. no. 2 (1974), 49-56.

Gross, O. [56], *A class of discrete type minimization problems*, Rand Corporation. Research Memorandum No. 1644, 1956.

Gross, O., and Johnson, S. [59], *Sequential minimax search for a zero of a convex function*, Math. Tables and Other Aids to Computation **13** no. 65 (1959), 44-51.

Gupal, A.M. [79], *Stochastic Methods of Solving Non-smooth Extremal Problems*, [Russian], Nauk. Dumka, Kiev, USSR, 1979.

Gurin, L.S., Dymarskii, Yu.A., and Merkulov, A.D. [68], *Problems and Methods of Optimal Resource Allocation*, [Russian], Sov. Radio, Moscow, USSR, 1968.

Gvishiani, D.M., and Emel'yanov, S.V. [78], *Decision Problems* (D.M. Gvishiani, and S.V. Emel'yanov,, eds.), [Russian], Mashinostroyenie, Moscow, USSR, 1978.

Haber, S. [75], *Adaptive integration and improper integrals*, Math. Comp. **29** no. 131 (1975), 806-809.

Hansen, E.R. [79], *Global optimization using interval analysis: The one-dimensional case*, J. Optimization Theory and Appl. **29** no. 3 (1979), 331-344.

[80], *Global optimization using interval analysis: The two-dimensional case*, Numer. Math. **34** no. 3 (1980), 247-270.

[84], *Global optimization with data perturbations*, Comput. and Oper. Res. **11** no. 2 (1984), 97-104.

Heindl, G. [82], *Optimal quadrature of convex functions*, Numerical integration. International series of numerical mathematics, v.57, Birkhauser Verlag, Basel, 1982, pp. 138-147.

Heyman, M. [68], *Optimal simultaneous search for the maximum by the principle of statistical information*, Operations Res. **16** no. 6 (1968), 1194-1205.

Himmelblau, D. [71], *Applied Nonlinear Programming*, McGraw-Hill, New York, 1971.

Hofman, K.L. [81], *A method for globally minimizing concave functions over convex sets*, Math. Programming **20** no. 1 (1981), 22-32.

Hyafil, L. [77], *Optimal search for the zero of the* $(n-1)st$ *derivative*, IRIA/LABORIA Rep. No. 247 (1977).

Ivanov, V.V. [72a], *On optimal algorithms for numerical solution of integral equations*, [Russian], Continuum Mechanics and Related Problems of Analysis, Nauka, Moscow, USSR, 1972, pp. 209-219.

[72b], *On optimal algorithms for computing singular integrals*, [Russian], Dokl. Akad. Nauk SSSR **204** no. 1 (1972), 21-24.

[72c], *On optimal algorithms for minimizing functions of certain classes*, [Russian], Kibernetika no. 4 (1972), 81-94.

[75], *Algorithms of optimal accuracy for the approximate solution of operator equations of the first kind*, [Russian], Zh. Vychisl. Mat. Mat. Fiz. **15** no. 1 (1975), 3-11; USSR Comput. Math. Math. Phys. **15** no. 1 (1975), 1-9, [English transl.].

[77], *Algorithms of optimal accuracy for approximating functions from certain classes*, [Russian], Theory of Function Approximation, Nauka, Moscow, USSR, 1977, pp. 195-200.

[78], *On minimization of the number of operations for ill-posed linear problems*, [Russian], Izv. Vuzov, Ser. Matematika no. 11 (1978), 47-54.

[86], *Methods of Computations using Computers. Reference Book*, [Russian], Nauk. Dumka, Kiev, USSR, 1986.

Ivanov, V.V., and Korzhova, V.N. [83], *Optimization of algorithms for estimating certain probability characteristics*, [Russian], Kibernetika no. 3 (1983), 94-102.

Ivanov, V.V., Ludvichenko, V.A., Mikhalevich, V.S., Trubin, V.A., and Shor, N.Z. [79], *Problems of Improving Efficiency of Algorithms for Minimization of Functions and Mathematical Programming,*, [Russian], Rep. 79-59, Institute of Cybernetics, Kiev, USSR, 1979.

Ivanov, V.V., Vasin, V.V., and Tanana, V.P. [78], *Theory of Ill-posed Linear Problems and Its Applications*, [Russian], Nauka, Moscow, USSR, 1978.

Ivanov, V.V., and Zadiraka, V.K. [79], *Problems of Optimization of Computations*, [Russian], Ob-vo "Znanie", Kiev, USSR, 1979.

Johnson, S.J., *Best exploration for maximum is Fibonaccian*, Rand Corporation Research Memorandum RM 1590, Santa-Monica, 1955.

Kacewicz, B.Z. [82], *On the optimal error of algorithms for solving a scalar autonomous ODE*, BIT **22** no. 4 (1982), 503-518.

[83], *Optimality of Euler-integral information for solving a scalar autonomous ODE*, BIT **23** no. 2 (1983), 217-230.

[84], *How to increase the order to get minimal-error algorithms for systems of ODE*, Num. Math. **45** no. 1 (1984), 93-104.

Kahaner, D.K. [71], *Comparison of Numerical Quadrature Formulas*, Mathematical Software, Academic Press, New York, London, 1971, pp. 229-259.

Kalinin, I.N. [84], *On investigation and comparison of optimization algorithms*, [Russian], Kibernetika no. 1 (1984), 77-80.

Karlin, S. [59], *Mathematical Methods and Theory in Games, Programming and Economics*, Pergamon Press, London, 1959.

Karmanov, V.G. [86], *Mathematical Programming*, [Russian], Nauka, Moscow, USSR, 1986.

Karp, R.M., and Miranker, W.L. [68], *Parallel minimax search for a maximum*, J. Comb. Th. **4** no. 1 (1968), 19-35.

Kaupe, A.F. [64], *On optimal search techniques*, Comm. Assoc. Comput. Mach. **7** no. 1 (1964), 38.

Keeney, R.L., Raiffa, H. [76], *Decisions with Multiple Objectives: Preferences and Value Tradeoffs*, John Wiley & Sons, New York, 1976.

Kiefer, J. [53], *Sequential minimax search for a maximum*, Proc. Amer. Math. Soc. **4** no. 3 (1953), 502-506.

[57], *Optimum sequential search and approximation methods under minimum regularity assumptions*, J. Soc. Indust. Appl. Math. **5** no. 3 (1957), 105-136.

Kolmogorov, A.N., and Fomin, S.B. [76], *Elements of Function Theory and Functional Analysis*, [Russian], Nauka, Moscow, USSR, 1976.

Kolmogorov, A.N., and Tikhomirov, V.M. [59], *The ε-entropy and ε-capacity of sets in functional spaces*, [Russian], UMN **14** no. 2 (1959), 3-80.

Kononov, V.A., and Biryukova, T.L. [75], *A method for seeking the extrema for a certain class of univariate functions*, [Russian], Computational Mathematics, Institute of Cybernetics, Kiev, USSR, 1975, pp. 89-98.

Korchanov, S.V. [84], *On optimal methods for integrating unimodal functions*, [Russian], Vestn. Mosk. Un-ta, Ser. Vychisl. Matem. i Kibern., no. 3 (1984), 38-43.

[86a], *On optimal algorithms for integrating functions with a given number of extrema*, [Russian], Software and Models in Operations Research, Izd-vo Mosk. Un-ta, Moscow, USSR, 1986, pp. 177-185.

[86b], *On algorithms for recovery of an iterated functional that are optimal in the classes of products of lattices*, [Russian], Vestn. Mosk. Un-ta, Ser. Vychisl. Matem. i Kibern., no. 4 (1986), 25-30.

[86c], *On optimal algorithms for recovery of an iterated functional*, [Russian], VINITI, No. 2489–B86, April 8, 1986.

[86d], *One-step optimal algorithms for integrating twice differentiable functions*, [Russian], VINITI, No. 6970–B86, Feb. 10, 1986.

Korneichuk, N.P. [68], *Best cubature formulas for some classes of functions of many variables*, [Russian], Mat. Zametki **3** no. 5 (1968), 565-576; Math. Notes **3** no. 5 (1968), 360-367, [English transl.].

[76], *Extremal Problems of Approximation Theory*, [Russian], Nauka, Moscow, USSR, 1976.

[84], *Splines in Approximation Theory*, [Russian], Nauka, Moscow, USSR, 1984.

[86], *Optimal methods for coding and recovering functions*, [Russian], Optimal Algorithms, BAN, Sofia, 1986, pp. 157-171.

Korotchenko, A.G. [78], *An algorithm for seeking the maximum value of univariate functions*, [Russian], Zh. Vychisl. Mat. Mat. Fiz. **18** no. 3 (1978), 563-573; USSR Comput. Math. Math. Phys. **18** no. 3 (1978), 34-45, [English transl.].

[79], *On search for the minimum of several unimodal functions*, [Russian], Zh. Vychisl. Mat. Mat. Fiz. **19** no. 5 (1979), 1337-1340; USSR Comput. Math. Math. Phys. **19** no. 5 (1979), [English transl.].

Krasnoshchekov, P.S. [84], *Mathematical Models in Operations Research*, [Russian], Znanie, Moscow, USSR, 1984.

Krasnoshchekov, P.S., Morozov, V.V., and Fedorov, V.V. [79a], *Decomposition in design problems*, [Russian], Izv. AN SSSR, Ser. Tekhn. Kibern., no. 2 (1979), 7-17.

[79b], *Successive aggregation in problems of internal design of technological Systems*, [Russian], Izv. AN SSSR, Ser. Tekhn. Kibern., no. 5 (1979), 5-12.

Krasnoshchekov, P.S., and Petrov, A.A. [83], *Principles of Constructing Models*, [Russian], Izd-vo Mosk. Un-ta, Moscow, USSR, 1983.

Krasnoshchekov, P.S., Petrov, A.A., and Fedorov, V.V. [86], *Computer Science and Design*, [Russian], Znanie, Moscow, USSR, 1986.

Krolak, P.D. [66], *A property of the Klorak-Cooper extension of Fibonaccian search*, SIAM Review **8** no. 4 (1966), 510-517.

[68], *Further extensions of Fibonaccian search to nonlinear programming problems*, SIAM J. Control **6** no. 2 (1968), 258-265.

Krolak, P.D., and Cooper, L. [63], *An extension of Fibonaccian search to several variables*, Comm. Assoc. Comput. Mach. **6** no. 10 (1963), 639-641.

Krotov, V.F., and Piyavskii, S.A. [68], *Sufficient conditions of optimality in optimal covering problems*, [Russian], Izv. AN SSSR, Ser. Tekhn. Kibern., no. 2 (1968), 10-17.

Kruger, A.Ya. [75], *On optimal search for the root of a function*, [Russian], Vestn. Belor. Un-ta, Ser. I, no. 3 (1975), 3-4.

[76], *On the problem of finding the root of a function*, [Russian], Vestn. Belor. Un-ta, Ser. I, no. 1 (1976), 3-6.

Kuhn, H.W. [53], *Extensive games and the problem of information*, Contributions to the Theory of Games, vol. II, Princeton, 1953, pp. 193-216.

Kuritskii, A.M. [71], *Energetics of mechanical balance trigger regulators*, [Russian], Proceedings of NIIChasprom, No. 8, Izd-vo NIIChasprom, Moscow, USSR, 1971, pp. 66-81.

Kurzhanskii, A.B. [71], *Control and Supervision in Uncertainty Conditions*, [Russian], Nauka, Moscow, USSR, 1977.

Kushner, H. [62], *A versatile stochastic model of a function of unknown and time varying form*, J. Math. Anal. and Appl. **5** no. 1 (1962), 150-167.

[64], *A new method of locating the maximum point of an arbitrary multipeak curve in the presence of noise*, Trans. ASME. Ser. D. J. Basic Eng. **86** no. 1 (1964), 97-105.

Kuzovkin, A.I., and Tikhomirov, V.M. [67], *On the number of computations for finding the minimum of a convex function*, [Russian], Ekonom. i Mat. Metody **3** no. 1 (1967), 95-103.

Larichev, O.I., and Nikiforov, A.D. [86], *Analytical survey of procedures for solving multi-criterion problems of mathematical programming*, [Russian], Ekonom. i Mat. Metody **22** no. 3 (1986), 508-523.

Lbov, G.S., and Grunov, A.A. [86], *An algorithm for seeking the global extremum of a function*, [Russian], Computational Systems, No. 67, Nauka, Novosibirsk, USSR, 1976, pp. 69-76.

Leonov, V.V. [70], *A covering method for finding the global maximum of a multivariate function*, [Russian], Studies in Cybernetics, Sov. Radio, Moscow, USSR, 1970, pp. 41-52.

Levin, A.Yu. [65], *On an algorithm for the minimization of convex functions*, [Russian], Dokl. Akad. Nauk SSSR **160** no. 6 (1965), 1244-1247; Soviet Math. Dokl. **6** (1965), 286-289, [English transl.].

Lootsma, F. [85], *Comparative performance evaluation, experimental design, and generation of test problems in nonlinear optimization*, Computational mathematical programming, Springer-Verlag, Berlin, 1985, pp. 249-260.

Luce, R., and Raiffa, H. [57], *Games and Decisions. Introduction and Critical Survey*, John Wiley & Sons, New York, 1957.

Luzin, N.N. [48], *Theory of Functions of a Real Variable*, [Russian], Uchpedgiz, Moscow, USSR, 1948.

Lyness, J.N. [69], *Notes on adaptive Simpson quadrature*, J. Assoc. Comput. Mach. **16** no. 3 (1969), 483-495.

Lyness, J.N., and Kaganove, J.J. [76], *Comments on the nature of automatic quadrature routines*, ACM Trans. Math. Software **2** no. 1 (1976), 65-81.

Madsen, K. [75], *Minimax solution of non-linear equations without calculating derivatives*, Nondifferentiable optimization, Mathematical Programming Study, 3, North-Holland, Amsterdam, 1975, pp. 110-126.

Maistrovskii, G.D. [72], *On the optimality of Newton's method*, [Russian], Dokl. Akad. Nauk SSSR **204** no. 6 (1972), 1313-1315; Soviet Math. Dokl. **13** (1972), 838-840, [English transl.].

Malcolm, M.A., and Simpson, R.B. [75], *Local versus global strategies for adaptive quadrature*, ACM Trans. Math. Software **1** no. 2 (1975), 129-146.

Mancini, L., and McCormick, G.P. [76], *Bounding minima*, Math. Oper. Res. **1** no. 1 (1976), 50-53.

[79], *Bounding global minima with interval arithmetic*, Operations Research **26** no. 4 (1979), 743-754.

Marchuk, A.G. [76], *Optimal error methods for solving linear recovery problems*, [Russian], Rep. VTs SO AN SSSR, Novosibirsk, USSR, 1976.

Marchuk, A.G., and Osipenko, K.Yu. [75], *Best approximation of functions specified with an error at a finite number of points*, [Russian], Mat. Zametki **17** no. 3 (1975), 359-368; Math. Notes **17** no. 3 (1975), 207-212, [English transl.].

Marchuk, G.I. [80], *Methods of Computational Mathematics*, [Russian], Nauka, Moscow, USSR, 1980.

Marchuk, G.I., and Kuznetsov, Yu.A. [68], *On optimality of iterative processes*, [Russian], Dokl. Akad. Nauk SSSR **181** no. 6 (1968), 1331-1334.

Markin, D.L., and Strongin, R.G. [87], *A method for solving multi-extremum problems with non-convex constraints using a priori information on estimates of the optimum*, [Russian], Zh. Vychisl. Mat. Mat. Fiz. **27** no. 1 (1987), 52-62; USSR Comput. Math. Math. Phys. **27** no. 1 (1987), [English transl.].

Maung Cho Niun, and Sharygin, I.F. [71], *Optimal cubature formulas for the classes $D_2^{1,c}$ and D_2^{1,l_1}*, [Russian], Issues of Computational and Applied Mathematics, No. 5, Institute of Cybernetics, An UzSSR, Tashkent, USSR, 1971, pp. 22-27.

Mayurova, I.V., and Strongin, R.G. [84], *Minimization of a multi-extremum function with a discontinuity*, [Russian], Zh. Vychisl. Mat. Mat. Fiz. **24** no. 12 (1984), 1789-1798; USSR Comput. Math. Math. Phys. **24** no. 6 (1984), 121-126, [English transl.].

McCormick, G.P. [72], *Attempts to calculate global solutions of problems that may have local minima*, Numerical methods for non-linear optimization, Academic Press, London, New York, 1972, pp. 209-221.

[80], *Locating an isolated global minimizer of a constrained nonconvex programm*, Math. Oper. Res. **5** no. 3 (1980), 435-443.

[83], *Nonlinear Programming. Theory, Algorithms and Applications*, Wiley, New York, 1983.

Melkman, A.A., and Micchelli, C.A. [79], *Optimal estimation of linear operators in Hilbert spaces from inaccurate data*, SIAM J. Numer. Anal. **16** no. 1 (1979), 87-105.

Micchelli, C.A. [78], *Optimal estimation of smooth functions from inaccurate data*, IBM Research Report 7024, Yorktown Heights, New York, 1978.

[84], *Orthogonal projections are optimal algorithms*, J. Approx. Theory **40** no. 2 (1984), 101-110.

Micchelli, C.A., and Miranker, W.L. [75], *High order search methods for finding roots*, J. Assoc. Comput. Mach. **22** no. 1 (1975), 51-60.

Micchelli, C.A., and Rivlin, T.J. [77], *A survey of optimal recovery*, Optimal estimation in approximation theory, Plenum Press, New York, London, 1977, pp. 1-54.

Micchelli, C.A., Rivlin, T.J., and Winograd, S. [76], *The optimal recovery of smooth functions*, Numer. Math. **260** no. 2 (1976), 191-200.

Mikhailov, G.A. [80], *Variance of vector Monte Carlo algorithms*, [Russian], Dokl. Akad. Nauk SSSR **263** no. 5 (1980), 1047-1050.

[81], *Optimization of vector Monte Carlo algorithms*, [Russian], Dokl. Akad. Nauk SSSR **260** no. 1 (1981), 26-31.

[84], *Minimax theory of weighted Monte Carlo methods*, [Russian], Zh. Vychisl. Mat. Mat. Fiz. **24** no. 9 (1984), 1294-1302; USSR Comput. Math. Math. Phys. **24** no. 5 (1984), 8-13, [English transl.].

Mikhalevich, V.S. [65], *Sequential optimization algorithms and their application, I, II*, [Russian], Kibernetika no. 1 (1965), 45-56; no. 2 (1965), 85-88.

Mikhalevich, V.S., and Kuksa, A.I. [83], *Methods of Sequential Optimization*, [Russian], Nauka, Moscow, USSR, 1983.

Milanese, M., Tempo, R. [85], *Optimal algorithms theory for robust estimation and prediction*, IEEE Trans. Automat. Contr. **AC-30** no. 8 (1985), 730-738.

Milne-Thompson, L.M. [33], *The Calculus of Finite Differences*, MacMillan, London, 1933.

Miranker, W.L. [69], *Parallel methods of approximating the root of a function*, IBM J. of Research and Development **13** no. 3 (1969), 297-301.

[77], *Parallel methods for solving equations*, IBM T. J. Watson Research Center Rep. 6545, 1977.

Mjelde, K.M. [83], *Methods of the Allocation of Limited Resources*, Wiley, New York, 1983.

Mockus, J. [67], *Multi-extremum Problems in Design*, [Russian], Nauka, Moscow, USSR, 1967.

[72], *On Bayesian methods of seeking the extremum*, [Russian], Avtomatika i Vychisl. Tekhn. no. 3 (1972), 53-62.

[77], *On Bayesian methods of seeking the extremum and their applications*, Information processing 77, North-Holland, Amsterdam, 1977, pp. 195-200.

[80], *The simple Bayesian algorithm for the multidimensional global optimization*, Numerical techniques for stochastic systems, North-Holland, Amsterdam, 1980, pp. 369-377.

Moiseyev, N.N. [75], *Elements of Theory of Optimal Systems*, [Russian], Nauka, Moscow, USSR, 1975.

[79], *Present State of the Theory of Operations Reaserch* (N.N. Moiseyev,, ed.), [Russian], Nauka, Moscow, USSR, 1979.

[81], *Mathematical Problems of Systems Analysis*, [Russian], Nauka, Moscow, USSR, 1981.

Moiseyev, N.N., Ivanilov, Yu.P., and Stolyarova, E.M. [78], *Optimization Methods*, [Russian], Nauka, Moscow, USSR, 1978.

Molchanova, E.S. [86], *Construction of optimal numerical integration algorithms for a certain functional class*, [Russian], VINITI, No. 104-B, Jan. 3, 1986.

Molodtsov, D.A. [78], *Regularization of a set of Pareto points*, [Russian], Zh. Vychisl. Mat. Mat. Fiz. **18** no. 3 (1978), 597-602; USSR Comput. Math. Math. Phys. **18** no. 3 (1978), 68-74, [English transl.].

[80], *Stability and regularization of optimality principles*, [Russian], Zh. Vychisl. Mat. Mat. Fiz. **20** no. 5 (1980), 1115-1129; USSR Comput. Math. Math. Phys. **20** no. 5 (1980), [English transl.].

Morozov, V.A. [74], *An optimality principle for the error when solving approximately equations with non-linear operators*, [Russian], Zh. Vychisl. Mat. Mat. Fiz. **14** no. 4 (1974), 819-827; USSR Comput. Math. Math. Phys. **14** no. 4 (1974), 1-9, [English transl.].

Morozov, V.A., and Grebennikov, A.I. [75], *On optimal approximation of operators*, [Russian], Dokl. Akad. Nauk SSSR **223** no. 6 (1975), 1307-1310.

Morozov, V.A., Sukharev, A.G., and Fedorov, V.V. [86], *Operations Research, Problems and Drills*, [Russian], Vysshaya Shkola, Moscow, USSR, 1986.

Mustăta, C. [77], *Best approximation and unique extension of Lipschitz functions*, J. Approx. Theory **19** no. 3 (1977), 222-230.

Nalimov, V.V., and Chernova, N.A. [65], *Statistical Methods for Extremal Experiments Design*, [Russian], Nauka, Moscow, USSR, 1965.

Nefedov, V.N. [84], *On the approximation of a Pareto set*, [Russian], Zh. Vychisl. Mat. Mat. Fiz. **24** no. 7 (1984), 993-1007; USSR Comput. Math. Math. Phys. **24** no. 4 (1984), 19-28, [English transl.].

[87], *The search for a global maximum of a function of several variables in a set specified by constraints of the inequality type*, [Russian], Zh. Vychisl. Mat. Mat. Fiz. **27** no. 1 (1987), 35-51; USSR Comput. Math. Math. Phys. **27** no. 1 (1987), 23-32, [English transl.].

Neimark, Yu.I., and Strongin, R.G. [66], *Informational approach to the problem of search for the extremum of a function*, [Russian], Izv. AN SSSR, Ser. Tekhn. Kibern., no. 1 (1966), 17-26.

Nemirovsky, A.S., and Nesterov, Yu.E. [85], *Optimal methods of smooth convex minimization*, [Russian], Zh. Vychisl. Mat. Mat. Fiz. **25** no. 3 (1985), 356-369; USSR Comput. Math. Math. Phys. **25** no. 2 (1985), 21-30, [English transl.].

Nemirovsky, A.S., and Yudin, D.B. [79], *Problem Complexity and Method Efficiency in Optimization*, [Russian], Nauka, Moscow, USSR, 1979; Wiley-Interscience, New York, [English transl.].

[83], *Informational complexity of mathematical programming*, [Russian], Izv. AN SSSR, Ser. Tekhn. Kibern., no. 1 (1983), 88-117.

Neuman, P. [81], *An asymptotically optimal procedure for searching a zero or an extremum of a function if a priori distribution of its location is known*, Proc. of a second Prague symposium on asymp. statistics, Academia, Prague, 1981, pp. 291-302.

Newman, D.J. [65], *Location of the maximum on unimodal surfaces*, J. Assoc. Comput. Mach. **12** no. 3 (1965), 395-398.

Niederreiter, H, and McCurley, K. [79], *Optimization of functions by quasy-random search methods*, Computing **22** no. 2 (1979), 119-123.

Niederreiter, H, and Peart, P. [86], *Localization of search in quasi Monte Carlo methods for global optimization*, SIAM J. Sci. Stat. Comput. **7** no. 2 (1986), 660-664.

Nikolskii, S.M. [50], *On estimating accuracy of approximation by quadrature formulae*, [Russian], UMN **5** no. 2 (1950), 165-177.

[79], *Quadrature Formulae*, [Russian], Nauka, Moscow, USSR, 1979.

Oliver, L.T., and Wilde, D.J. [64], *Symmetric sequential minimax search for a maximum*, Fibonacci Quart **2** no. 1 (1964), 169-175.

Ortega, J., and Rheinboldt, W. [70], *Iterative Solution of Nonlinear Equations in Several Variables*, McGraw-Hill, New York, 1970.

Osipenko, K.Yu. [72], *Optimal interpolation of analytic functions*, [Russian], Mat. Zametki **12** no. 4 (1972), 465-476; Math. Notes **12** no. 4 (1972), 712-719, [English transl.].

[76], *Best approximation of analytic functions from information about their values at a finite number of points*, [Russian], Mat. Zametki **19** no. 1 (1976), 29-40; Math. Notes **19** no. 1 (1976), 17-23, [English transl.].

Ostapenko, O.S. [83], *On optimal algorithms for minimization of functions from the classes* $C^n_{1,L,N}$, $C^n_{1,L,N,\varepsilon}$, $\widetilde{C}^n_{1,L,N,\delta}$, [Russian], Kibernetika no. 5 (1983), 88-95.

Ostrovski, A.M. [63], *Solution of Equations and Systems of Equations*, IL, Moscow, USSR, 1963, [Russian]; Academic Press, New York, [English transl.].

Papadimitriou, Ch., and Steiglitz, K. [82], *Combinatorial Optimization: Algorithms and Complexity*, Prentice-Hall, Englewood Cliffs (USA), 1982.

Pevnyi, A.B. [82], *On optimal search strategies for the maximum of a function with bounded highest derivative*, [Russian], Zh. Vychisl. Mat. Mat. Fiz. **22** no. 5 (1982), 1061-1066; USSR Comput. Math. Math. Phys. **22** no. 5 (1982), 38-44, [English transl.].

Piyavskii, S.A. [67], *Algorithms for finding the absolute minimum of a function*, [Russian], Seminar on the Theory of Optimal Solutions, No. 2, Izd-vo IK AN USSR, Kiev, USSR, 1967, pp. 13-24.

[68], *Optimization of nets*, [Russian], Izv. AN SSSR, Ser. Tekhn. Kibern., no. 1 (1968), 68-80.

[72], *An algorithm for finding the absolute extremum of a function*, [Russian], Zh. Vychisl. Mat. Mat. Fiz. **12** no. 4 (1972), 888-896; USSR Comput. Math. Math. Phys. **12** no. 4 (1972), 57-67, [English transl.].

Plaskota, L. [86], *Optimal linear information in the problem of reconstructing the global maximum of a real function*, [Russian], Zh. Vychisl. Mat. Mat. Fiz. **26** no. 6 (1986), 934-938; USSR Comput. Math. Math. Phys. **26** no. 3 (1986), 186-190, [English transl.].

Podinovskii, V.V., and Gavrilov, V.M. [75], *Optimization Using Successively Applied Criteria*, [Russian], Sov. Radio, Moscow, USSR, 1975.

Podinovskii, V.V., and Nogin, V.D. [82], *Pareto-optimal methods for solving multicriterion problems*, [Russian], Nauka, Moscow, USSR, 1982.

Podobedov, V.E. [87], *An optimal algorithm for seeking the extremum of a Lipschitz function using arbitrary linear information*, [Russian], System Programming and Issues of Optimization, Izd-vo Mosk. Un-ta, Moscow, USSR, 1987, pp. 180-187.

Polyak, B.T. [83], *Introduction to Optimization*, [Russian], Nauka, Moscow, USSR, 1983.

Polyak, B.T., Tsypkin, Ya.Z. [80], *Optimal pseudo-gradient adaptation algorithms*, [Russian], Avtomatika i Telemekhanika no. 8 (1980), 74-84.

Popov, N.M. [80], *Approximation of the set of semi-effective points in decomposition of design problems*, [Russian], Vestn. Mosk. Un-ta Ser. Vychisl. Matem. i Kibern., no. 4 (1980), 43-48.

[86], *Approximate solution of multicriterion problems with functional limitations*, [Russian], Zh. Vychisl. Mat. Mat. Fiz. **26** no. 10 (1986), 1468-1481; USSR Comput. Math. Math. Phys. **26** no. 5 (1986), 125-134, [English transl.].

[87], *Numerical methods of multicriterion optimization*, [Russian], System Programming and Issues of Optimization, Izd-vo Mosk. Un-ta, Moscow, USSR, 1987, pp. 155-168.

Pourciau, B.H. [77], *Analysis and optimization of Lipschitz continuous mappings*, J. Optimization Theory and Appl. **22** no. 3 (1977), 331-351.

Poznyak, A.S., and Tsypkin, Ya.Z. [84], *Class-wise optimal algorithms for optimization in correlated-noise conditions*, [Russian], Zh. Vychisl. Mat. Mat. Fiz. **24** no. 6 (1984), 806-822; USSR Comput. Math. Math. Phys. **24** no. 3 (1984), 112-122, [English transl.].

Pshenichnyi, B.N. [69], *Necessary Conditions of Extremum*, [Russian], Nauka, Moscow, USSR, 1969.

[80], *Convex Analysis and Extremal Problems*, [Russian], Nauka, Moscow, USSR, 1980.

Pshenichnyi, B.N., and Danilin, Yu.M. [75], *Numerical Methods for Extremal Problems*, [Russian], Nauka, Moscow, USSR, 1975.

Rastrigin, L.A. [68], *Statistical Search Methods*, [Russian], Nauka, Moscow, USSR, 1968.

[73], *Random search: problems, trends, prospects*, [Russian], Issues of Cybernetics. Problems of Random Search, Nauchn. Soviet AN SSSR po Kompl. Probl. "Kibernetika", Moscow, USSR, 1973, pp. 3-17.

Renegar, J. [85a], *On the complexity of a piecewise linear algorithms for approximating roots of complex polynomials*, Math. Programming **32** no. 3 (1985), 301-318.

[85b], *On the cost of approximating all roots of a complex polynomial*, Math. Programming **32** no. 3 (1985), 319-336.

[85c], *On the efficiency of Newton's method in approximating all zeros of a system of complex polynomials*, Berkeley Mathematical Sciences Research Institute Report, MSRI 00318-86, 1985.

[86], *Rudiments of an average case complexity theory for piecewise-linear path following algorithms*, Berkeley Mathematical Sciences Research Institute Report, MSRI 03418-86, 1986.

Rice, J.R. [74], *Parallel algorithms for adaptive quadrature – convergence*, Proc. IFIP Congress 74, Stokholm, North-Holland, Amsterdam, 1974, pp. 600-604.

[75], *A metaalgorithm for adaptive quadrature*, J. Assoc. Comput. Mach. **22** no. 1 (1975), 61-82.

[76a], *Parallel algorithms for adaptive quadrature II – metalogical correctness*, Acta Informatica **5** no. 4 (1976), 273-275.

[76b], *Parallel algorithms for adaptive quadrature III – program correctness*, ACM Trans. Math. Software **2** no. 1 (1976), 1-30.

[83], *Numerical Methods, Software and Analysis*, McGraw-Hill, New York, 1983.

Rinnooy Kan, A.H.G., Boender, C.G.E., and Timmer, G.T. [85], *A stochastic approach to global optimization*, Computational mathematical programming, Springer-Verlag, Berlin, 1985, pp. 281-305.

Rogers, C.A. [57], *A note on coverings*, Mathematika. A Journal of Pure and Applied Mathematics **4** Part 1, no. 7 (1957), 1-6.

[64], *Packing and Covering*, Cambridge University Press, Cambridge, 1964.

Rubal'skii, G.B. [82], *Search for an extremum of unimodal functions of one variable in an unbounded set*, [Russian], Zh. Vychisl. Mat. Mat. Fiz. **22** no. 1 (1982), 10-16; USSR Comput. Math. Math. Phys. **22** no. 1 (1982), 8-15, [English transl.].

Ryshkov, S.S. [67], *Effectivization of a certain Davenport method in covering theory*, [Russian], Dokl. Akad. Nauk SSSR **175** no. 2 (1967), 303-305.

Ryshkov, S.S., and Baranovskii, E.P. [76], *C-types of n-dimensional lattices and five-dimensional primitive parallelohedrons (with application to covering theory)*, [Russian], Tr. MIAN SSSR **137** (1976).

Saaty, T. [70], *Optimization in Integers and Related Extremal Problems*, McGraw-Hill, New York, 1970.

Sadovnikov, Yu. Yu. [86], *Operator method of global optimization*, [Russian], Kibernetika no. 2 (1986), 39-43.

Šaltenis, V. [71], *A method of multi-extremum optimization*, [Russian], Avtomatika i Vychisl. Techn. no. 3 (1971), 33-38.

Samarskii, A.A. [77], *Theory of Difference Schemes*, [Russian], Nauka, Moscow, USSR, 1977.

[82], *Introduction to Numerical Methods*, [Russian], Nauka, Moscow, USSR, 1982.

Sard, A. [49], *Best approximate integration formulas: best approximation formulas*, Amer. J. Math. **71** no. 1 (1949), 80-91.

Shapiro, H.D. [84], *Increasing robustness in global adaptive quadrature through interval selection heuristics*, ACM Trans. Math. Software **10** no. 2 (1984), 117-139.

Shapiro, R.D., and Wilde, D.J. [74a], *Optimal minimax search with unequal block sizes*, Stanford University, Dept. of Operations Res., Technical Report 74-16, 1974.

[74b], *Variable block search strategies with prior information*, Stanford University, Dept. of Operations Res., Technical Report 74-17, 1974.

Shih, W. [74], *A new application of incremental analysis in resource allocations*, Operational Research Quartely **25** no. 4 (1974), 587-597.

Shokin, Yu.I. [81], *Interval Analysis*, [Russian], Nauka, Novosibirsk, USSR, 1981.

Shor, N.Z. [79], *Methods for Minimization of Non-differentiable Functions and Their Applications*, [Russian], Nauk. Dumka, Kiev, USSR, 1979.

Shub, M., and Smale, S. [85], *Computational complexity: on the geometry of polynomials and a theory of cost, Part I*, Ann. scient. Éc. Norm. Sup. 4 ser. **18** no. 1 (1985), 107-142; *Part II*, SIAM J. Comput. **15** no. 1 (1986), 145-161.

[86], *On the existence of generally convergent algorithms*, J. Complexity **2** no. 1 (1986), 2-11.

Shubert, B.O. [72a], *A sequential method seeking the global maximum of a function*, SIAM J. Numer. Anal. **9** no. 3 (1972), 379-388.

[72b], *Sequential optimization of multimodal discrete function with bounded rate of change*, Manage. Sci. **18** no. 11 (1972), 687-693.

Sikorski, K. [82], *Bisection is optimal*, Numer. Math. **40** no. 1 (1982), 111-117.

[84], *Optimal solution of nonlinear equations satisfying a Lipschitz condition*, Num. Math. **43** no. 2 (1984), 225-240.

Smale, S. [81], *The fundamental theorem of algebra and complexity theory*, Bull. Amer. Math. Soc. (N.S.) **4** no. 1 (1981), 1-36.

Smolyak, S.A. [65], *On Optimal Restoration of Functions and Functionals of Them,*, [Russian], PhD Thesis, Moscow State University, 1965.

Sobol, I.M. [69], *Multidimensional Quadratures and Haar Functions*, [Russian], Nauka, Moscow, USSR, 1969.

[73], *Numerical Monte Carlo Methods*, [Russian], Nauka, Moscow, USSR, 1973.

[79], *On the systematic search in a hypercube*, SIAM J. Numer. Anal. **16** no. 5 (1979), 790-793.

[82], *On estimation of accuracy of the simplest multidimensional search*, [Russian], Dokl. Akad. Nauk SSSR **266** no. 3 (1982), 569-572.

Sobol, I.M., and Statnikov, R.B. [81], *Choice of the Optimal Parameters in Multicriterion Problems*, [Russian], Nauka, Moscow, USSR, 1981.

Sobolev, S.L. [74], *Introduction to the Theory of Cubatures*, [Russian], Nauka, Moscow, USSR, 1974.

[77], *Coefficients of optimal quadrature formulas*, [Russian], Dokl. Akad. Nauk SSSR **235** no. 1 (1977), 34-37.

Sonnevend, G. [77], *On optimization of algorithms for function minimization*, JVM and MF **17** no. 3 (1977), 591-609.

[78], *On optimization of adaptive algorithms*, Operations Research Verfahren, v.31, Athenäum, 1978, pp. 581-595.

[83a], *Optimal passive and sequential algorithms for the approximation of convex functions in* $L([0,1]^s)$, $p=1,\infty$, Constructive theory of functions 81, Sofia, 1983, pp. 535-542.

[83b], *An optimal sequential algorithm for the uniform approximation of convex functions on* $[0,1]^2$, Applied Mathematics and Optimization **10** no. 2 (1983), 127-142.

Sorokina, T.G., and Sorokin, I.M. [81], *Minimization of multi-extremum functions*, [Russian], Mathematical Methods of Optimization and Control of Complex Systems, Izd-vo KGU, Kalinin, USSR, 1981, pp. 60-68.

Strigul, O.I. [83], *Optimal algorithms for seeking minimax of functions from certain classes*, [Russian], Kibernetika no. 5 (1983), 126.

Strongin, R.G. [78], *Numerical Methods for Multi-exremum Problems*, [Russian], Nauka, Moscow, USSR, 1978.

Strongin, R.G., and Markin, D.L. [86], *Minimization of multi-extremum functions with non-convex constraints*, [Russian], Kibernetika no. 4 (1986), 64-69.

Sugie, N. [64], *An extension of Fibonaccian searching to multidimensional cases*, IEEE Trans. on Automatic Control. **9** no. 1 (1964), 105.

Sukharev, A.G. [71], *Optimal strategies of the search for an extremum*, [Russian], Zh. Vychisl. Mat. Mat. Fiz. **11** no. 4 (1971), 910-924; USSR Comput. Math. Math. Phys. **11** no. 4 (1971), 119-137, [English transl.].

[72], *Best sequential search strategies for finding an extremum*, [Russian], Zh. Vychisl. Mat. Mat. Fiz. **12** no. 1 (1972), 35-50; USSR Comput. Math. Math. Phys. **12** no. 1 (1972), 39-59, [English transl.].

[75], *Optimal Search for an Extremum*, [Russian], Izd-vo Mosk. Un-ta, Moscow, USSR, 1975.

[76a], *Optimal search for the roots of a function satisfying a Lipschitz condition*, [Russian], Zh. Vychisl. Mat. Mat. Fiz. **16** no. 1 (1976),

20-29; USSR Comput. Math. Math. Phys. **16** no. 1 (1976), 17-26, [English transl.].

[76b], *Optimal passive and sequential algorithms for constructing best approximations for functions satisfying a Lipschitz condition*, [Russian], Dokl. Akad. Nauk SSSR **231** no. 4 (1976), 814-817; Soviet Math. Dokl. **17** (1976), 1660-1664, [English transl.].

[77], *Solution of multistep antagonistic approximation game*, [Russian], Vestn. Mosk. Un-ta, Ser. Vychisl. Matem. i Kibern., no. 2 (1977), 43-57.

[78], *Optimal method of constructing best uniform approximation for functions of a certain class*, [Russian], Zh. Vychisl. Mat. Mat. Fiz. **18** no. 2 (1978), 302-313; USSR Comput. Math. Math. Phys. **18** no. 2 (1978), 21-31, [English transl.].

[79a], *Optimal numerical integration formulas for some classes of functions of several variables*, [Russian], Dokl. Akad. Nauk SSSR **246** no. 2 (1979), 282-285; Soviet Math. Dokl. **20** (1979), 472-475, [English transl.].

[79b], *A sequentially optimal algorithm for numerical integration*, J. Optimization Theory and Appl. **28** no. 3 (1979), 363-373.

[79c], *Optimal algorithms for iterated numerical integration*, J. Optimization Theory and Appl. **28** no. 3 (1979), 375-390.

[81a], *Global extremum and methods for finding it*, [Russian], Mathematical Methods in Operations Research, Izd-vo Mosk. Un-ta, Moscow, USSR, 1981, pp. 4-37.

[81b], *Optimal and sequentially optimal algorithms in problems of numerical analysis*, [Russian], Mathematical Methods in Operations Research, Izd-vo Mosk. Un-ta, Moscow, USSR, 1981, pp. 54-68.

[81c], *A stochastic algorithm for extremum search, optimal in one step*, [Russian], Zh. Vychisl. Mat. Mat. Fiz. **21** no. 6 (1981), 1385-1401; USSR Comput. Math. Math. Phys. **21** no. 6 (1981), 23-39, [English transl.].

[81d], *Method of investigation of the effectiveness of quadrature formulas*, [Russian], Kibernetika no. 6 (1981), 75-80.

[81e], *On optimal stochastic and deterministic search for global extremum*, Internationale Tagung Mathematische Optimierung – Theorie und Anwendungen, Technische Hochschule, DDR, Ilmenau, 1981, pp. 163-166.

[82a], *Problem of the construction of optimal quadratures for functions of several variables*, [Russian], Kibernetika no. 1 (1982), 7-11.

[82b], *On optimal methods for solving multi-criterion problems*, [Russian], Izv. AN SSSR, Ser. Tekhn. Kibern., no. 3 (1982), 67-73.

[84], *On optimal methods in numerical analysis*, Computational mathematics, Banach Center Publications, v.13, PWN – Polish Scientific Publishers, Warsaw, 1984, pp. 575-587.

[85], *On coincidence of errors in the classes of passive and sequential algorithms*, [Russian], Zh. Vychisl. Mat. Mat. Fiz. **25** no. 2 (1985), 295-298; USSR Comput. Math. Math. Phys. **25** no. 1 (1985), 193-195, [English transl.].

[86], *On the existence of optimal affine methods for approximating linear functionals*, J. Complexity **2** no. 4 (1986), 317-322.

Sukharev, A.G., and Chuyan, O.R. [84], *Application of the sufficient conditions of coincidence of the worst-case errors in the classes of adaptive and nonadaptive algorithms*, [Russian], VINITI, No. 3535-84, May 31, 1984.

Sukharev, A.G., and Fedorov, V.V. [79], *Minimax Problems and Minimax Algorithms*, [Russian], Izd-vo Mosk. Un-ta, Moscow, USSR, 1979.

[81], *Optimal search for the maximum of a minimum function in the case of coupled variables*, [Russian], Vestn. Mosk. Un-ta, Ser. Vychisl. Matem. i Kibern., no. 1 (1981), 45-50.

[85], *Optimal global search algorithms for minimax problems*, [Russian], Issues of Cybernetics. Models and Methods of Global Optimization, Nauchn. Soviet AN SSSR po Kompl. Probl. “Kibernetika”, 1985, pp. 70-79.

Sukharev, A.G., Timokhov, A.V., and Fedorov, V.V. [86], *Optimization Methods*, [Russian], Nauka, Moscow, USSR, 1986.

Sysoyev, V.V., and Petrov, V.A. [76], *Development of a program for optimization of multi-extremum functions using ψ-transform method*, [Russian], Econom. i Matem. Metody **12** no. 1 (1976), 178-184.

Tanana, V.P. [81], *Methods for Solving Operator Equations*, [Russian], Nauka, Moscow, USSR, 1981.

Tarasova, V.P. [78], *Optimal strategies for seeking the domain of greatest values for a class of functions*, [Russian], Zh. Vychisl. Mat. Mat. Fiz. **18** no. 4 (1978), 886-896; USSR Comput. Math. Math. Phys. **18** (1978), 886-896, [English transl.].

[81], *Optimal algorithms for seeking the interval of greatest values for a class of functions*, [Russian], Zh. Vychisl. Mat. Mat. Fiz. **21** no. 5 (1981), 1108-1115; USSR Comput. Math. Math. Phys. **21** no. 5 (1981), 33-41, [English transl.].

[84], *Optimal search for the extremum of a locally unimodal function*, [Russian], Kibernetika no. 1 (1984), 65-68.

Tikhomirov, V.M. [69], *Best methods for approximation and interpolation of differentiable functions in the space* $C[-1;1]$, [Russian], Mat. Sb. **80** no. 2 (1969), 290-304; Math. USSR Sb. **9** (1969), 275-289, [English transl.].

[76], *Some Problems of Approximation Theory*, [Russian], Izd-vo Mosk. Un-ta, Moscow, USSR, 1976.

Tikhonov, A.N. [63a], *On solution of ill-posed problems and a method of regularization*, [Russian], Dokl. Akad. Nauk SSSR **151** no. 3 (1963), 501-504.

[63b], *On regularization of ill-posed problems*, [Russian], Dokl. Akad. Nauk SSSR **153** no. 1 (1963), 49-52.

[65], *On non-linear equations of the first kind*, [Russian], Dokl. Akad. Nauk SSSR **161** no. 5 (1965), 1023-1026.

Tikhonov, A.N., and Arsenin, V.Ya. [79], *Methods of Solving Ill-posed Problems*, [Russian], Nauka, Moscow, USSR, 1979.

Tikhonov, A.N., and Gaisaryan, S.S. [69], *The choice of optimum networks in the approximate calculation of quadratures*, [Russian], Zh. Vychisl. Mat. Mat. Fiz. **9** no. 5 (1969), 1170-1176; USSR Comput. Math. Math. Phys. **9** no. 5 (1969), 252-262, [English transl.].

Timonov, L.N. [77], *An algorithm for seeking the global extremum*, [Russian], Izv. AN SSSR, Ser. Tekhn. Kibern., no. 3 (1977), 53-60.

Todd, M.J. [76], *Optimal dissection of simplices*, Cornell University, Dept. of Operations Res. Report, 1976.

Torokhtii, A.P. [83], *On mesh-class-optimal methods of computing quadratures*, [Russian], Zh. Vychisl. Mat. Mat. Fiz. **23** no. 1 (1983), 29-38; USSR Comput. Math. Math. Phys. **23** no. 1 (1983), 20-26, [English transl.].

Traub, J.F. [82], *Iterative Methods for the Solution of Equations*, Chelsea, New York, 1982.

[85], *Complexity of approximately solved problems*, J. Complexity **1** no. 1 (1985), 3-10.

Traub, J.F., Wasilkovski, G.M., and Woźniakowski, H. [83], *Information, Uncertainty, Complexity*, [Russian], Addison-Wesley, Reading, Massachusetts, 1983.

[84], *Average case optimality for linear problems*, Theoret. Comp. Sci. **29** no. 1 (1984), 1-25.

Traub, J.F., and Woźniakowski, H. [80], *A General Theory of Optimal Algorithms*, [Russian], Academic Press, New York, 1980.

[84a], *Information and computation*, Advances in computers, v.23, Academic Press, New York, London, 1984, pp. 35-92.

[84b], *On the optimal solution of large linear systems*, J. Assoc. Comput. Mach. **31** no. 3 (1984), 545-559.

Ust'uzhaninov, V.G. [78], *On random search for an optimal system of criteria*, [Russian], Computational Systems, No. 77. Computer-aided Design in Microelectronics. Theory, Methods, Algorithms, Nauka, Novosibirsk, 1978, pp. 42-52.

[80a], *Limiting abilities of random search algorithms*, [Russian], Avtomatika i Vychisl. Tekhn. no. 1 (1980), 71-77.

[80b], *On efficiency of random search*, [Russian], Avtomatika i Vychisl. Tekhn. no. 3 (1980), 70-74.

[80c], *Search for the maximum of a function satisfying the Lipschitz restriction*, [Russian], Problems of Random Search, No. 8, Zinatne, Riga, 1980, pp. 135-148.

[81], *Existence and efficiency of optimal random search algorithms*, [Russian], Problems of Random Search, No. 9, Zinatne, Riga, 1981, pp. 9-29.
[83], *Abilities of random search in solving discrete extremal problems*, [Russian], Kibernetika no. 2 (1983), 64-77.
[85], *Random search in continuous problems of global optimization*, [Russian], Issues of Cybernetics. Models and Methods of Global Optimization, Nauchn. Soviet AN SSSR po Kompl. Probl. "Kibernetika", 1985, pp. 37-45.
Vasil'ev, F.P. [80], *Numerical Methods of Solving Extremal Problems*, [Russian], Nauka, Moscow, USSR, 1980.
[81], *Methods of Solving Extremal Problems*, [Russian], Nauka, Moscow, USSR, 1981.
Vasil'ev, N.S. [83], *Methods of finding the global minimum of a quasi-concave function*, [Russian], Zh. Vychisl. Mat. Mat. Fiz. **23** (1983), 307-313; USSR Comput. Math. Math. Phys. **23** (1983), 31-35, [English transl.].
[84], *An active method of searching for the global minimum of a concave function*, [Russian], Zh. Vychisl. Mat. Mat. Fiz. **24** (1984), 152-156; USSR Comput. Math. Math. Phys. **24** (1984), 96-100, [English transl.].
[85], *Minimum search in concave problems using the sufficient condition for a global extremum*, [Russian], Zh. Vychisl. Mat. Mat. Fiz. **25** (1985), 190-199; USSR Comput. Math. Math. Phys. **25** (1985), 123-129, [English transl.].
[86], *Approximation of convex sets by polyhedrons and its application to search for global extremum*, [Russian], Tekhn. Kibernetika no. 1 (1986), 27-36.
Vasil'ev, P.P. [83a], *Optimal search for zero of an approximately evaluated function*, [Russian], Mathematical Methods of Optimization and Control of Complex Systems, Izd-vo KGU, Kalinin, 1983, pp. 19-25.
[83b], *Optimal search for the root of an approximately evaluated function*, [Russian], VINITI No. 3486-83, June 28, 1983.
[84], *On one-step optimal stochastic search for the root of a function satisfying the Lipschitz condition*, [Russian], VINITI No. 827-84, Feb. 10, 1984.
[85], *On optimal passive search for the root of a monotonic function*, [Russian], Issues of Cybernetics. Models and Methods of Global Optimazation, Nauchn. Soviet AN SSSR po Kompl. Probl. "Kibernetika", Moscow, USSR, 1985, pp. 98-111.
Vasil'ev, P.P., and Sukharev, A.G. [83], *Program for numerical solution of a system of nonlinear equations using Brown's method with preliminary selection of the initial approximation*, GosFAP, No. П006441,

Kalinin, (Annotation in: *Algorithms and Programs* no. 5 (1983), p. 20 [Russian]).

Vasil'ev, S.B., and Ganshin, G.S. [82], *A sequential algorithm of search for the maximum of a twice differentiable function*, [Russian], Mat. Zametki **31** (1982), 613-618; Math. Notes **31** (1982), [English transl.].

Vasil'eva, L.G., et al. [72], *Algol Procedures for Numerical Integration*, [Russian], Izd-vo Mosk. Un-ta, Moscow, USSR, 1972.

Veinott, A.F. [66], *Production planning with convex costs: a parametric study*, Manage. Sci. **12** no. 11 (1966), 745-777.

Vitushkin, A.G. [59], *Estimation of Complexity of Tabulation Problems*, [Russian], Fizmatgiz, Moscow, USSR, 1959; *Theory of the Transmission and Processing Information*, Pergamon Press, Oxford, 1961, [English transl.].

Volkov, Ye.A. [74], *On search for the maximum of a function and approximate global solution of a system of nonlinear equations*, [Russian], Tr. MIAN SSSR **131** (1974), 64-80.

Vorob'yov, N.N., and Vrublevskaya, I.N. [67], [Russian], *Extensive games* (N.N. Vorob'yov and I.N. Vrublevskaya, eds.), Nauka, Moscow, USSR, 1967.

Vysotskaya, I.N., and Strongin, R.G. [83], *A method of solving nonlinear equations using a priori probability estimates of the roots*, [Russian], Zh. Vychisl. Mat. Mat. Fiz. **23** (1983), 3-12; USSR Comput. Math. Math. Phys. **23** (1983), 1-7, [English transl.].

Wald, A. [47], *Sequential Analysis*, John Wiley & Sons, New York, Chapman & Hall, London, 1947.

[67], *Statistical Desision Functions*, John Wiley & Sons, New York, Chapman & Hall, London.

Wasilkowski, G.W. [84], *Some nonlinear problems are as easy as the approximation problem*, Comp. and Maths. with Appls. **10** no. 4/5 (1984), 351-363.

[85], *Average case optimality*, J. Complexity **1** no. 1 (1985), 107-117.

Wasilkowski, G.W., and Woźniakowski, H. [84], *Can adaptation help on the average?*, Num. Math. **44** no. 2 (1984), 169-190.

Werschulz, A.G. [79], *Optimal order for approximation of derivatives*, J. Comp. Syst. Sci. **18** no. 3 (1979), 213-217.

[80], *Computational complexity of one-step methods for systems of differential equations*, Math. Comp. **34** no. 149 (1980), 155-174.

[81], *On maximal order for local and global problems*, J. Comp. Syst. Sci. **23** no. 1 (1981), 38-48.

[84], *Does increased regularity lower complexity?*, Math. Comp. **42** no. 165 (1984), 69-93.

[86], *Optimal algorithms for a problem of optimal control*, Optimal algorithms, BAS, Sofia, 1986, pp. 228-234.

Wiener, N. [63], *Fourier Integral and Some of Its Applications*, Fizmatgiz, Moscow, USSR, 1963, [Russian].

Wilde, D.J. [64], *Optimum Seeking Methods*, Prentice-Hall, Englewood Cliffs, New Jersey, 1964, [Russian].

[65], *A multivariable dichotomous optimum-seeking method*, IEEE Trans. on Automatic Control **10** no. 1 (1965), 85-87.

Wilde, D.J., and Beightler, C.S. [67], *Foundations of optimization*, Prentice-Hall, Englewood Cliffs, New Jersey, 1967.

Wilde, D.J., and Sanchez-Anton, J.M. [71a], *Discrete optimization on a multivariable Boolean lattice*, Math. Programming **1** no. 3 (1971), 301-306.

[71b], *Multivariable monotonic optimization over multivalued logics and rectangular design lattices*, Discrete Maths. **1** no. 3 (1971), 277-294.

Witzgall, C. [72], *Fibonacci search with arbitrary first evaluation*, Fibonacci Quart **10** no. 2 (1972), 113-134.

Woźniakowski, H. [85], *A survey of information-based complexity*, J. Complexity **1** no. 1 (1985), 11-44.

[86], *Complexity of integration in different settings*, Optimal algorithms, BAS, Sofia, 1986, pp. 235-240.

Yermakov, S.M. [71], *Monte Carlo Method and Related Problems*, [Russian], Nauka, Moscow, USSR, 1971.

Yermol'ev, Yu.M. [78], *Stochastic Programming Methods*, [Russian], Nauka, Moscow, USSR, 1978.

Yudin, D.B. [83], *Computational methods of multi-criterion optimization*, [Russian], Izv. AN SSSR, Ser. Tekhn. Kibern., no. 4 (1983), 3-16.

Zadiraka, V.K. [83], *Theory of Computation of the Fourier Transform*, [Russian], Nauk. Dumka, Kiev, USSR, 1983.

Zadiraka, V.K., and Igisinov, K. [73], *Analysis of accuracy and efficiency of fast Fourier transform algorithms and some of their applications*, [Russian], Kibernetika no. 6 (1973), 57-61.

Zadiraka, V.K., and Ivanov, V.V. [79], *Problems of Optimization of Computations (Numerical Integration and Differentiation)*, [Russian], Ob-vo "Znanie", Kiev, USSR, 1979.

Zaguskin, V.L. [60], *Reference Book on Numerical Methods for Solving Algebraic and Transcendental Equations*, [Russian], Fizmatgiz, Moscow, USSR, 1960.

Zaliznyak, N.F., and Ligun, A.A. [78], *Optimal strategies for seeking the global maximum of a function*, [Russian], Zh. Vychisl. Mat. Mat. Fiz. **18** no. 2 (1978), 314-321; USSR Comput. Math. Math. Phys. **18** no. 2 (1978), 31-38, [English transl.].

Zermelo, E. [13], *Über eine Anwendung der Mengenlehre auf die Theorie des Schachspiels*, Proceedings of the Fifth International Congress of Mathematicians (Cambridge, 1912), Cambridge University Press, 1913, pp. 501-504.

Zhensykbayev, A.A. [81], *Monosplines with the minimum norm and the best quadrature formulas*, [Russian], UMN **36** (1981), 107-159.

Zhidkov, N.P. [77], *Linear Approximations of Functionals*, [Russian], Izd-vo Mosk. Un-ta, Moscow, USSR, 1977.

Zhigliavskii, A.A. [85], *Mathematical Theory of Global Random Search*, [Russian], Izd-vo Leningr. Un-ta, Leningrad, USSR, 1985.

Zhileikin, Ya.M., and Kukarkin, A.B. [78], *Optimal evaluation of integrals of rapidly oscillating functions*, [Russian], Zh. Vychisl. Mat. Mat. Fiz. **18** no. 2 (1978), 294-301; USSR Comput. Math. Math. Phys. **18** no. 2 (1978), 15-21, [English transl.].

Zhirov, V.S. [85], *Search for the global extremum of a polynomial on a parallelepiped*, [Russian], Zh. Vychisl. Mat. Mat. Fiz. **25** no. 2 (1985), 163-180; USSR Comput. Math. Math. Phys. **25** no. 1 (1985), 105-116, [English transl.].

Žilinskas, A. [75], *One-step Bayesian method for seeking the extremum of a univariate function*, [Russian], Kibernetika no. 1 (1975), 139-144.

[76], *A method for one-dimensional multi-extremum minimization*, [Russian], Izv. AN SSSR, Ser. Tekhn. Kibern., no. 4 (1976), 71-76.

[86], *Global Optimization. Axiomatics of Statistical Models, Algorithms, Applications*, [Russian], Mokslas, Vilnius, 1986.

AUTHOR INDEX

W

Y

Z

SUBJECT INDEX

THEORY AND DECISION LIBRARY

SERIES B: MATHEMATICAL AND STATISTICAL METHODS
Editor: H. J. Skala, *University of Paderborn, Germany*

1. D. Rasch and M.L. Tiku (eds.): *Robustness of Statistical Methods and Nonparametric Statistics*. 1984 ISBN 90-277-2076-2
2. J.K. Sengupta: *Stochastic Optimization and Economic Models*. 1986 ISBN 90-277-2301-X
3. J. Aczél: *A Short Course on Functional Equations*. Based upon Recent Applications to the Social Behavioral Sciences. 1987 ISBN Hb 90-277-2376-1; Pb 90-277-2377-X
4. J. Kacprzyk and S.A. Orlovski (eds.): *Optimization Models Using Fuzzy Sets and Possibility Theory*. 1987 ISBN 90-277-2492-X
5. A.K. Gupta (ed.): *Advances in Multivariate Statistical Analysis*. Pillai Memorial Volume. 1987 ISBN 90-277-2531-4
6. R. Kruse and K.D. Meyer: *Statistics with Vague Data*. 1987 ISBN 90-277-2562-4
7. J.K. Sengupta: *Applied Mathematics for Economics*. 1987 ISBN 90-277-2588-8
8. H. Bozdogan and A.K. Gupta (eds.): *Multivariate Statistical Modeling and Data Analysis*. 1987 ISBN 90-277-2592-6
9. B.R. Munier (ed.): *Risk, Decision and Rationality*. 1988 ISBN 90-277-2624-8
10. F. Seo and M. Sakawa: *Multiple Criteria Decision Analysis in Regional Planning*. Concepts, Methods and Applications. 1988 ISBN 90-277-2641-8
11. I. Vajda: *Theory of Statistical Inference and Information*. 1989 ISBN 90-277-2781-3
12. J.K. Sengupta: *Efficiency Analysis by Production Frontiers*. The Non-parametric Approach. 1989 ISBN 0-7923-0028-9
13. A. Chikán (ed.): *Progress in Decision, Utility and Risk Theory*. 1991 ISBN 0-7923-1211-2
14. S.E. Rodabaugh, E.P. Klement and U. Höhle (eds.): *Applications of Category Theory to Fuzzy Subsets*. 1992 ISBN 0-7923-1511-1
15. A. Rapoport: *Decision Theory and Decision Behaviour*. Normative and Descriptive Approaches. 1989 ISBN 0-7923-0297-4
16. A. Chikán (ed.): *Inventory Models*. 1990 ISBN 0-7923-0494-2
17. T. Bromek and E. Pleszczyńska (eds.): *Statistical Inference*. Theory and Practice. 1991 ISBN 0-7923-0718-6

18. J. Kacprzyk and M. Fedrizzi (eds.): *Multiperson Decision Making Models Using Fuzzy Sets and Possibility Theory*. 1990 ISBN 0-7923-0884-0
19. G.L. Gómez M.: *Dynamic Probabilistic Models and Social Structure*. Essays on Socioeconomic Continuity. 1992 ISBN 0-7923-1713-0
20. H. Bandemer and W. Näther: *Fuzzy Data Analysis*. 1992 ISBN 0-7923-1772-6
21. A.G. Sukharev: *Minimax Models in the Theory of Numerical Methods*. 1992 ISBN 0-7923-1821-8

KLUWER ACADEMIC PUBLISHERS – DORDRECHT / BOSTON / LONDON

The manufacturer's authorised representative in the EU is Springer
Nature Customer Service Centre GmbH, Europaplatz 3, 69115 Heidelberg,
Germany. If you have any concerns regarding our products, please
contact ProductSafety@springernature.com

Printed and bound by CPI Group (UK) Ltd, Croydon, CR0 4YY
23/04/2026
02095625-0012